Volcanism

Hans-Ulrich Schmincke

Volcanism

With 401 Figures, 396 in Color

 Springer

Author

Professor Hans-Ulrich Schmincke

Dept. of Volcanology and Petrology
GEOMAR Research Center
University of Kiel (Germany)

E-mail: *hschmincke@geomar.de*

All photographs by the author unless otherwise noted

Cover illustrations by:
Nagasaki Photo Service (front cover);
H.-U. Schmincke and M. Sumita (back cover).

ISBN 3-540-43650-2 Springer Berlin Heidelberg New York
1st ed. 2004. Corr. 2nd printing

Library of Congress Cataloging-in-Publication Data applied for

Bibliograhic information published by Die Deutsche Bibliothek
Die Deutsche Bibliothek lists this publication in the Deutsche Nationalbibliografie; detailed bibliographic data is available in the Internet at <http://dnb.ddb.de>.

Springer-Verlag is a part of Springer Science + Business Media
springeronline.com
© Springer-Verlag Berlin Heidelberg 2004, 2006
Printed in Germany

© 1998 (2.überarb. Auflage) by Wissenschaftliche Buchgesellschaft, Darmstadt.

Typesetting: schreiberVIS, Seeheim, Germany
Illustrations: www.schreibervis.de in cooperation with
Elke Göpfert, Mörlenbach-Weiher,
Marion Mayer, Darmstadt.
Dr. Mari Sumita, Ascheberg

Cover design: E. Kirchner, Heidelberg
Printing and binding: Stürtz AG, Würzburg

Printed on acid-free paper   32/3141/as   5 4 3 2 1 0

Preface

The reason why I wrote a book on volcanism in German (second edition published in 2000) rather than in English, the *lingua franca* of science, was simple. The intended users of this book were not only scientists and beginning students in the Earth sciences, but also school teachers and lay people interested in volcanoes. The updated English translation likewise attempts to strike a balance between providing enough science and a few up-to-date references for those who want to dig deeper, while at the same time remaining digestible for those who simply want to know more about volcanoes. Translation of the text by the author had the advantage that it allowed rewriting and updating of major sections of several chapters, within the constraints of the publication schedule.

The first edition of this book appeared at a time (1986) when a certain plateau had been reached in volcanology, following an almost explosive growth phase in the study of volcanic rocks and processes. The basic motivation to study volcanoes is the sustained curiosity about what happens beneath volcanoes, at the source of magmas, during their rise, their vesiculation, their many ways to erupt explosively or quietly and the transport of hot and cold materials along the Earth's surface or through the atmosphere. This book attempts to point out some of the major current issues in volcanology. Two fundamental goals for motivating research have recently been reinforced in national and international research programs. One goal is that of mitigating volcanic disasters. The other is concerned with the interaction of volcanic processes with the environment including the solid earth, groundwater and surface waters, as well as the atmosphere.

This book shares the fate of all books attempting a broad overview of a subject. Experts will not be content with the superficial account of their specialized fields. Yet there is a great need for broad-brush treatments in the face of ever-growing specialization. I have listed a few more specialized books below, some of which contain extensive lists of references for those who want to delve deeper into a subject.

The first chapters (Chaps. 1–9) are more general in nature and thus do not require detailed references. Most recent studies quoted are the sources of the figures. Hence, the list of papers quoted is in no way representative. In view of the large amount of literature consulted when writing this volume, I have not been able to trace back some information to its primary source. I apologize to those colleagues whose paper I have failed to quote. The 2000 German edition was also conceived to summarize some aspects of the research of our group and to represent a symbolic thanks to several research agencies that had generously funded our work for more than four decades. Many figures are sourced in these publications, which are thus greatly overrepresented in the list of references. The present text has been updated and some parts were rewritten to provide a more logical flow and to make reference to important recent eruptions. Some 35 graphs and 85 photos were added to the English-language edition or exchanged with previous illustrations.

About half of the roughly 450 references – generally indicated in the text by a number in parenthesis (the number of the chapter in which they are quoted is given in the list of references) but listed alphabetically at the end – were published during the past six years. Recent books that treat some aspects of volcanism in more detail include Francis (1993), Wohletz and Heiken (1992), McGuire (1995), Scarpa and Tilling (1996), Sparks et al. (1997), Gilbert and Sparks (1998), Heiken and Wohletz (1998), and Freundt and Rosi (1998). The monumental *Encyclopaedia of Volcanology* (Sigurdsson et al. 2000), published after the second German edition of *Vulkanismus* appeared, will remain the standard source of information for many years. Two older classic textbooks in volcanology, by Macdonald (1972) and Williams and McBirney (1979), are still informative introductions to the subject. Blong (1984) provides a detailed and comprehensive overview of the effects of volcanic eruptions. There are also many popular accounts of volcanology, particularly readable books being those by Decker and Decker (1994), of which there are several editions. Woods and Kienle (1998) provide a guide to many areas of volcanic interest in the United States. Volcanic

A volcano is not made on purpose to frighten superstitious people into fits of piety and devotion; nor to overwhelm devoted cities with destruction; a volcano should be considered as a spiracle to the subterranean furnace, in order to prevent the unnecessary elevation of land, and letal effects of earthquakes.

James Hutton, Theory of the Earth, Codicote, 1795

national parks in the United States are succinctly discussed and supplied with illustrative maps by Decker and Decker (2002). Planetary volcanology, a subject not treated in this book for simple space reasons, is discussed in several dedicated books such as Cattermole (1996). The history of man's attempt to understand volcanoes is discussed in some detail by Sigurdsson (2000).

There are two journals in volcanology: *Bulletin of Volcanology* and *Journal of Volcanology and Geothermal Research*. Publications treating various aspects of volcanology also appear in journals in geology, petrology, geophysics, geochemistry and geomorphology.

Details of current eruptions can be obtained from the Internet, especially www.volcano.si.edu/gvp, the Global Volcanism Network of the Smithsonian Institution (Washington, D.C.) and the US Geological Survey. They contain weekly and monthly activity reports where details of current eruptions and background data are summarized, as well as data on volcanoes of the world and various links to other volcano sites. The Internet site of IAVCEI, the International Association of Volcanology and Chemistry of the Earth's Interior (www.iavcei.org) also contains much useful information on volcano observatories, current courses and links to various volcano-related sites.

A list of common abbreviations and physical units is given at the end of this book along with SI units. Leading journals continue to use some of the old units. I have followed this practice and still use kb (kilobar) instead of Gpa (giga pascal). For density I mostly use g/cm^3 which is easier to visualize than kg/m^3.

Several colleagues were kind enough to take out time from their busy schedule to critically scan earlier versions of one or more chapters and make helpful suggestions. These include Colin Devey, Wendell Duffield, Jim Gardner, Martina Halmer, Jason Phipps Morgan, Chris Newhall, Jon Snow, Roland von Huene and, for the German edition, Armin Freundt and Matthias Hort. I am especially grateful to Shane Cronin, a volcanologist as well as a native English speaker. He did a trail-blazing review of first drafts of all chapters chopping overlong Germanic sentences and making the text easier to follow. Kathy Cashman did a final review of several chapters pointing out logical flaws and other inconsistencies. I owe much to Tad Ui, my host at Hokkaido University where the final drafts of the translation were prepared. He also looked over several chapters and made many useful suggestions. Several colleagues kindly provided me with files of figures from their publications. I was lucky to have been associated with many able students and some of our joint work is quoted throughout the book. I was advised on graphics by Conny Park. Much of the book owes its appearance to the graphic skills of Mari Sumita who prepared preliminary versions of many graphs and graciously helped completing the book in many other ways. Joachim Schreiber and colleagues prepared the final figures and mastered the layout. I am very grateful to Wolfgang Engel (Springer Verlag) who patiently paved the way for the English edition. Luisa Tonarelli (Springer Verlag) took care of the manuscript during the final stages and saw the book to the press. Many thanks to all of you. All the faults remaining are my own responsibility, and there are undoubtedly many. I encourage readers to point these out and also suggest more appropriate references or figures for future revisions.

Lisch, March 2003   *Hans-Ulrich Schmincke*

Contents

Introduction

On 14 November 1964, a new volcanic island rose out of the sea, about 30 km south of Iceland. Its complex volcanic evolution, with major changes in eruptive mechanisms and constant erosion by the waves (which continuously threatened to destroy the entire island) came to an end in June 1967. The island was called *Surtsey* by Icelandic volcanologists in reference to the mythical fire giant Surtur. The development of the island Surtsey has been well-documented (376), often under harsh conditions. The nature, however, of the physical processes that take place when water and magma meet below the water surface as well as in the interface between water and land were not of major interest to scientists in the 1960s.

Tragic was the death of 31 scientists and crew members when the research vessel No.5 *Kaiyo-maru* approached the emerging submarine dacitic volcano Myojinsho, located some 420 km south of Tokyo, during an apparent lull in activity. During the 1952–1953 370-day-long eruptive phase, the volcano had breached sea level repeatedly, forming temporary islands each time followed by their own explosive destruction. An answer for the disappearance of the ship was found when wooden planks were inspected, the only remains of the research boat. Pumice forcefully implanted into the wood left no doubt that a powerful explosive blast spreading laterally from the hot lava dome interacting with seawater had sunk the ship. In contrast to Surtsey, Myojinsho did not succeed in becoming an island.

The fact that a volcano can grow from below the waves to become an island, is, in mythological terms, a victory of the element fire over water. Indeed, questions concerning the nature of fire and water have been asked for thousands of years. Fire and water have been traditional symbols for fundamentally different powers of nature from the occident and orient, as in Chinese medicine. The antagonism between fire and water, however, has also been a major topic in the early decades when the scientific discipline of volcanology came of age, but then was largely neglected until quite recently.

Below, I will briefly review early ideas on the causes of volcanism ending in the early nineteenth century when the field of volcanology became firmly established. The subsequent section on how volcanoes work and how they are studied is intended to provide a snapshot of current views of pro-cesses and evolution of volcano-magma systems and major methods employed by volcanologists in their work. This discussion introduces the way the book is organized.

Early Perception of Volcanoes and Volcanic Actions

Volcanic eruptions have always filled humankind with fear and anxiety. Questions as to the nature and the roots of these natural phenomena have been asked for thousands of years. Myths about demons and gods in the interior of the Earth have developed in many countries from the Pacific to the ancient cultures of the Mediterranean (27, 399). The oriental traditions of a cleansing fire at depth are also reflected in the Christian view of purgatory in the Middle Ages. However, the beneficial sides of volcanic powers have also been the subject of some classic myths. According to Greek mythology, Prometheus provided the basis for human existence by presenting the fire he had stolen from the god Hephaistos from inside the Earth (a deed for which he was punished cruelly). Religious worship of volcanoes is still widespread, even in technically highly advanced societies as in Japan, where it is based on the Shintoist perception of animated natural objects (Fig. 1.1). The interior of mighty Mt. Shasta in northern California is believed by some to be the home of exotic communities and bizarre creatures. Volcanoes, the most dramatic natural phenomenon, appeal to all our senses. They always fascinate man, and emotions also come into play when analyzing volcanoes to understand their origins and workings.

When ancient Greek natural philosophers started to look rationally into nature, they understandably included volcanoes in their curiosity, the two most active ones in Europe being Etna volcano on Sicily – where some of the pre-Socratic philosophers lived – and nearly permanently active Stromboli Island in the Mediterranean. Etna, the most active large volcano in Europe, so fascinated the Greek philosopher Empedocles that he is said to have ended his life by jumping into one of its

The object of the following essay is to throw some light on those phenomena which consist in the development of subterranean activity in the form of Volcanoes and Earthquakes, the investigation of which appears to me of primary importance to the progress of Geological science.

G. Poulett Scrope, Considerations on Volcanoes, the probable causes of their phenomena, their laws which determine their march, the description of their products, and their connection with the present state and past history of the Globe, London, 1825

▶ Fig. 1.1. Pilgrims' trails and stations on the flanks of sacred Mt. Fuji

▶ Fig. 1.2. Early ideas on the causes of volcanic activity: heat panned by winds and sulfur or organic substances as fuel for the fire. These ideas developed by the early Greek naturalist-philosophers remained popular through the late 18th century. The background portrait shows Pliny the Elder based on an etching by A. Thenet (1684) reproduced in (334) (306)

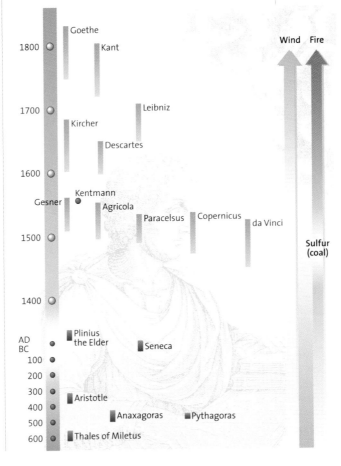

craters. Early ideas and perceptions of volcanic processes, developed by Greek and Roman philosophers and scientists, survived until the seventeenth and, in part, late eighteenth century (Fig. 1.2). The main elements of these long-held views of the underworld include *winds* panning the flames thought to be raging beneath the volcanic valves and *sulfur* as the burning substance. Various materials invoked later included coal or other bituminous substances. These views peaked in two masterly cross sections of the Earth by Athanasius Kircher, one of which is shown in Fig. 1.3.

Another milestone illustration reflecting early perceptions was the drawing of basaltic columns at Stolpen near Dresden (Saxony) (Fig. 1.4), picturing the most widely portrayed and debated lava columns in the history of earth science. The columns from this locality were mentioned as early as the first half of the sixteenth century by Agricola in his famous volume on mining. Stolpen also represents a pivotal locality for the Neptunist's views (see be-

▶ Fig. 1.3. Cross section through the Earth by Athanasius Kircher (1664). Water in the chambers (*hydrophylacia*, light blue) is panned by the central fire (light red) and small fire hearths (*pyrophylacia*), is heated, evaporates and recondenses in springs (176)

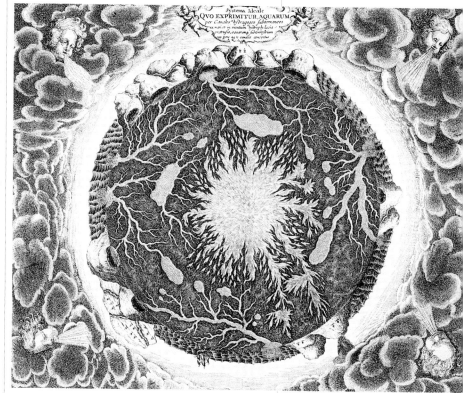

low). The rendition of the columns was masterly in objectivity in several respects. The columns were shown to not only be bordered by six, but also five or seven sides – as is common in nature. Secondly, the horizontal stripes of the columns may not be purely ornamental. Such ledges commonly surround basalt columns and are now known to be caused by the episodic advance of cooling fractures from the cooler brittle into the hotter ductile part of cooling lava flows (Fig. 1.5). This was excellent descriptive science.

The other side of the coin – and indeed typical of scientific endeavor in general – was the interpretation. This illustration manifested the old idea that basalt columns actually represent crystals grown from a solution – an idea that was to play a decisive role in the Neptunist-Volcanist debate. The philosopher Gesner – or his draftsman Kentmann – expressed this interpretation by the imaginary pyramids added to the top of the columns in the otherwise naturalistic drawing from the field. To make his point, the artist also showed a garnet crystal at the base of the group of columns, perhaps to drive the idea home of columns actually being crystals.

Neptunists, Volcanists and Plutonists

In the scientific discipline of geology, two major debates have raged over decades. The fundamental topic of discussion, when earth science came of age at the end of the eighteenth century, was the question whether or not columnar basalt had crystallized at low temperature out of water or had formed from hot molten rock welled out onto the surface from the interior of the Earth. The focus of this debate was the true nature of the roots of volcanoes. The protagonists and followers were named after the Roman gods, understandable in view of the classicist culture of the times. In the outgoing eighteenth century, the *Neptunists* strictly adhered to the idea of basalt columns

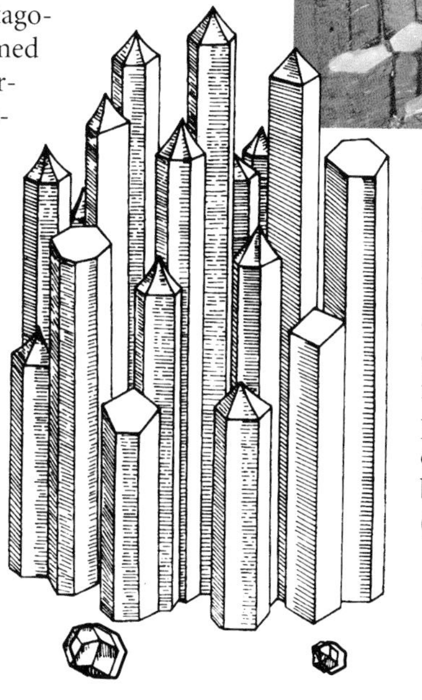

▶ Fig. 1.4. Drawing by Kentmann (1565) in (117) of basalt columns at Stolpen (Saxony), topped by imaginary prisms on most columns. Two garnet dodecahedra just in front of the columns

as having crystallized from water at low temperature. Their most prominent defender was Abraham Gottlob Werner, teaching at the mining academy of Freiberg (Saxony). Werner's dominating influence was based on his fame as the most inspiring geology professor of his time and in having developed the first system of minerals based on their external characteristics (416). The Neptunists also believed

▲ Fig. 1.5. Columnar basalt from the eroded lower part of a Tertiary volcano near Kamenický Šenov (Czech Republic) with horizontal ledges reflecting episodic advance of fractures during cooling

granite to represent the primeval rock type forming the basement on which all other rocks of the Earth's crust were deposited.

Werner's dogmatic approach in explaining the nature of volcanoes contrasted strongly with the empirical deductive methodology applied by his contemporary Sir William Hamilton (1730–1803), then the British ambassador in Naples (138). Vesuvius, which is sleeping at present, towering over the Bay of Naples, was unusually active during the second half of the eighteenth century. Hamilton was fascinated by the explosions in the crater and lava flows emitted from the volcano. He provided careful documentation of frequent changes in the crater region of Vesuvius (Fig. 1.6), as well as balanced and well-reasoned deductions as to the age of volcanism in this Campanian volcanic field of Italy. Hamilton became the father of modern volcanological scientific analysis and his beautifully illustrated masterpiece, *Campi Phlegraei*, is an impressive monument of early volcanological studies (138). Hamilton is better known to the general public on account of the famous love affair between his wife Emma and Lord Nelson (350).

The *Volcanists* were more pragmatic by deriving their interpretation from actual field observations. In 1765, Nicolas Desmarest, one of the pioneers, found that columnar prismatic basalt in the Auvergne (France) not only graded into the scoriaceous top of a lava flow but also noted baked soil beneath (Fig. 7.2). He therefore interpreted basalt as having formed by solidification of lava erupted on the Earth's surface and not by crystallization out of water (81, 416). His countryman, Jean-Etienne Guettard, had earlier (132) identified the youthful conical mountains near Volvic in central France as actual volcanic edifices associated with lava flows (Fig. 1.7). It is often overlooked that these two discoveries were made at a time when A.G. Werner (1749–1817) was still a young boy. The Neptunists' view of the world, that had a long history and was based on empirical evidence by Guettard and Desmarest, began to be doubted, Werner merely being by far the most influential protagonist. Both Neptunists and Volcanists, however, explained the fire (i.e., the high temperature of volcanoes) by oxidizing sulfide deposits, or burning coal seams, located not far below the surface of the Earth (Fig. 1.8). Volcanoes were also commonly regarded as rather superficial features, having developed very late during Earth's history.

The breakthrough in global understanding of magma emplacement came with James Hutton (1795, "no beginning, no end") who proposed a steady-state model of the Earth (164, 165). He thought that, in contrast to the prevailing dogma (favoring the unidirectional evolution of the Earth), basaltic magmas owed their origin to heat in the interior of the Earth and could rise at any time during the evolution of the Earth. This mechanism also held for granites. Hutton would have been delighted to know that granites may be as young as 340 000 years and more than 500 °C hot, as recently recovered in a geothermal bore hole at a depth of 2 860 m in Kakkonda (Japan) (293). Huttons followers were called *Plutonists*, alluding to the fact that they placed the roots of volcanoes much deeper in the Earth. Although history proved the plutonist's position right, their victory generated another dogma. Apart from the fact that Hutton also explained the cementation

▶ Fig. 1.6. Growth stages in the central crater of Vesuvius between July 8 to October 29, 1767 originally published by Hamilton (138) and redrawn in an 19th century engraving (authors collection)

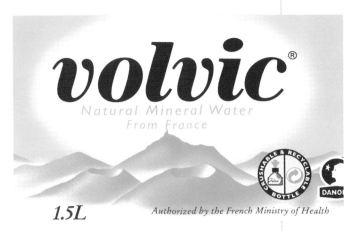

of sediments by heat, the explosivity of volcanic eruptions was henceforth thought to be due entirely to endogenic magmatic processes, especially degassing of magma during its rise to the surface.

The Neptunist view of volcanoes and their origin was practically dead by 1810 (save for Robert Jameson and his followers in Edinburgh). Volcanology was firmly established as a subdiscipline of earth science in the first quarter of the nineteenth century by Leopold von Buch, Alexander von Humboldt and Poulett Scrope who wrote the first textbook on volcanology (319). I will return to Scrope's excellent interpretation of the formation of basalt columns in Chapter 9. Here, I present his view of the roots of volcanoes, the role of updoming and fractures to serve as channels for the rising magma (Fig. 1.9). A recent example of drastic updoming was demonstrated by a rising probably dacitic dome during the spectacular April 2000 eruptions within the Usu volcanic complex (Hokkaido, Japan), shown in Figs. 1.10 a, b.

How and Why Do Volcanoes Work?

Volcanoes not only exist and erupt to provide jobs for volcanologists. They are the most visible proof that planet Earth is alive and well. This statement seems trivial to a scientist. The sensual appeal of volcanoes and volcanic eruptions to people in general is, however, fundamental to appreciate that we live on a planet that is very hot and dynamic in its interior. This planet constantly feeds material to the Earth's surface and the atmosphere that was generated by degassing of the Earth in the first place.

Present views of how a volcano works naturally depend on our constantly evolving understanding of the processes in the interior of a volcano. Beyond this truism, I want to stress three fundamental aspects of volcanism that form the matrix of many chapters in this book. A volcano is merely representing – albeit often impressively –

▲ Fig. 1.7. Quaternary scoria cones and trachyte dome Puy de Dome in the Auvergne (central France)

▼ Fig. 1.9 a,b. Cross section through three volcanoes. a) The area abcd below volcano e has lost heat during an eruption so that further heat flows into this area (*arrows*). Moreover, heat is released from deeper areas (*C – D*) more rapidly than prior to the eruption. When the vent is closed, the upper crust expands (*A – B*) so that new lava can be erupted through the fractures so formed. If a new vent opens nearby (e. g. f), heat will be removed from the older eruptive site. The term heat is used by Scrope synonymously with magma. b) Cross section through the Earth´s crust and magma chamber. Fractures open upwards at (*A*) and (*B*) during doming and tilting of larger crustal portions. In between (*C*) fractures will open downward so that magma can accumulate in the reservoir so formed and rise along fractures to the Earth's surface. After Scrope (1825, Figs. 9 and 33) redrawn in (306)

▲ Fig. 1.8. Miocene lake sediments, heated by burning coal seams. Some sediments have become partially melted and form porcellanite (Dobrčíče, Czech Republic)

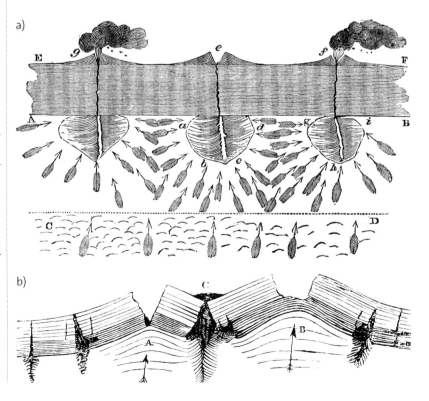

▲◢ Fig. 1.10 a, b. Updoming of a 10 km² area in the Usu volcanic complex (Hokkaido, Japan) caused by an intrusion during the 2000 eruption. Total uplift by 70 m was accompanied by drastic faulting and horst and graben formation. ▲ a) Shows a hill at a place where the road climbed gently prior to April 2000 (graph by Usu Prefecture). ◢ b) Displacements in the uplifted road (Fig. 1.10a) are the surface expressions of faults bordering the center of the uplifted horst

one of several arrested stages in the flux of matter and energy from the Earth's interior to its surface. Any attempt to understand the causes and consequences of volcanic eruptions must therefore consider the entire volcano-magma system, including the deep-reaching, much larger root system that lies beneath a volcano, as well as the atmosphere above (Fig. 1.11). Secondly, volcanology has fundamentally benefited from the models of global tectonics more than any field of the Earth sciences, driving home the point that we can only hope to understand why volcanoes work if we look at them in a planetary context. Thirdly, there is growing recognition that many processes within volcanic edifices and in the eruptive systems are governed not only by the input from greater depth (mass, composition and physical properties of rising magma), but also by interaction with the surface or near-surface environment.

The Volcano-Magma System
The magma-volcano system is divided here into four depth zones for simplicity (Fig. 1.11). In the root zones, magma is generated by partial melting of pre-existing older rocks (Fig. 1.12), notwithstanding the widespread belief occasionally taught in some schools that magma is siphoned off a liquid layer circling the globe beneath the crust. Parent rocks in the root zones vary widely in geological history and present composition. Both the contrast in source rocks as well as in the mechanisms of magma generation are best understood within their specific global tectonic framework. The composition of magmas, of the xenoliths brought up from depth and remote seismic sensing of the Earth's interior, are fundamental to pro-

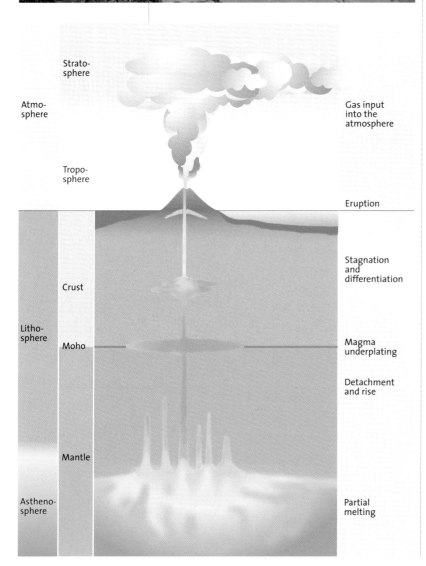

◢ Fig. 1.11. Schematic volcano-magma system (303)

vide answers on the nature of, and processes within, root zones. Taking account of processes within the root zones can help explain why a volcano forms at a particular site on Earth, at a specific time and with a characteristic magma composition, or how it erupts, quietly, effusively or highly explosively. The deep roots of volcanoes are treated in Chapter 3, in which the origin, composition and evolution of magmas and their crystallization are discussed.

Whether a magma erupts effusively as lava flow (Fig. 1.13), or dome extrusion, or explosively, depends on parameters such as composition, viscosity and rise speed, interaction with external water and, in particular, on the formation, expansion and bursting of bubbles formed when a magma becomes saturated with volatile compounds. All active volcanoes emit gases, sometimes tens of thousands of tons a day without erupting explosively. Obviously, degassing of a bubbly magma can take place several kilometers below the surface without fragmenting and erupting explosively. High ascent rates may be especially important prerequisites for an explosive eruption. Lava domes that ooze out slowly over years or decades, such as at Merapi volcano (Indonesia) or Santiaguito (Guatemala), appear to have lost most of their volatiles by the time the magma appears at the surface. At the other extreme, expanded lava particles and gases may rise as high as 40 km into the atmosphere in powerful explosive eruptions and gases can rise even higher. Magmatic gases, rheological properties of magmas and the eruptive behavior of volcanoes are discussed in Chapters 4 and 10–12.

The Global Framework of Volcanism

Convincing interpretations of volcanic processes can only be developed in the framework of broad hypotheses, generalizing theories being indispensable for the growth of science. Volcanic eruptions were responsible for the creation of the first crust on our planet about 4.6 billion years ago. This volcanic crust was subsequently modified by erosion, covered by sediments, folded and buckled by mountain-building, and transformed through metamorphism. This has been common knowledge in Earth science for some time. The fact that processes such as crust formation occur on a grand scale day by day is a relatively new development in geology, springing from the second major controversy in the history of Earth science. Why Surtsey, and indeed the entire micro-continent of Iceland, formed at that particular spot was not so

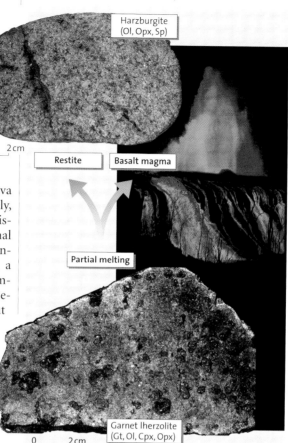

Harzburgite (Ol, Opx, Sp)

Restite

Basalt magma

Partial melting

Garnet lherzolite (Gt, Ol, Cpx, Opx)

0   2 cm

Fig. 1.12. Partial melting of mantle rock (garnet lherzolite) generates basaltic magma erupted in lava fountains at Kilauea volcano (Hawaii). Crystalline residue (restite) represented by xenoliths (olivine nodule) found in volcanoes (306). Photo of lherzolite rock slab from (438).
Abbreviations:
Ol = olivine, Opx = orthopyroxene, Sp = spinel, Cpx = clinopyroxene, Gt = garnet

Fig. 1.13. Road on the south flank of Kilauea volcano (Hawaii National Park), flooded by lava flows erupted by Pu'u 'O'o volcano, active since 1983 (commercial postcard)

a)

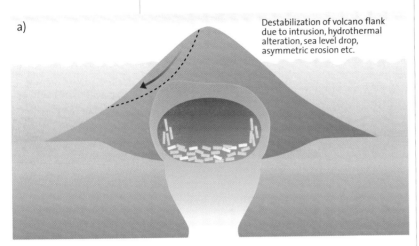

Destabilization of volcano flank
due to intrusion, hydrothermal
alteration, sea level drop,
asymmetric erosion etc.

b)

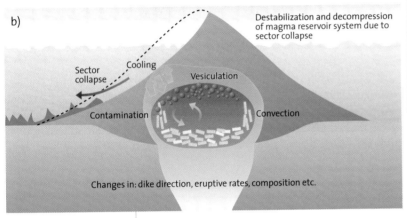

Destabilization and decompression
of magma reservoir system due to
sector collapse

Cooling

Sector
collapse

Vesiculation

Contamination

Convection

Changes in: dike direction, eruptive rates, composition etc.

▲ Fig. 1.14 a, b. Schematic of volcano flank and magma chamber showing a) likely trace of future flank collapse. A sector collapse b) affects the entire volcano-magma system including asymmetric cooling, increasing vesiculation, increased rates of crystallization due to degassing, changes in eruptive patterns and composition of lavas erupted, rates of eruption etc

evident in the mid-1960s. It was only obvious that two of the most active volcanic fields on Earth, Iceland and Hawaii, developed in the Atlantic and Pacific Ocean basins, reminders that magma-forming processes may be especially powerful beneath oceanic crust. To understand the origin of large composite volcanic massifs, such as Iceland or Hawaii, Vesuvius, Mt. Fuji or Mount St. Helens, volcanoes had to be viewed in their overall global tectonic framework.

Alfred Wegener's fundamentally new view of "drifting" continents, followed by Arthur Holme's powerful models of thermal convection and decompression melting beneath the oceanic crust – expanded and superseded by seafloor spreading, plate tectonic and plume models of the 1960s, provided the basic context in which to explain global volcanism. Many fundamental aspects of volcanism such as the composition of magmas, their formation, production and eruption rate, gas inventory, eruptive mechanisms, and so on suddenly became amenable to rational and more realistic analysis. Early on in the plate tectonic debate, three major volcano-tectonic settings could be distinguished: mid-ocean ridges, subduction zones and intraplate areas. I will discuss plate tec-

tonics in more detail in Chapter 2 and the three main types of global or volcano-tectonic environments in Chapters 5 – 8.

Nature – Nurture

In psychology and anthropology, the relative influences of *nature* (genetic makeup or, in short, heredity) and *nurture* (family-societal and other educational and environmental factors) in determining the character of a person have been debated for ages. In volcanology – as understood since the beginning of the nineteenth century – the spectacular power of volcanic eruptions as well as the complex evolution of volcanic edifices in time, space and composition were traditionally thought to be governed entirely from within. There is growing evidence, however, that external factors play a major role in many processes during the evolution of a volcano-magma system. For example, the formation of a free gas phase depends critically on pressure exerted by the overburden of a water or rock column. This is the reason why eruptions on the seafloor at water depths greater than say 2 000 m, in shallow water, or on land result in very different products. Volcanoes evolve by long periods of construction, punctuated by short and sometimes violent destructive events, during which major segments of a volcanic edifice may collapse catastrophically within a few minutes. Not only may huge volumes of groundwater rush into the hot interior of the decapitated volcano. The rapid decompression of a high-level magma reservoir during sudden unloading of a major flank segment may lead to increased vesiculation, cooling and crystallization as well as interaction of magma with seawater and so on (Fig. 1.14). Thus, lavas may erupt following a sector collapse that otherwise might have evolved differently or have never erupted. Indeed, the entire stress field of a volcano will have to adjust to the new load distribution and dike/rift directions and eruptive frequencies and erupted volumes may change drastically following major flank collapse.

Flank collapses themselves may be facilitated by asymmetric erosion of large volcanic complexes. For example, there is abundant evidence on land as well as in drill cores through the submarine volcanic apron that the large volcanic island of Gran Canaria (Canary Islands) was eroded preferentially in the north from its basaltic shield stage from the middle Miocene to the present. The reason is the apparent constancy of the trade wind directions from at least some 15 million years ago to the present (Fig. 1.15). The prevalence of sector collapses on the northern flanks of the Canary Islands may be due to repeated oversteepening and thus destabilization of the flanks because of the asymmetric erosional removal. Even the change in major dike swarm directions on Gran Canaria from a radial pattern in the Miocene shield stage to northwest-southeast directions from the late Pliocene to the Holocene roughly parallel to the strike of the eroded flank of the volcano may thus be ultimately climate-controlled. Such an interpretation is highly speculative since both external and internal forcing or positive or negative feedback loops may determine the stress field of a volcano. This unorthodox interpretation of the possible major influence of climate on the structural evolution of major volcanic edifices is intended to emphasize the importance of constant interplay between internal and external forcing mechanisms in volcanoes although we still know very little about the causes and negative and positive feedback loops.

The most obvious and nowadays accepted interaction of external and internal factors is the ubiquitous encounter of rising magma with water. This may be groundwater and flow of surface/groundwater into partially collapsed conduit systems/magma chambers during the course of an eruption, eruption of a volcano in a lake and so on, as detailed in Chapter 12. Glacier-clad volcanoes are particularly dangerous because of the potential for quickly generating floods of water. This was demonstrated tragically by the eruption of Nevado del Ruiz whose eruption on 13 November 1985 caused the extinction of a town and death of 23 000 people (Chap. 13). Many phenomena in active volcanoes are now known to be influenced by high rates of precipitation. The fact that climate can be modulated by major Plinian eruptions has been a major insight, especially following the eruption of Pinatubo in 1991. Even partial melting in the mantle, the source regions of the basaltic magmas, may be influenced by major unloading of the lithosphere as during melting of major ice sheets.

Clearly, the interdependency of external and internal processes in volcano-magma systems although little appreciated at present is likely to develop into a major research direction in the future. I will return to the topic at the end of Chapter 4.

How Do Volcanologists Work and Why Do They Work on Volcanoes?

Volcanologists are people whose job and/or passion is to work on volcanoes. Most of them work in universities, other research institutions and in observatories. Many are motivated – or hired by an institution – because of the need to prevent disasters, others by a strong interest in the most visible manifestation of a dynamic Earth. For geo-

morphologists, the definition of a volcano will be quite different than for a petrologist, and a geophysicist has a different mental picture of the interior and workings of a volcano than a geochemist. Consequently, the methods employed to study volcanoes vary widely. Most volcanologists study on-land volcanoes, but the study and monitoring of underwater volcanoes is also an increasing field of inquiry. The form of volcanoes is not a major topic in present-day research – except for the scars on volcanoes, where huge flanks have slipped off – but the morphology of a volcano tells us a lot about the physical properties of the erupting magma and specific eruptive processes. For this reason, and in addition to Chapters 5–8, the main types of volcanoes and their explosive and effusive deposits are introduced in Chapter 9 as an observational framework for Chapters 10–12. Planetary volcanology has developed as a field of its own with many lessons learnt from active volcanism on Io, a moon of the planet Jupiter, but also by studying volcanic features on Mars and the Moon.

Methods and emphasis of volcanological research have evolved over time, sometimes marked by sudden changes in paradigm. Major increases

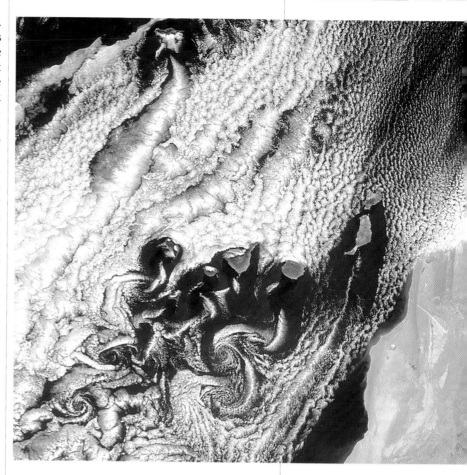

▲ Fig. 1.15. Satellite view of cloud pattern around the Canary Islands, clouds piling up on the steep northern flanks due to the prevailing north-northeast directions of the trade winds. Dominant wind directions appear to have been constant since the middle Miocene at least for Gran Canaria. The higher erosional rates on the northern flanks resulted in steeper slope gradients and higher susceptibility for flank collapse. Image provided by SeaWiFS Project, NASA/Goddard Space Center, and ORBIMAGE

in our understanding of how volcanoes work are most commonly triggered by large and well-studied eruptions. Some of these eruptions produced previously unknown or greatly underestimated physical effects, some of them being discussed in Chapters 10–14. Examples include the eruption of Mt. Pelée (Martinique, 1902; nuées ardentes, Chap. 11), Taal volcano (Philippines, 1965, magma-water contact, base surges, Chap. 12), Mount St. Helens (Washington, USA, 18 May 1980 lasting until 1986; sector collapse, lateral blasts, eruption forecast Chap. 10), El Chichón (Mexico, 1982; sulfuric acid aerosols, climate impact, Fig. 1.16 Chap. 14), Pinatubo (Philippines, 15 June 1991; another milestone in eruption prediction and effective mitigation and in understanding volcanic climate forcing Chap. 13), Montserrat (Lesser Antilles, 1995 to present; Fig. 1.17; mechanisms of dome growth, pyroclastic density currents), Usu volcano (Japan, March 31–August 2000; phreatic eruptions and major ground deformation, Fig. 1.10, Chap. 12) and Miyakejima (Japan, July-August, 2000, lateral magma withdrawal, caldera collapse and ensuing phreatomagmatic eruption Chap. 9).

Major advances have also been made through long-term pioneering studies of active or dormant volcanoes over several decades with a variety of methods resulting in a steady stream of papers on subaerial and submarine processes taking place in large active volcanoes. The foremost example is Kilauea volcano on the island of Hawaii that has no equal on Earth with respect to its voluminous activity (Chap. 6). Its study has also spawned more new insight into the architecture and dynamics of active volcanoes, flow and crystallization of lava and gas evolution than any other volcano on the globe. An example of a dormant volcano that has been studied for several decades is the Laacher See volcano in the Eifel (Chap. 11).

Other rapid changes have come with the development of new analytical tools such as mass spectrometers, their employment being vastly accelerated by the necessity to carefully analyze lunar basaltic rock samples, electron microprobe, broad-band seismometers, remote sensing using various types of instruments on satellite platforms, global positioning system (GPS), high resolution aerial laser scanning and computer power, to name a few. These advances in technology help to accelerate our ability to look into basic processes operating in volcano-magma systems in more detail than before. Some important methods have been with us for several decades. Newer methods include ion-probe, single crystal dating, and analytical probing into crystals with a variety of methods to determine their trace element and isotopic signatures as well as focused study of glass and fluid inclusions. These latter techniques have the potential to yield results that will enable us to better understand the origin of magmas and their evolution prior to eruption.

New ideas and creative research strategies have often helped to advance our understanding of volcanic processes. Traditional work focused on the study of the *structure*, *stratigraphy* and *lithology* of volcanoes and their rocks, and *observation* and *description of volcanic eruptions* and their *impact on the environment*. These studies still represent the backbone of current volcanological work for the simple reason that the history of a volcano and its age pattern and types of eruptive behavior continue to offer the best clues to forecasting its future behavior. Objects of study are lavas, intrusions, pyroclastic rocks, and volcanic gases and their sublimates. For example, answers to the questions as to the nature of processes determining the generation of volcanic islands, such as Surtsey, or volcanoes in general, still need to be based to a large extent on precise descriptions and interpretations of the deposits resulting from eruptions. Examination of the tuffs, breccias, dikes and lava flows is a form of empirical inductive research pioneered by Hamilton in the outgoing eighteenth century.

The roots of volcanoes and their proverbial engines or hearts, the magma chambers, are located too deeply in the Earth to be observed directly. The understandable dream of some scientists not only to observe volcanoes from a safe distance, but to travel with the magma from the source regions in the Earth's mantle to the surface is best treated by novelists (397). *Remote sensing* has therefore become a major approach in studying active volcanoes. From the Earth's surface it is carried out by employing seismic, gravimetric, magnetic, electric, and other methods. From space, volcanoes are observed with a variety of instruments on satellite platforms providing us with important tools to study the dynamic behavior of volcanoes even in remote areas. Observations that allow us to detect and quantify volcanic processes and help to predict volcanic eruptions (Chap. 13) include: tracking of airborne ash clouds, deformational evolution of volcanic edifices by radar,

▼ Fig. 1.16. Title page of *Geophysical Research Letters* reporting on the climatic effects of volcano El Chichón (Mexico) erupted in 1982 based on sampling of aerosols by high flying aircraft and satellite monitoring (TOMS, housed on Nimbus platform). With permission by American Geophysical Union

Geophysical Research Letters

Climatic Effects of the Eruption of El Chichón

volume 10 number 11

NOVEMBER 1983

infrared radiation of higher temperature areas on active volcanoes, quantitative detection of gas emissions especially SO_2 using the total ozone mapping spectrometer (TOMS), mapping the generation and evolution of aerosol clouds resulting from major Plinian eruptions, and mapping the surface of volcanoes with a spatial resolution of better than 10 m.

Today, empirical research into the nature of volcanic systems is increasingly accompanied by theoretical and experimental volcanology, both of which also trace their beginnings back some 200 years. This type of work includes *experimental simulations and numerical modeling* to determine physical and thermodynamic properties of magmas, lavas and pyroclastic systems. Experiments to understand volcanic processes and the generation of melts and crystallization of melts were conducted as early as the nineteenth century, while experiments by studying phase equilibria only started in the beginning of the twentieth century. The past two decades have seen an increase in analog experimental studies focusing on a large variety of volcanic processes. These include dynamic processes in magma reservoirs, injection of magma into volcanic edifices, transport of lava flows or clastic systems under different conditions on dry surfaces or under water, bubble formation and fragmentation processes in eruption columns and pyroclastic flows and volcano-tectonic processes such as caldera and sector collapse. The analog materials used by experimental volcanologists range widely from flour to natural pine resin, gum resin, polyethylene glycol, gelatine, and some scientists work with volcanic ash, solidified lava or hot silicate liquid in their experiments.

Volcanoes, volcanic eruptions, volcanic impacts on the solid Earth, the hydrosphere and the atmosphere are exceedingly complex and contain a huge number of variables. It is thus impossible to provide a concise definition of volcanoes or present a robust model of the processes, say within a moving pyroclastic flow. Interdisciplinary work is the most appropriate way to study volcanic processes. In reality, however, this has proven to be difficult. During the late eighteenth century, science and humanities were very close and poets and scientists "spoke the same language". Some even excelled in both science and humanities, as, for instance, the famous German poet and scientist Johann Wolfgang von Goethe. Today, the modelers, experimentalists, material analysts, remote sensers, field geologists and so on in a single research field such as volcanology often find it difficult to productively communicate, since their scientific approach relies on contrasting ways to search for the most convincing answers. Modelers may look down on field-oriented volcanology as merely descriptive and of local relevance only. Field volcanologists may complain that the work of some modelers or experimentalists, whose work by necessity includes a limited number of variables that can be handled even by supercomputers, is far removed from the real world. Needless to say, both approaches are indispensable when searching for robust answers. Most subaerial volcanoes erupt predominantly explosively. The analysis of explosive volcanic eruptions by examining their deposits has thus been a major endeavor of volcanological research for the past three decades, combining both empirical field and laboratory data and pioneering modeling. I have devoted four chapters to the discussion of this topic: one general (Chap. 4) and three more specific (Chaps. 10–12), treating the most common types of explosive eruptions and the resulting products.

Examples of interdisciplinary studies with exciting results include the impact of slow or highly explosive degassing of volcanoes and the release of vast volumes of gas into the troposphere or even the stratosphere where they have a major effect on climate and the stability of ozone (Chap. 14). Large-scale tectonic earthquakes are generally not thought to be directly associated with volcanic eruptions, even though this is a common question volcanologists are asked by the media. Recent work shows that the spatial, temporal, and causal relationships are probably much more closely connected than previously thought, although at the moment there are just a few glimpses of the complex relationship with these two major dynamic paroxysms of the solid Earth. I will return to the subject when briefly discussing different triggering mechanisms of volcanic eruption (Chap. 4). Similarly, physical processes that could explain the wide variety of so-called volcanic earthquakes are still poorly understood. Recent interdisciplinary work may reveal more convincing solutions to the problem.

The Impact of Volcanic Activity on the Environment and on Society

The last three chapters concern the direct and indirect impact of volcanoes on the environment and on society. In the media, volcanic eruptions are reported only when people or buildings are harmed. Media interest is also high when political or social problems develop that result from the increasing activity of a volcano or possibly from the necessity of evacuation. Long-term evacuation is sometimes unavoidable, despite the hardship for the people concerned, such as during the eruption of Soufrière Hills volcano on the island

▲ Fig. 1.17. Volcano Soufrière Hills on the Caribbean island of Montserrat active since 1995. Small pyroclastic flows descend the slope. Two months after the photograph was taken (May 1997), the entire area in the foreground was swept by pyroclastic flows (see also Fig. 11.26)

One major task of hazard-focused work is the assemblage of hazard maps. This includes the analysis and mapping of (1) products of previous volcanic eruptions, (2) modern theoretical insight into transport mechanisms, (3) the energy involved in the various types of eruptive mechanisms and transport processes, (4) the ability to forecast likely energies released, and (5) pathways based on an analysis of older deposits. These hazard maps are increasingly enhanced by computer-generated maps showing the probable location of the anticipated layers in reference to infrastructure, population, or economic facilities exposed to the various levels of hazard.

Public education has become another branch of hazard-focused volcanology to allow people to prepare for a possible eruption including the preparation for evacuation. This is commonly a major problem because people are reluctant to be evacuated, unless they have been convinced by widespread information – or by a drastic order. Volcanologists working in areas with active volcanoes also have to deal with a wide range of administrative civil servants and people with a political mandate, a job quite different from the painstaking work of a scientist investigating the nature of eruptive processes.

Concerning such issues as global warming and greenhouse effect, volcanologists are often consulted on the possible climatic impact of volcanic eruptions. If our Earth had not degassed since its formation 4.6 billion years ago, there would be no atmosphere, no water and, thus, no life on Earth. Our climate repeatedly changed drastically during the course of Earth history and its possible future changes have unforeseeable consequences for mankind. Are these changes also caused by volcanic eruptions? Hence, the study of volcanic forcing of climate is of increasing interest for assessing the causes of climatic change. This is relevant where anthropogenic short-term climate change has become a major concern of mankind. The relationship of climate, ozone, and volcanoes is treated in Chapter 14.

A seeming paradox, however, is that people have always benefited from volcanoes much more than they have suffered from their eruptions. Five of the major benefits of volcanoes are treated very briefly in the final Chapter 15: (1) geothermal energy; (2) ore deposits; (3) volcanic soils; (4) volcanic raw materials and
(5) their "beauty".

of Montserrat since 1995, or on Miyakejima since September 2000. In the latter case, a return of the population is impossible because of continuing massive SO_2 emissions (Fig. 4.25). The most common types of volcanic hazard, the chief monitoring methods, and success in prediction and evacuation are the main topics of Chapter 13.

The number of scientists concerned with the mitigation of volcanic disasters or who work full-time in universities, observatories or other governmental or nongovernmental institutions on hazard aspects has increased significantly in recent years. One reason for this new emphasis is the recognition that so-called natural disasters are mainly due to the inability of a community or larger segment of society to protect themselves against a natural hazard. Moreover, the awareness is increasing that modern societies are much more vulnerable to natural hazards. Insight is growing in some societies that people have a right to be informed of potential natural hazards enabling them to plan more prudently for the future or prepare themselves mentally for the possibility of evacuation in the case of a crisis. This growing acceptance of responsible authorities to fully inform the public represents a significant departure from traditional policy attempting to retain hazard information under the guise of preventing uncontrolled and disastrous reactions.

Plate Tectonics

In 1912, Alfred Wegener took the scientific community by surprise in postulating that the present continents had once been united in a mega-continent named *Pangea* that broke up about 200 million years ago into smaller fragments – the present continents –, which drifted apart and still do so today. This went against all contemporary scientific belief of the permanence of ocean basins and continents. A fundamental argument was the very good morphological and structural fit of South America and Africa and many other geological similarities between the southern continents. Wegener's revolutionary vision of a mobile dynamic Earth continued to be opposed by most scientists for almost fifty years. His ideas were dramatically revived during the 1960s. The hypothesis of *continental drift* was then greatly expanded and superseded by the models of *seafloor spreading* and *plate tectonics*, both of which have come to revolutionize all of Earth sciences. This new view of a dynamic planet also heralded a new phase in our understanding of the location and origin of volcanoes. The striking difference in their eruptive behavior could now be related to different magma sources, processes of magma generation and thus magma compositions. The time was passed to view volcanoes such as Surtsey or Mauna Loa as isolated local phenomena.

Alfred Wegener did not regard volcanoes, their composition and the origin of magmas as fundamental to his model. While his conception of a mobile Earth turned out to be correct, the models of seafloor spreading and plate tectonics are based on geological and geophysical observations in the present ocean basins, where the crust is entirely volcanic (Fig. 2.1). These hypotheses constitute the most important change in geological paradigm of the last 100 years. The story of the development of these ideas has been recounted numerous times and is part of basic general geological education. I will here focus on those aspects relevant for understanding the origin and evolution of magmas beneath volcanoes –

Don't be alarmed at the heat, my boy.
Jules Verne, Journey to the center of the Earth, 1864
Translated from the French

▼ Fig. 2.1. Major lithospheric plates and global distribution of some active and dormant volcanoes

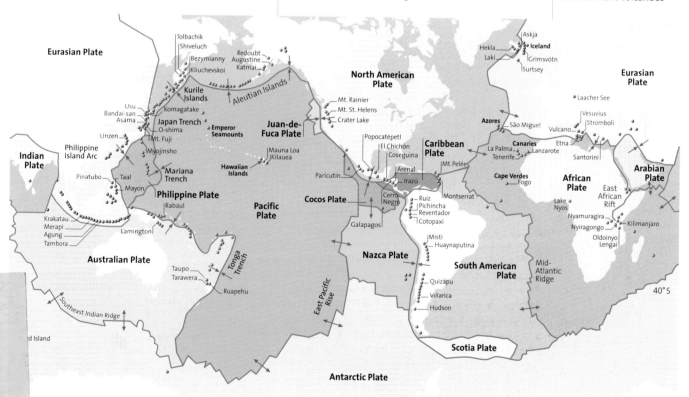

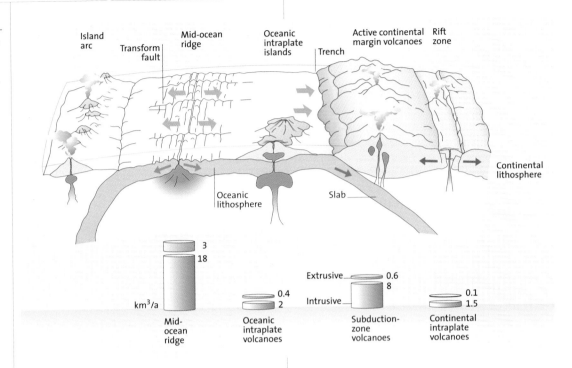

► Fig. 2.2. Divergent (mid-ocean ridges) and convergent (subduction or Wadati-Benioff zones) plate margins (after 269). Annual magma production rate (km³/a) after several sources (303).

processes that occupy a central position in the new global view of Earth science.

Below, I will briefly review fundamental geodynamic relationships between volcanoes and the structure of the Earth, as well as dynamic processes in the Earth's crust and the mantle, prior to discussing more specific aspects of volcanoes and volcanic eruptions. Causes of volcanic eruptions and volcanic processes in general cannot be understood without a rudimentary understanding of the composition of magmas and gases and the distribution of volcanoes on the planet. In other words the entire framework of the magma-volcano system needs to be laid out before treating the volcano itself.

Fundamental is the observation that not only the ca. 550 active subaerial volcanoes (340), but all Quaternary, Tertiary, and older volcanoes and volcanic fields are not distributed haphazardly on the surface of the Earth.

The Conveyor Belt of the Mid-Ocean Ridges

One of the breakthrough discoveries made during the 1960s was the recognition that changes in the orientation of the magnetic field of the Earth occurred every few hundred thousand to million years. In other words, the *magnetic* North Pole (to which compass needles currently point) changes position from the *geographical* North Pole (or close to it) to the *geographical* South Pole. These opposite orientations, occupied by the magnetic field of the Earth in irregular intervals of its history, can be well-documented because they are, so

to speak, frozen into lava flows. When hot lava flows cool, magnetic minerals crystallize and the submicroscopic magnetic domains in the mineral phases align along the direction of the prevailing orientation of the magnetic field.

When magnetic orientations were measured in volcanic and plutonic rocks of the ocean floors, scientists were struck by a major puzzle; the oceanic crust turned out to be characterized by stripes with alternating opposite magnetic orientations. These stripes of normal and reverse polarity paralleled the mid-ocean ridges (Fig. 2.2). Why should the ocean floor show parallel belts of opposite orientation, when, according to the prevailing idea, the ocean floor was very old? Frederick Vine and Drummond Matthews from England discovered in 1963 that the magnetic zebra patterns on the sea floor on either side of a mid-ocean ridge are symmetrical, with the mid-ocean ridge representing the mirror plane. When comparing the relative widths of these normally and reversely magnetized stripes, it became apparent that relatively narrow stripes corresponded to short time intervals and larger ones to longer periods of constant orientation of the Earth's magnetic field.

These discoveries provided compelling evidence that the oceanic crust of the deep sea (three quarters of the entire Earth's surface) could not be billions of years old; it must have formed in the recent geological past. And this crust was apparently not formed where it is today but rather along welts in the middle of the ocean basins. Hence it had to migrate away from this zone, as on a constantly moving conveyor belt. Moreover, it soon became apparent that the contrasting widths of these stripes corresponded to periods of normal and reverse polarity preserved in deposits on land as discovered a few years earlier. All this combined to show that the Mid-Atlantic Ridge (roughly central in the Atlantic basin) and the asymmetrically located East Pacific Rise, are young geological features where new oceanic crust is being formed day by day. These revolutionary concepts necessitated an expansion and major modification of the conventional view of a layered Earth.

The Layered Earth

The velocity of seismic waves depends mainly on the density of a rock. Acoustic compressional waves, generated during earthquakes or artificial explosions, and which migrate through the Earth, increase their velocity with depth in a stepwise fashion. These abrupt changes in velocity allowed geophysicists to subdivide the Earth into three main layers or shells of different density (Fig. 2.3):

- The Earth's crust, the thin outer rind of the Earth with a mean density of $2.67\,g/cm^3$, averages about 25 km thick in the continents, but can reach up to 70 km beneath some mountain ranges. The mainly basaltic crust in the ocean basins is only 5 to 7 km thick, with a mean density of $2.8\,g/cm^3$. The crust is distinguished from the underlying mantle along the so-called *Mohorovičić discontinuity* (Moho), a surface defined by a sudden increase in wave velocity from about 7 to over 8 km/s.
- The Earth's mantle underlying the crust consists of Fe- and Mg-rich silicates, chiefly olivine in the upper 660 km. At greater depth the high pressure forms of olivine with closer packing of the atoms dominate. The mantle is about 2 870 km thick and has a mean density of $4.6\,g/cm^3$.
- The core, probably consisting largely of nickel and iron, has a radius of 3 480 km and a mean density of $10.6\,g/cm^3$. Its outer shell is liquid, its interior core solid.

How have these layers of different density developed? Current consensus is that the Earth originally aggregated from cold cosmic dust, about 4.6 billion years ago. This protoplanet was melted by meteorite impacts, radioactive decay and gravitational energy, causing differentiation of the Earth's core. The end result was a more or less homogeneous planet. In the gravity field of these masses, the shells probably developed by density separation. This was via the process of partial melting, which was not only active in the early history of the Earth, but is continuing today, as shown by many volcanic eruptions each year.

Because the Earth is very much hotter in its core than the outer mantle or the crust, heat migrates along thermal gradients to the Earth's surface and is then radiated into space. The thermal gradient does not, however, increase linearly downward. If the measured crustal temperature increase of about 3 °C per 100 m would be constant to the center of the Earth, the core would have the unbelievably high temperatures of almost 200 000 °C, instead of the likely 5 000 to 6 000 °C. For comparison, the surface temperature of the Sun is estimated to be about 5 500 °C.

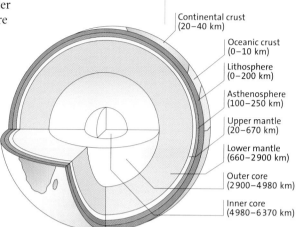

Dynamic Subdivision of the Earth

When looking at a globe, one is struck by the contrast between continents and ocean basins. The belts of seismic and volcanic activity stand out strongly. Jason Morgan (241) subdivided the outer rind of the Earth into a number of "plates", whose boundaries are characterized by earthquakes and often also by volcanic activity. These plates are constantly moving against one another (Fig. 2.1). Many plate boundaries do not coincide with the boundaries between continental and oceanic crusts. Some, like the Mid-Atlantic Ridge, are roughly parallel to the continent-ocean boundaries, while others do not, such as the East Pacific Rise, which impinges on the North American continent at an acute angle.

Because the Earth does not expand, the constant production of new crust must be compensated by destruction of crust at some other place. Wadati-Benioff zones (Chap. 8), recognized since the 1930s, are characterized by planes of earthquake foci (hypocenters), dipping beneath continents or island arcs (such as the American continent and the Japanese Islands, respectively). These Wadati-Benioff zones became recognized as counterparts to the axial zones. Ocean plates, composed of igneous ocean crust and underlying mantle, dip downward along these zones characterized morphologically by deep-sea trenches alongside convergent continental margins or island arcs. A major portion of the sediments deposited on top of the igneous crust in the ocean and shed from land is also dragged down into the mantle.

► Fig. 2.4. Distribution of P (*red*) and S (*blue*) velocities of seismic waves and density (*violet*) and a cross-section through the Earth (*left*). Subdivision of the upper mantle by shear wave velocity at *right* (269)

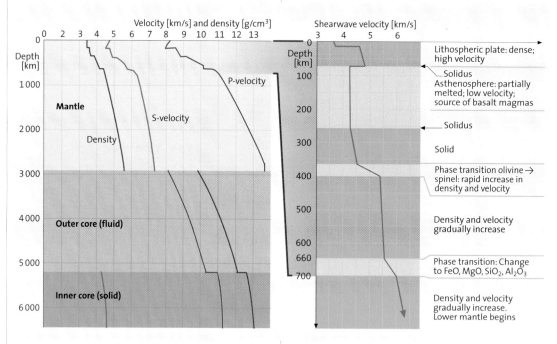

These new concepts necessitated the introduction of a more dynamic subdivision of the outer rind of the Earth (Figs. 2.3–2.5). The crust and the uppermost part of the mantle are mechanically coupled and are termed *lithosphere*. The lithospheric plates of plate tectonics are visualized as relatively cool rigid, highly viscous outer rinds of the Earth, being underlain by the more mobile, weaker *asthenosphere*. There are several different definitions of lithosphere, however (11).

The *seismic lithosphere* is defined by a zone of higher seismic velocity, contrasting with the underlying *low velocity zone* (LVZ). Its thickness increases in the ocean basins from less than twenty km at the mid-ocean ridges, to more than sixty km along passive continental margins (the oldest oceanic lithosphere being about 180 Ma old). Beneath the so-called *cratons* (the old shields of the continents) the lithosphere reaches thicknesses in excess of 200 km. The temperature at the base of the lithosphere is estimated as 600–650 °C.

The *elastic lithosphere* is derived from the measured rise and sinking of the Earth's surface when subjected to loading and unloading (such as by huge volcanic edifices as in Hawaii – Chap. 6 – or by mountain ranges). The elastic lithosphere in the ocean basins is similar in thickness to that of the seismic lithosphere, but is a little thinner (30–40 km) under continents. In the ocean basins, the base of the elastic lithosphere roughly corresponds to a temperature of about 500 °C for dry olivine rheology (Chap. 4). The lower sharp boundaries of the seismic and elastic lithosphere are

probably due to a change in the mineral composition of mantle rocks and/or due to a phase change.

The *thermal lithosphere*, i.e. the cool outer rind or thermal boundary layer (TBL), is determined by a conductive gradient with a basal temperature of about 1 280 °C. In other words, the thermal lithosphere is about twice as thick as the seismic and elastic lithosphere. In the ocean basins, its thickness increases from a few kilometers in the center to about 100 km at the margins, thickening to about 150 km beneath the continents.

Plates along active continental margins that descend beneath continents or island arcs are called *slabs*.

A *low velocity zone* (LVZ) is almost universally recognized to be characterized by up to 2 % wave velocity decrease beneath the lithosphere. Beneath cratons, it begins at about 150 km and locally as deep as 400 km. The LVZ is roughly equivalent to the asthenosphere, the "soft" layer beneath the rigid lithospheric plate. The low seismic velocity is commonly explained by an increase in temperature approaching the melting point of mantle peridotite. Hence, partial melts (magma) may be present in this zone, although recently it is also interpreted as a decrease in H_2O concentrations, caused by partial melting (e.g. 169). The asthenosphere is widely thought to be the source region for geochemically depleted basaltic magma, which intrudes and erupts along mid-ocean ridges (Chap. 5).

A second layer that has recently attracted much scientific attention is the D-region – called

D double prime (D'') – at the base of the mantle above the outer core. This layer, typically 200 – 300 km thick, is characterized by enhanced scattering of seismic waves. Locally the layer may be very hot and even contain traces of magma.

The lithosphere consists of about 16 larger and many smaller, relatively rigid plates, which move against each other. Most large plates, such as the North American and the Eurasian plates, comprise continental as well as oceanic lithosphere (Fig. 2.1).

These discoveries provided robust evidence that continents and ocean basins were not formed at one time, remaining stationary for billions of years. Continents have drifted apart, became welded together and broken up repeatedly. The large mountain chains, such as the Alps in Europe or the Himalayas, are the crumple zones between giant colliding plates. Energy accumulating during such collisions is released in earthquakes, as in Afghanistan, Armenia, Iran, India or Turkey, sometimes with catastrophic results. The present Atlantic Ocean basin, which started to form only about 200 million years ago when a large super-continent called *Gondwana* broke up, widens by about 2 cm/year. The migration of the lithospheric plates, their constant formation along mid-ocean ridges and their plunge into the mantle along Wadati-Benioff zones, are governed by deep-reaching motions in the Earth's interior (Chaps. 6 – 8).

Distribution of Volcanoes on the Earth's Surface

Alexander von Humboldt (163) noted that geologically young Quaternary and Tertiary volcanoes and volcanic fields are not distributed evenly or randomly on the surface of the Earth. He considered that the concentration of volcanoes was due to deep-reaching causes, an idea that strongly contrasted with that of his teacher, Abraham Gottlob Werner, and disciples of the Neptunist school, who thought that volcanoes had shallow sources and only grew late during Earth history.

Humboldt's view has been confirmed time and time again. In some areas, such as geologically old shields or *cratons*, in which the crust is maybe more than 1 billion years old (e.g. Central Canada, Sweden, Finland and Western Australia), almost no volcanic eruptions have occurred in the recent geological past. Young volcanoes preferably occur in specific zones. These are foremost island arcs or

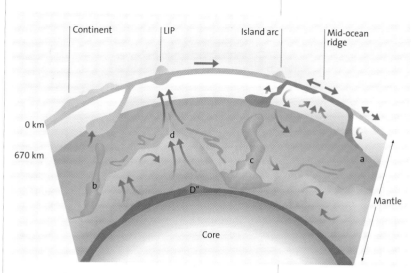

▶ Fig. 2.6. Comparison of the number of historic eruptions and mass of annual global volcanic production in three different plate tectonic settings: subduction zones, intraplate environments and rift zones, including mid-ocean ridges (339)

convergent continental margins, such as around the Pacific Ocean or in the Mediterranean, in rift zones and their uplifted shoulders, such as the Rio Grande Rift, the Rhine Graben or the East African Rift or, by far most abundantly, in the ocean basins. Today, this geographic subdivision can be interpreted in a geodynamic framework, which greatly helps us to better understand eruptive rates, chemical compositions of magmas, or contrasting explosivity in volcanoes.

Diverging or constructive plate margins are the oceanic axial zones with the highest production and eruption of magmas on Earth and the largest number (although unknown precisely) of active submarine volcanoes (Fig. 2.6). Iceland is a mid-

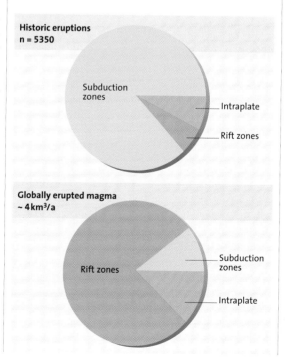

▲ Fig. 2.5. Model of plate dynamics and magma sources. Volcanoes erupted along mid-ocean ridges are fed by magmas from the asthenosphere (depleted mantle). Volcanoes above subduction zones derive their magma chiefly from the overlying mantle wedge, melting being triggered by hydrous fluids released from the descending slab. Magmas in oceanic and continental intraplate volcanoes (LIP) may be generated in rising mantle diapirs (plumes). Some downgoing lithospheric slabs may become subducted to the boundary between upper and lower mantle at 670 km (a). Others may penetrate deep into the thick intrinsically dense layer in the lower mantle (b) or to the core-mantle boundary (c), eventually forming a major magma source for intra-plate magma systems (oceanic volcanic islands, LIPs, continental intraplate volcanic fields). Plumes are thought to start at high spots (d) of the dense layer in the lower mantle and consist of mantle peridotite variably mixed with recycled lithosphere. The core is separated from the mantle by the D'' layer that may be the starting point for plumes (395)

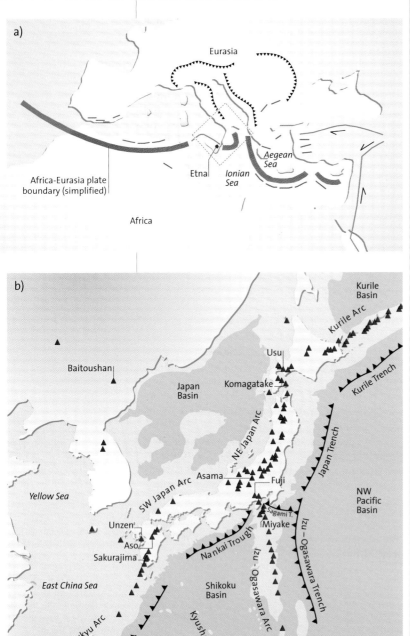

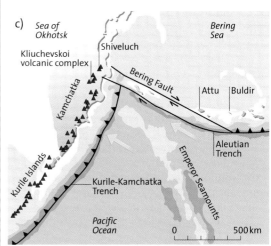

▲ ◄ ▲ Fig. 2.7. Three examples of highly productive basaltic magma – volcano systems in *slab-edge tectonic settings*. a) Simplified tectonic schemes of Mt. Etna in the Mediterranean (135). b) Distribution of major volcanoes in Japan, eight very active and/or famous volcanoes being highlighted, and the trench position along the Wadati-Benioff Zone in Japan and adjoining arcs (several sources). The location of Mt. Fuji at the triple junction between three plates is shown in detail in Fig. 8.4. c) Kliuchevskoi/Tolbachik volcanic complex (Kamchatka) close to the edge of the subduction zone where it is cut off by the complex Bering fault zone. Note oblique sub-duction (yellow arrow) along the Aleutian Trench. The island of Adak, type locality for adakite magmas inter-preted as slab melts, just outside frame (Smithsonian). Frame in a shows area detailed in Figs. 2.8 and 2.9

into the Earth's mantle. Most presently active *sub-aerial* volcanoes are formed above these *sub-duction zones*. Even though eruptive volumes of magmas along subduction zones comprise less than 10% of the global production or about 4 km³ per year, they represent more than 80% of the roughly 5 350 eruptions recorded in historic times.

Intraplate volcanoes or volcanic fields are sub-divided into oceanic ones, such as Hawaii, and continental ones, such as the volcanic fields of Michoacán-Guanajuato in Mexico, San Francisco in Arizona (USA), Eifel in Germany, or Chaîne des Puy in France.

Hybrid Tectonic Settings

As can be expected, not all volcanoes fit neatly into the plate tectonic pigeonholes. For example, some particularly productive basaltic volcanoes occur at complex junctions between subduction zones, or at or near the edge of a downgoing slab (Figs. 2.7 – 2.9). Prominent examples are Mt. Etna, the largest active volcano in Europe rising to 3 350 m a.s.l., Mt. Fuji, the famous landmark tow-ering over wide stretches of Honshu (3 776 m

ocean ridge volcano, in which particularly high magma production and lava eruption rates above a plume (Chap. 6) have generated a large volcanic island.

Convergent or destructive plate margins are zones along which oceanic lithosphere descends

a.s.l.) and the more remote Kliuchevskoi volcanic complex, the highest (4 835 m a.s.l.) and most active volcano on Kamchatka. The chemical signatures of the mafic magmas of some of these *slab-edge volcanoes* are most similar to intraplate magmas but they also show some characteristics of subduction zone-related magmas, Mt. Etna being a prominent example of this hybrid magma source. Etna and Mt. Fuji volcanoes appear to have developed in a tensional environment between subducting plates. Mt. Fuji has developed at a triple junction at the southern end of the downgoing Pacific Plate (Fig. 8.4) and the Kliuchevskoi/Tolbachik complex at the northern end of the Honshu-Hokkaido-Kurile-Kamchatka arc system (Fig. 8.17). Here, a 1 000 km long strike slip fault borders the downgoing slab to the north connecting this region to the Aleutian Arc. A major reason for foci of high mafic magma production rates in slab-edge settings may be the fact that fertile asthenosphere may be able – or is forced – to rise sideways of the edges of downgoing slabs, partially melting as it rises and decompresses. The magmas of both Mt. Fuji and the big northern Kamchatka magmas show subduction signatures, however. Magma genesis above downgoing slabs is discussed in more detail in Chap. 8. Examples of volcanoes in other plate tectonic settings that do not fit easily into the simple three-fold plate tectonic subdivision are back arc magma-volcano systems (Chap. 8) and ocean islands or island groups such as Iceland and the Azores showing weak to strong intraplate magma compositional characteristics but have grown above or along the Mid-Atlantic Ridge.

Summary

The volumes, heights and forms of volcanoes mainly depend on the physical properties and composition of the magma and thus processes in, and composition of, the root zones of volcanoes, whose dynamics are governed by their plate tectonic setting. Hence, a single volcano does not only contain information about its local origin. An andesitic stratocone is typical for a subduction zone tectonic setting. Morphology and architecture of a volcano by itself is not diagnostic, however. For example, caldera volcanoes can form in very different types of tectonic environments. Increasing numbers of calderas are now recognized to occur even in submarine volcanoes both, on seamounts as well as along mid-ocean ridges. Some types of volcanoes, such as maars, whose form is basically governed by near-surface processes (e.g. the interaction of magma and water) (Chap. 12), are particularly unsuitable to infer a particular tectonic setting.

Most volcanoes on Earth form either along divergent or convergent plate margins or in the continental or oceanic plate interiors. The divergent and convergent plate margins and the plate interiors can be characterized by a number of physical parameters (crustal and plate thickness, heat flow, stress field, earthquakes, etc). The mag-

▼ Fig. 2.8. Map of the south Tyrrhenian subduction zone with north and south reversed for easier visualization. The Ionian microplate(*green area*) descends northward under the Tyrrhenian microplate, contours in km (5 – 450) representing the depth to the top of the downgoing slab. The Aeolian islands (*dots*) form a volcanic arc (134)

▼ Fig. 2.9. Three-dimensional model of the south Tyrrhenian subduction zone looking southeast from a point above the Tyrrhenian Sea (see Fig. 2.8). *Red lines* represent magma pathways from the top of the slab to the arc. *Arrows* represent local patterns of asthenospheric mantle flow driven by slab motion. Etna volcano is located outside the Aeolian arc and above asthenospheric mantle rising sideways from beneath the African plate. The Adriatic microplate dipping steeply west of Vesuvius is thought to be broken.

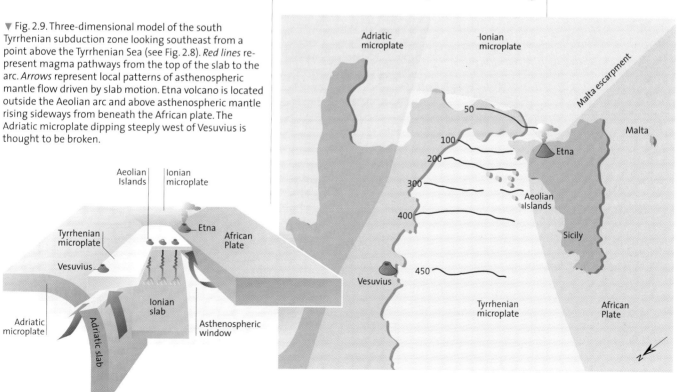

mas of volcanoes in each of these major plate tectonic setting are characterized by specific chemical compositions comprising major and trace elements, isotope ratios and composition of mineral phases. Different volatile contents of the magmas are particularly well reflected in contrasting modes of eruptions. For example, volcanoes grown above subduction zones where water-rich sediments and ocean crust are dehydrated at depth are typically highly explosive because the processes of magma formation in subduction zones are strongly governed by fluid release from the subducted slab. Magma compositions and types of volcanoes in hybrid plate tectonic settings commonly show more complex characteristics.

Soon after the theory of plate tectonics revolutionized Earth science, attempts increased to infer paleotectonic settings based on the composition of older volcanic rocks. This is sometimes problematic because volcanic rocks change their composition during alteration. However, the number of publications that attempt reconstructions of paleotectonic settings based on a comparison with modern plate tectonic characteristics still increases.

Before I discuss the main plate tectonic settings of volcanic activity on Earth in detail (Chaps. 5–8), I will treat the material of volcanoes, the magma (i.e. the melt) and the volcanic rocks formed by cooling of the magma in more detail.

Magma

Magma is an impressive word. Little wonder then, that a pop band, now forgotten, once called itself Magma. But the mysterious and fearsome associations evoked by the word magma have far deeper roots. Magma loses its mythical aura once it appears at the surface of the Earth; it is consequently given a different name: erupted magma is called lava.

What is magma? What is the source material for magma? At what depths and by which processes does magma form? Why does magma rise upward from depth into the Earth's crust or even to the Earth's surface? How can the bewildering types of eruptions and forms of volcanoes be explained? By different magma compositions or different eruptive conditions? Why is magma erupted explosively in some volcanoes, but pours out of a conduit quietly in others? Why do some volcanoes spew out only a few tens of thousands m^3 of lava, while others emit lava in colossal volumes of more than one thousand km^3 within the incredibly short time of a few months or less?

Magma is generated by partial melting of rocks in the Earth's mantle or, in much smaller amounts, in the lower crust. Volcanoes are features on the Earth's surface where magma is erupted, either effusively as lava flows or explosively as a mixture of fragmented magma, country rock fragments and gases. Volcanoes, however, erupt only a minute fraction of the melts that rise from greater depths in the Earth. The bulk of magma generated in the source areas is trapped during ascent and congeals; in other words, it crystallizes to become a coarse-grained plutonic rock. If we want to understand the origin of magma and of volcanoes, we have to take a closer look at the interior of the Earth beneath volcanoes.

Until a few years ago, the processes by which magma is generated were treated only briefly in textbooks on volcanology. This was because the relationship between volcanic eruptions, generation of magmas, and tectonic framework have only become reasonably well understood after acceptance of plate tectonics as the theoretical framework of Earth science. For example, why are most volcanoes basaltic, but only few of rhyolitic composition? Why do volcanoes of rhyolitic composition grow preferentially above subduction zones? Satisfying answers to questions concerning the origin and evolution of magmas and volcanoes in different tectonic settings can only be obtained by using experimental petrological, geophysical and chemical data (major and trace elements, radiogenic and stable isotope ratios), as well as volcanological field data. Nevertheless, we are far from being able to document different stages in the evolution of magmas to the degree of fully understanding the causes and processes of magma formation and evolution.

The sinking of crystals through a viscid substance like molten rock, as is unequivocally shown to have been the case in the experiments of M. Drée, is worthy of further consideration, as throwing light on the separation of the trachytic and basaltic series of lavas.

Charles Darwin, Geological observations on the volcanic islands, London, 1844

What Is Magma?

Magma is molten matter of silicate composition. Silicon (Si) is, apart from oxygen (O), the main constituent of most minerals and rocks in the Earth's crust, as well as in the underlying mantle. When one analyzes igneous rocks that contain glass and several types of minerals, practically all contain between 40–75 wt % SiO_2 (Fig. 3.1). The dominant volcanic rocks on Earth (and on most planets) are basaltic lavas such as those of the ocean crust. They contain about 50 wt % SiO_2. Granites, at the other end of the chemical spectrum of igneous rocks, contain between 70–75 wt % SiO_2. There are also unusual non-silicate magmas, such as the *carbonatites* that erupt in the Tanzanian volcano Oldoinyo Lengai and *sulfide melts* that ooze out of some fumaroles, but both are rare. Most lavas erupted on the Earth's surface contain crystals that have grown in the magma during its slow cooling in magma reservoirs or en-route during its migration through the lithosphere. The crystals visible to the naked eye are called *phenocrysts*. Zones of different chemical compositions and inclusions of glass and smaller crystals within them reflect complex conditions during their growth (Fig. 3.2). Successive crystallization of different types of phenocrysts changes the chemical composition of the remaining liquid, a topic discussed more fully at the end of this chapter.

Classification of Igneous Rocks

Igneous rocks can be simply classified based on their grain size, which reflects contrasting cooling histories, and their geological occurrence. When a

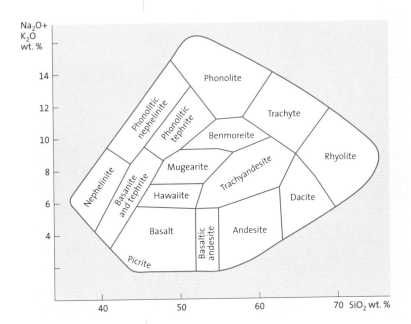

magma cools slowly, early crystals become larger and new ones form until no melt remains; the coarse-grained, entirely crystallized rocks are called *plutonic* (Fig. 3.3). Most plutons in the continental crust are of granitic to granodioritic composition. Magmas that rise very close to the Earth's surface but do not erupt, cool more quickly; these finer-grained rocks are called *subvolcanic*. Lava erupted onto the Earth's surface cools very quickly (Fig. 3.4). The groundmass of the resulting *volcanic rocks* is thus very fine-grained or even glassy. The different constituents can generally only be distinguished under the microscope.

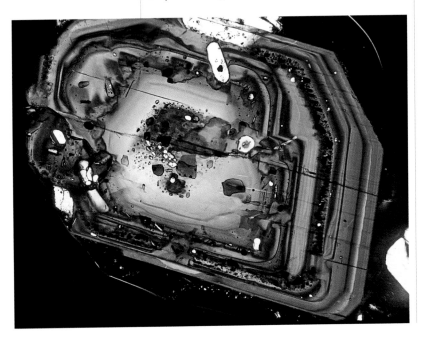

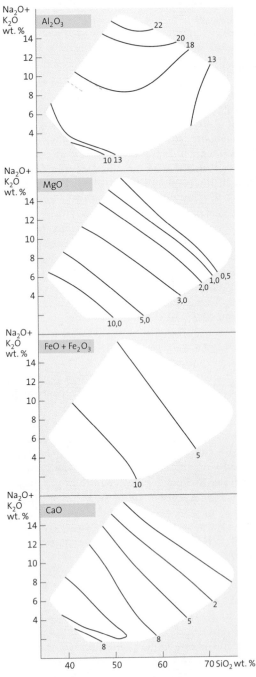

▲ Fig. 3.1. Chemical classification of major volcanic rock types (67)

◄ Fig. 3.2. Photomicrograph of a plagioclase crystal from andesitic ash erupted in 1976 from Augustine volcano (Alaska). The polarized light shows zones of different calcium concentrations. Compositional zonations reflect changes in the physico-chemical conditions in the magma reservoir such as increases and decreases in the partial pressure of H_2O. The inclusions are described in Fig. 3.19 (the same crystal in non-polarized light). Diameter of the crystal is 1 cm

Plutonic rocks can be classified relatively easily based on different volumes of the constituent minerals. Volcanic rocks, on the other hand, are best classified using their chemical composition (Fig. 3.1). One can classify volcanic rocks into (1) dark, *mafic* or *primitive*, mostly basaltic volcanic rocks; (2) lighter-colored, *intermediate* rocks, such as andesite and (3) light-colored, highly differentiated, *felsic* rocks, such as *rhyolite* or *phonolite*, sometimes also called acid rocks. Phonolites, which are common in some intraplate volcanic fields such as the Eifel in Germany (Laacher See phonolite) or the Canary Islands (Fig. 6.27), are the end product of magmatic differentiation of very alkaline basaltic parent magmas. Their SiO_2 concentration is around 55%, similar to that of basaltic andesites in subduction zones, a reason why the term acidic is quite inappropriate for such rocks. The high degree of differentiation in phonolites is expressed in the extremely high concentration of Al_2O_3, alkali and *incompatible elements* (that is, in those trace elements concentrated in the remaining melts, such as Zr, Nb, La, Th, Rb).

There are two main groups of basaltic magmas. The first and by far most common group is that of the *tholeiitic* or *subalkaline* basaltic magmas, which are poor in K, P, Ti, Rb, Zr, Nb, U and Th. They are typical of the ocean crust, flood basalts, some island arcs and especially productive volcanic islands such as Hawaii and Iceland. The second group is that of the *alkaline basalts*, which are enriched in K and Na and the trace elements listed above but have a lower concentration of SiO_2. They occur more commonly in relatively small volcanoes, in continents, or in many oceanic islands.

Basaltic magmas are the *parent liquid* for the intermediate and more strongly differentiated magmas. Many granitic magmas (chemically identical to rhyolite) also form by partial melting of the lower crust. When we want to better understand the complex processes by which the primary magmas are generated, we have to consider the layered structure of the Earth.

Where Are Magmas Generated?

Most magmas erupted on the Earth's surface are of basaltic composition and have temperatures at eruption of 1100 to 1250 °C. For this reason alone they could not have been generated in the Earth's crust. For example, the temperature of the lower crust in central Europe is estimated to be around 500–600 °C. In addition, the Earth's core can be excluded as a source of silicate magmas because it is mostly composed of iron and nickel. Alkaline basaltic magmas (which are common on oceanic islands and in continents) often bring up

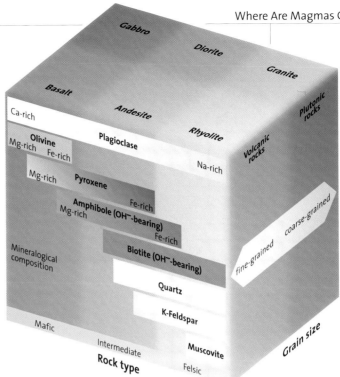

fragments from depth of a rock called *peridotite*. Peridotites consist mainly of olivine (Ol) and the fragments are commonly known as *olivine nodules* (Fig. 3.5). Other minerals common in peridotite fragments include orthopyroxene (Opx), much smaller amounts of clinopyroxene (Cpx) and the alumina-rich phase chrome spinel (Sp). When a peridotite contains Ol + Opx + Cpx + Sp, it is called *spinel-lherzolite*, based on the locality Lherz in the Pyrenees. It is called *harzburgite* when Ol dominates with minor Opx and Sp. High-pressure

▲ Fig. 3.3. Mineralogical composition of the major groups of igneous rocks (269)

▼ Fig. 3.4. Pahoehoe lava breaking through the crust of already cooled lava flows. Width of red lava tongue is about 4 m. Pu'u 'O'o eruption on the east flank of Kilauea volcano (Hawaii)

▶ Fig. 3.5. Olivine nodule (harzburgite) from the phreatomagmatic deposits of the Late Quaternary Dreiser Weiher maar (Eifel, Germany). The light-green crystals are olivine, the dark-green pyroxene and the black crystals chromium spinel

experiments have shown that lherzolite changes its mineralogical composition at a depth of more than 80 km (i.e., at pressures over ca. 25 kb) into garnet-bearing lherzolite. Garnet, like spinel, is the main carrier of aluminum in lherzolite, but has a greater density than spinel. At depths less than 30 km, *plagioclase lherzolite* would be stable. We therefore have reason to assume that basaltic magmas are formed in the upper Earth's mantle, under mid-oceanic ridges (MOR), at a depth of about 20–70 km. Under continents and many oceanic islands (beneath which the lithosphere is colder

▶ Fig. 3.6. Three modes of generation of basaltic magma by partial melting of peridotite.
a Increasing temperature;
b decreasing pressure;
c wet melting curve, lowering of the melting point (curve) by addition of H_2O and CO_2. Oceanic crustal conditions are assumed

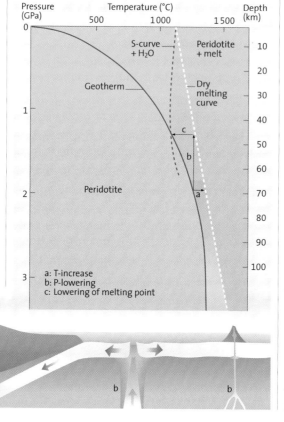

and thicker compared to mid-ocean ridges) alkaline basaltic magmas form at depths of 80–150 km.

Very different source-rock compositions must be assumed for the light-colored, rhyolitic to *dacitic* magmas. Based on their chemical composition, they cannot be derived from the Earth's mantle, but quite plausibly derive from the lower crust. The erupted temperature of these magmas (about 700 °C for H_2O-rich granitic and up to 1 000 °C for more mafic magmas of granodioritic composition) are higher than those normally occurring in the Earth's crust. These high temperatures are thought to result from very hot basaltic magma *underplating* the crust and inducing its partial melting. The lower crust is chemically and mineralogically heterogeneous and may therefore generate a range of magma compositions.

Much of the evidence for generation of evolved magma by partial melting is based on geochemical arguments discussed fully in textbooks of petrology. Sometimes, pieces of partially melted source rocks are brought up by the magmas themselves. The dacitic pumice which forms the base of the famous thick basaltic Plinian tephra deposit erupted in December 1707 from a crater on the flanks of Fuji volcano (Fig. 10.10) may have been generated by partial melting when fresh basaltic magma intruded an older magma reservoir filled with crystallized basaltic magma now present as olivine gabbro (295). Silicic magma can also form by differentiation of basaltic magmas (as will be discussed at the end of this chapter).

How Are Magmas Generated?

The fact that shear waves, which cannot propagate in fluids, are transmitted through the potential source areas of magmas (albeit with reduced velocity) is robust evidence that magmas form by partial melting of crystalline rocks. They are not siphoned off from a liquid magma layer below the brittle Earth's crust, as was once assumed and is often believed by lay people. The process of partial melting can be illustrated in a schematic cross section through the lithosphere. Figure 3.6 shows three curves. The geotherm represents the temperature increase from the Earth's surface to greater depths. This increase is not linear but instead decreases quickly below the first few km. Also shown is the melting curve for peridotite under two conditions: (1) dry at higher temperature and (2) rich in volatiles, resulting in melting at lower temperature. When a peridotite is melted at the Earth's surface (atmospheric pressure) the melting point is about 1 100 °C. Scientists in the first half of the nineteenth century showed that the melting temperature increases with pressure. In other words, the deeper one goes into the

Earth, the higher the melting temperature of peridotite. The temperature of all changes in the physical state of matter that involve an increase in volume, such as the melting of crystals to form a magma, increases with increasing pressure. This behavior is known as a *positive melting curve*. The fundamental condition for the normal state of the lithosphere is that the temperature of the melting curve lies above that of the geotherm. In other words, rocks of the lithosphere are hot but crystalline and therefore solid. A melt forms only when geotherm and melting curves intersect.

We know three changes in the basic parameters P (pressure), T (temperature) and X (chemical composition) through which solid rocks can partially melt:

- Increase in T (P and X constant),
- Lowering of P (T and X constant),
- Change of X (P and T constant).

More than one of these can occur simultaneously in nature.

Increase in Temperature

The process of partial melting most easily understood from everyday experience is an increase in temperature. We can melt materials with low melting points, such as wax and lead, by heating them on a normal kitchen oven. When we heat a rock to its melting point at constant pressure, the temperature will not increase further by extra addition of heat. The reason is that the melting points of different mineral phases that make up a rock can differ appreciably from each other. Additional heat energy is required to continue melting the different mineral components until the rock has been completely melted. The melting temperature for some mineral combinations is lower, however, than for the pure mineral phases.

What are the sources of heat in the Earth and are they sufficient to melt a rock? The radioactive decay of the elements U, Th and K, which have become enriched in the Earth's crust during the 4.6 billion years of Earth history, generates heat. This heat escapes constantly by migrating to the cold surface of the Earth and radiating into space. The concentration of radioactive elements in the Earth's mantle is much too low, however, to heat rock to its melting point, even when no heat is removed.

Some scientists have postulated that sufficient heat can be generated by mechanical friction, thereby facilitating partial melting, e.g. at the base of the moving lithosphere or along Wadati-Benioff zones. Robust proof for the generation of significant amounts of magma by shear melting is, however, lacking.

Transport of heat through a solid body, or *heat conduction*, is a very slow process. Heat can be transported much faster in fluid matter; this is called *convective heat transport*. When large volumes of basaltic magmas at more than 1 200 °C rise from the mantle to the crust and pond at the crust-mantle boundary, crustal rocks can partially melt. When the basaltic magma starts to crystallize during slow cooling, it generates further heat, so-called *latent heat* or *heat of crystallization*. Many granitic (rhyolitic) magmas are probably generated in the lower Earth's crust by this process of partial melting, as evidenced by their chemical composition. Most diagnostic are the ratios of radiogenic isotopes, in which daughter products (e. g. ^{87}Sr) are strongly enriched at the expense of their parent isotopes (in this case ^{87}Rb), which were accumulated in the crust from the beginning of the Earth (i. e., billions of years ago). The most plausible process to generate enough heat to partially melt lower crust is thus the addition of large volumes of basaltic magma (150). This process is also plausible as intermediate density basaltic magmas are likely to accumulate at the boundary between Earth's dense mantle and low density crust, the *level of neutral buoyancy* (LNB).

Decompression

When the temperature of a given volume of mantle rock is held constant, decompression of that rock can cause melting (path b in Fig. 3.6). In other words, the internal heat of the rock volume moved up is sufficient to trigger the melting process. This is strictly true only for dry i.e., fluid-absent, systems (see below).

The process of decompression of rising mantle material is probably the most important mechanism for generating partial melts, i.e., basaltic magma, as the entire hypothesis of plate tectonics is based on the assumption of large-scale motions within the Earth's mantle. Along the diverging plate margins, i.e., beneath mid-ocean ridges or underneath hot spots (described in Chap. 6), mantle material rises convectively so fast that heat loss to the surroundings is suppressed. This is called *adiabatic rise*. To compensate for rising mantle flow, colder lithosphere enters the mantle along subduction zones. That rocks of lower density can flow, rise and break through the Earth's surface is a well-known phenomenon, as exemplified by salt diapirs that can rise through less mobile rocks in the Earth's crust and even pierce the surface.

Addition of Fluid Phases

Pressure (P) and temperature (T) are both held constant in the third possible mechanism for generating magma, but the melting curve is shifted to lower temperature by the addition of fluid phases (mainly H_2O and CO_2). In other words, fluid

▲ Fig. 3.7. Mauna Ulu lava lake with 10 m-high lava fountain (Kilauea volcano, Hawaii)

▼ Fig. 3.8. Aloi Pit crater along the East Rift Zone of Kilauea volcano one month after it had become filled on 12/29/1969. The walls are about 15 m high. The solidified crust of the ca. 20-m-deep lava lake has been broken up into plates. The decrease in volume as a result of cooling is also reflected in subsidence of the central part of the lava lake and the margins, which dip into the lake. Diameters of individual plates about 5 m

phases lower the melting point of a rock. How much the melting point can be lowered depends not only on the amount of a fluid phase but also on the ratio of fluid compounds, e.g., $H_2O/H_2O + CO_2$. A mantle or crustal rock will thus melt at a lower temperature when it is "wet". Because most basaltic magmas contain between about 0.1 wt % (tholeiites) and 1.5 wt % (alkaline basalts) primary H_2O (and other volatiles such as CO_2, SO_2, etc.) (Chap. 4), the Earth's mantle is not "dry". These fluid compounds are probably contained in mineral phases such as mica (phlogopite, H_2O), amphibole (H_2O) or dolomite (CO_2), and are released during the early stages of partial melting. Partial melting through lowering the melting point of peridotite by adding fluid phases may be the most important mechanism for generating magmas along a heated subducting slab. Slabs are composed of subducted sediments and about 1.5 km basaltic volcanic crust that overlies plutonic crust and the uppermost mantle (which may be partially serpentinized by seawater that has penetrated through large fractures). Compounds such as H_2O, CO_2, and other elements released into fluids during *slab heating* (such as K, Rb, Br, Cs, Sr), migrate into the overlying mantle wedge and trigger the generation of H_2O-rich basaltic or even andesitic magmas. Fluids probably also play an important role by migrating into the melting domains in other tectonic environments (such as hot spots), thus modifying the mantle material metasomatically and thereby facilitating partial melting.

Why Do Magmas Rise?

The physical reason for the ascent of basaltic magmas from the Earth's mantle is their lower density compared to surrounding rocks (i.e., their buoyancy). Because the boundary between mantle and crust is defined by a density jump from 3.3 to ca. 2.8 g/cm³ at the so-called *Mohorovičić discontinuity*, or *Moho* for short, the positive buoyancy of many basaltic magmas is strongly reduced at this level. The growth of the Earth's crust from below by intrusion of basaltic magmas is called *underplating*. The density difference is a necessary but insufficient condition for the rise of a magma. Dynamic triggers are likely to be involved as well, especially the pressure due to rising mantle plumes.

Cooling and Crystallization of Lava Lakes

The comparison between the Earth's mantle and wax or lead, is, like all comparisons, not strictly correct because lead and wax melt at well-defined temperatures. Peridotite, however, melts over a large temperature interval because it consists of several minerals, each of which has a different melting point.

Cooling behavior also differs between simple substances and melts/rocks. If liquid wax is cooled to its melting temperature, it becomes hard or solidifies. In contrast, a basaltic magma at the surface is completely liquid at about 1 200 (its *liquidus temperature*) but solidifies completely only when it has cooled by 200 °C to about 1 000 °C (its *solidus temperature*).

The formation of different mineral phases during cooling of a basaltic magma and the change in chemical composition of the remaining silicate liquid have been studied during cooling of a *lava lake* in Kilauea volcano (Hawaii) under near-laboratory conditions. Lava lakes form when lava erupts from a fissure or conduit and flows into older sub-circular collapse craters (*pit craters*), sometimes filling them to a depth of more than 100 m (Figs. 3.7–3.11). Analyses from four such lava lakes (Kilauea Iki, Makaopuhi, Aloi, Alae) have substantially improved our understanding of the physical and chemical processes that take place during cooling of a lava (e. g. 258; 436). Geologists of the US Geological Survey were able to drill through the crust of these lava lakes, measure the temperature of the crystal-liquid mixture in situ, and take samples in bore holes at different depths representing different degrees of

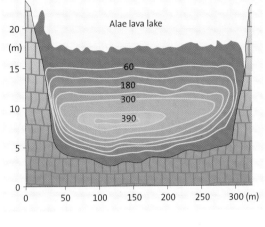

crystallization. These samples were analyzed microscopically and chemically to reconstruct crystallization processes during cooling (Figs. 3.10, 3.11).

Figure 3.9 shows in cross section the cooling of a lava lake, starting from a completely liquid state (liquidus temperature) to its almost complete consolidation (solidus temperature). Heat above a lava lake is removed rapidly from its surface by both radiation and convection of hot air from the surface. Heat migrates very slowly, however, from the lava lake into its surrounding rock, because conduction of heat is a slow process. The

◀ Fig. 3.9. Schematic cross section through Alae lava lake (East Rift Zone, Kilauea volcano, Hawaii). The *curves* and *numbers* indicate the migration of the 1 000 °C isotherm in days after formation of the lava lake (259)

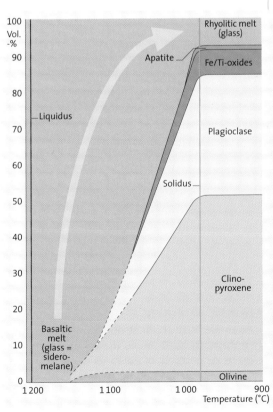

a) Photomicrograph of lava droplets (Pelées tears) containing abundant olivine crystals and few bubbles. The glass beads (sideromelane) are 2–4 mm in diameter. Kilauea volcano.

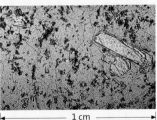

b) Photomicrograph of a basalt sample taken at 1 120 °C in drill holes in Makaopuhi lava lake. Large crystal: olivine; small crystals: clinopyroxene; yellow groundmass: basaltic glass (sideromelane)

c) Photomicrograph of basalt sample taken at 1 065 °C. White needles: plagioclase; round light crystals: clinopyroxene; dark crystals: ilmenite; brown groundmass: glass; large light-colored crystal: olivine.

◀ Fig. 3.10. Crystallization of a Hawaiian basaltic magma, derived from samples of drill cores in Alae lava lake. The relative volumes of the crystallizing mineral phases with decreasing cooling are accompanied by a decreasing amount of the not-yet crystallized melt, which would solidify to glass during quenching. The yellow arrow shows the change in chemical composition of the melt starting initially with a basaltic and developing to a highly evolved rhyolitic melt. Liquidus temperature 1 200 °C; solidus temperature 980 °C (modified after 258). Photomicrographs b and c courtesy TL Wright (US Geological Survey)

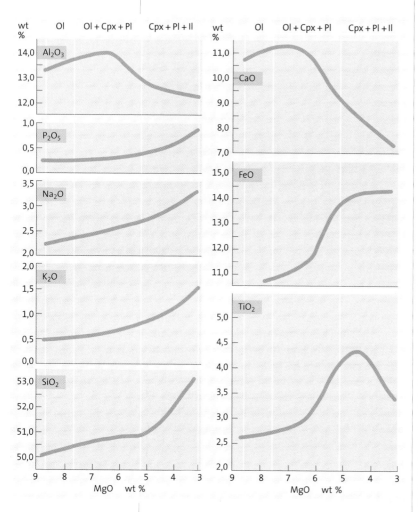

▲ Fig. 3.11. MgO-variation diagram (schematic) of the chemical composition of rocks drilled in Makaopuhi lava lake (Kilauea volcano, Hawaii), showing basaltic and more evolved rocks or the intergranular melts. The melt starts to crystallize at a temperature of about 1 200 °C with olivine (Ol). In magmas with less than about 7.7 wt% MgO, olivine stops crystallizing, while ilmenite (Il) appears as new opaque phase (436)

pyroxenes and Ca-Na-Al-bearing plagioclase, to crystallize. Pyroxene and plagioclase are the two main constituents of the rock type basalt and its intrusive equivalent gabbro. At about 1 070° C the melt has crystallized to about 50% and has become so enriched in iron and titanium that the opaque oxides (ore minerals) magnetite and/or ilmenite begin to form (Fig. 3.10c). When the melt is 75% crystallized the concentration of phosphorus remaining is sufficiently high that the common phosphate mineral apatite can crystallize. Apatite also forms part of our bones with a somewhat different composition.

When a lava flow or lava lake cools relatively quickly, the remaining liquid (about 8 vol% of the original liquid) can congeal near the solidus temperature (1 000) to glass. This glass has a composition that is vastly different from its parent liquid because the elements Mg, Fe, Cr, Ni, Ca, Ti and P have been depleted by crystallization of the minerals Ol, Px, Pl, Fe/Ti oxides and Ap. On the other hand, it has become enriched in the elements Si, Al, K, Na and many other trace elements, such as Rb, Y, Zr, Nb, U, Th and Ba. This remaining glass is of rhyolitic composition and if cooled slowly it would crystallize to quartz and alkali feldspar. Thus the glass represents the formation of the extreme derivative magma rhyolite. In summary, the chemical and mineralogical changes in such a lava lake show graphically how the very wide variety of igneous rocks can form in nature by cooling and differentiation via crystal fractionation of a single mafic parent magma.

temperature maximum thus migrates to the lower third of the lava body. The lava lake shown in Fig. 3.9 solidified after about 400 days, i.e., the temperature maximum in the lower part of the cooling mass reached about 1 000 °C.

The mineral olivine starts to crystallize a little below the liquidus temperature of a basaltic magma. In Hawaii, olivine commonly starts to crystallize in the slowly cooling magma reservoir at depth. Olivine is then erupted as a phenocryst together with the melt (Figs. 3.10, 3.11). Olivine has a composition of about 39 wt% SiO$_2$, 42 wt% MgO and 19 wt% FeO, different from that of basaltic magma with 50 wt% SiO$_2$, 8 wt% MgO, 12 wt% FeO and many other elements. Hence, the melt is impoverished in the element magnesium by the crystallization of olivine. Also depleted are the trace elements nickel (Ni), which substitutes for Mg in the olivine lattice, and chromium (Cr) contained in inclusions of the mineral chrome spinel. The remaining melt is enriched in all other elements. Further lowering of temperature causes other minerals, such as the Ca-Mg-Fe-bearing

Magma Chambers

Magma reservoirs are not only temporary holding reservoirs for magmas beneath volcanoes but are also the final site of crystallization of many magmas where they solidify to *plutons*. Granitic magmas cool and crystallize mainly within the upper crust. Magma reservoirs have existed since the formation of the Earth or at least since the beginning of formation of the Earth's crust. Bodies of plutonic rocks, now eroded, are robust evidence that large volumes of magmas can accumulate in the Earth's crust. Many more-or-less liquid magma reservoirs also exist in the crust at present. Nevertheless, to unequivocally identify stationary magma chambers by geophysical methods, e.g., in geothermally active areas with hot springs, fumaroles, etc. (The Geysers, Yellowstone, Iceland, Larderello, Wairakei) is difficult. One magma chamber that has been identified in the crust by employing a variety of methods is that which lies beneath Kilauea volcano on Hawaii (Chap. 6).

Magma reservoirs can be subdivided into rheologically different portions depending on the temperature, amount of crystals and magma rheology. Completely molten areas (above the liquidus temperature) have low viscosities and can convect, provided that the viscosity is sufficiently low, as in basaltic magmas. The degree of crystallinity increases with falling temperature. At crystallinities of about 50–60 %, the magma consists of a crystal mush and behaves rheologically like a solid body. This boundary has also been called *rigidus* (208). The temperature difference between liquidus and solidus, for example within tholeiitic magma, is about 200, that between liquidus and rigidus 100.

The term *differentiation* is used for processes by which chemically and mineralogically different types of derivative magmas evolve from *primitive* parent magmas. Conventionally, differentiation is thought to take place in magma reservoirs in the upper 10 km of the Earth's crust, where they stagnate, cool slowly and crystallize. Fractional crystallization is probably the main mechanism of magmatic differentiation. Many near-surface magma reservoirs are strongly compositionally zoned, as shown by detailed mineralogical and chemical studies of Plinian ash falls and pyroclastic flow deposits (see below). Mounting mineralogical and chemical evidence suggests that magma can also differentiate substantially at much higher pressures, such as at the base of, or even beneath, the crust. Green, iron-rich cores of clinopyroxenes are examples of such high-pressure fractionation (Fig. 3.12).

During *fractional crystallization*, dense crystals sink in a less dense magma, or rise in a denser magma. Crystal sinking may lead to accumulation of dense crystals at the bottom of a magma chamber, forming *cumulates*. However, the viscosity (Chap. 4) of strongly differentiated SiO_2-rich magmas is so high that the sinking velocity of phenocrysts is sufficiently low to prevent extreme fractionation by gravitational removal of crystals. In the model of *convective fractionation*, crystallization on the side walls of a magma reservoir generates highly evolved melts of low density that can rise convectively toward the top of a magma body (18, 355). Convective mass movements will also be discussed for different reasons in some later chapters, such as the convective rise of heated air above cooling lava flows or ignimbrites (Chap. 11), convectively rising eruption columns (Chap. 10) or convectively rising mantle plumes (Chap. 6). Other processes, such as *magma mixing* and uptake, modification or complete melting of rock fragments from the walls of a magma chamber, and possibly also diffusion in the liquid state, can further significantly change the composition of a magma.

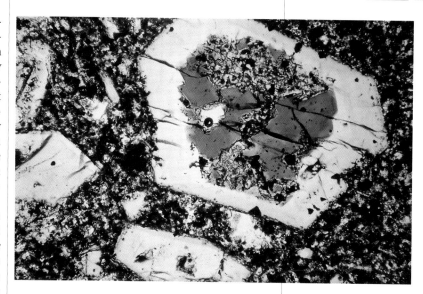

Zoning in Phenocryst Minerals

Phenocryst mineral phases in volcanic rocks are characteristically compositionally zoned, a nightmare to experimental petrologists trying to reconstruct equilibrium conditions in a magma body. There are several processes by which a mineral phase grows layers differing from each other in chemical composition, in some cases strikingly apparent under polarized light (Fig. 3.2). Magmas erupting relatively quietly in the form of lava flows, or, more commonly, explosively provide a snapshot of the conditions in a magma reservoir at a particular point in time as well as of processes that take place during slow or fast ascent in the conduit, or, in the case of long-lived magma effusion, during transport and cooling of lava flows.

Two main processes are commonly evoked to explain rapid chemical changes frozen in the growth rims of crystals. One is a new magma, commonly of more mafic composition, entering a magma reservoir and heating and/or mixing with the cooler resident magma. Such a *replenishment* will further add to the complexities of the dynamic processes in a magma reservoir such as convection, sidewall crystallization, etc. Another process that can strongly affect the equilibrium between a mineral surface growing in a melt is the episodic degassing of a magma, something that is very evident in many volcanoes that release magmatic gases without erupting their magmas from which the gases have been released. A classical case is the strong dependence of plagioclase composition on the partial pressure of H_2O in a magma, an effect well-supported by experimental evidence. The anorthite content of plagioclase increases strongly within increasing water content of a magma.

Depending on the composition of the primitive parent magma, several lines of differentiation

▲ Fig. 3.12. Photomicrograph (width 1 cm) of a leucite nephelinite lava from the Eifel (Germany) with large clinopyroxene phenocrysts. The greenish-brownish cores have already become partially decomposed and represent early crystallized phases of the magma at high pressure (probably boundary crust-mantle). The iron-rich green cores are overgrown by more magnesium-rich colorless to slightly brownish rims crystallized from a younger more primitive magma

► Fig. 3.13. Ignimbrite erupted at Mount Mazama about 7 700 years B.P. (Crater Lake volcano, Oregon, USA). Light-colored dacite in the lower part of the photograph (originally upper part of the magma column) and dark andesite with weathered-out degassing channels (fossil fumaroles) in the upper part reflect the chemical zonation of the magma reservoir in reverse order

can be identified. The two main trends are:

- Tholeiite → rhyolite;
- Alkaline basalt → phonolite (trachyte).

Several algorithms have been developed to unscramble the genetic relationships between rocks and their parent magmas. The best known model is the *AFC-model* (assimilation, fractionation, crystallization) (80). A sequence of cogenetic rocks can be analyzed using the AFC model to test the hypotheses that (1) the magma evolved in a closed system, or (2) that it exchanged some elements with country rocks, or (c) that larger masses of crustal rocks were partially melted and incorporated. In general, the chemical composition of a rock suite in a volcano or in a volcanic field is modified more by differentiation and magma mixing than by assimilation.

Compositionally Zoned Magma Reservoirs

The increasing attention paid by scientists to major pyroclastic deposits, such as huge ignimbrite sheets or voluminous Plinian pumice fallout deposits, has had a significant impact on our understanding of the volume, *compositional zonation*, and differentiation mechanisms of large felsic magma reservoirs. Almost all major tephra deposits show a systematic change in composition from highly evolved at the base of a deposit to more primitive mafic compositions at the top (Figs. 3.13–3.18, 10.10, 10.13). Commonly, this is seen in the field by a change in color from light-colored deposits at the base to darker ones at the top. The transition in some deposits is very sharp, while in others a whole gamut of dark- and light-colored banded pumice lapilli form a transition zone between felsic base and more mafic top (Fig. 3.13). Accompanying the compositional variation is another trend obvious to the naked eye and easily discernable with the hand lens, that of a major change in the content of visible crystals, or phenocrysts. The basal light-colored deposits commonly contain very few if any phenocrysts, while the uppermost darker deposits may contain as much as 50 % of crystals per volume. When the composition of whole rock samples or glass is analyzed (by methods such as the electron microprobe), chemical changes are apparent in major as well as trace elements and also in volatiles (specifically when glass inclusions in phenocrysts are analyzed). The earlier, most evolved subunits are commonly more widespread than the later, more mafic ones especially when these are represented by more crystal-rich and therefore more viscous magma, but there are major exceptions.

Let us look at a particularly instructive example that shows all these changes in a nutshell (Figs. 3.15–3.18). The Laacher See volcano in Germany erupted in the spring of 12 900 B.P. The structures and eruptive history of this deposit will be discussed in more detail in Chapter 11, the compositional zonation being treated below.

In the field, a change in color in the deposit is quite obvious, especially when approaching the eruptive center (which is a hole in the ground, slightly greater than 2 km in diameter, filled by a lake). The change in color of the pumice lapilli from light to dark is not gradual. There are three stratigraphic horizons in which the change is most pronounced. The change in color is associated with other changes such as chemical rock composition, volatile content and mineral types and proportions, degree of vesicularity of pumice and shape of vesicles and thus physical properties such as magma density. This correspondence well illustrates the fact that different magma properties strongly influence the eruptive behavior of a magma and thus the nature of the deposits laid down. Three major compositional changes occur at the stratigraphic boundaries (1) LLST (Lower Laacher See Tephra) to MLST (Middle Laacher See Tephra), (2) within MLST and (3) from MLST to ULST (Upper Laacher See Tephra). The pumice lapilli at boundary (3) also change from highly inflated to poorly vesiculated lapilli with vesicle volumes of 20% or less. These lapilli are angular, differing from the rounder and more highly inflated lapilli below. The lapilli in the ULST contrast with the lower ones by being larger, potato to cauliflower-shaped, very crystal-rich and dense, and containing abundant Devonian slate chips.

◀ Fig. 3.15. Basal fallout deposits of late Quaternary Laacher See Tephra (LST) (Germany) that erupted 12 900 years ago. The light-colored almost crystal-free pumice is of highly evolved phonolitic composition. The massive, 4-m thick layer above the bedded pumice and dark fine-grained tuff layers is an ignimbrite representing the Middle LST and is represented by a white massive layer in the lower right side of Fig. 3.16. Krufter Ofen

▲ Fig. 3.14. Extremely welded and plastically deformed high temperature Miocene ignimbrite VI (Gran Canaria) showing three sublayers of different composition as reflected in contrasting colors: peralkaline rhyolite (*light colored*), intermediate and mixed (*left side*) and dark flame-like trachyte. The layers have become sandwiched and folded by fast successive emplacement of flow units and delamination, the younger more mafic trachytic flow unit now occurring near the base of the cooling unit

The crystal content at boundary (2) changes from < 2 vol% to ca. 5 vol%, while the dense ULST lapilli generally contain > 40 vol% of crystals and also many more phenocryst species than in the lower parts of the deposit (Figs. 3.16–3.18). There are further changes: vesicles are typically elongate in LLST but become more spherical in MLST-C. Rock fragments in the LLST are fragments of near-surface deposits, such as Tertiary clays, quartzite pebbles from conglomerates and older Quaternary volcanics and Devonian slate. The relative proportion of volcanics to Devonian xenoliths changes repeatedly, reflecting alternating phases of lateral and downward enlargement of the conduit system. In the MLST, the ejected type of basement changes from slate to greenschist facies rocks, such as phyllites, reflecting the deepening of the conduit and downward evacuation of the magma column. Moreover, xenoliths of contact metamorphic rocks (buchites), slowly cooled plutonic rocks of phonolitic composition (syenites) and cumulates of mafic crystals, increase in abundance upward in the deposit. All these lithological, physical and compositional parameters indicate an extremely compositionally zoned pre-eruption magma reservoir that consisted of about four distinct layers, each zoned in itself and separated from the other layers by a relatively abrupt transition.

Compositionally zoned deposits such as the Laacher See Tephra pose several fundamental petrological and volcanological questions: first and foremost, how accurately does the stratigraphic zonation of a deposit reflect the zonation in the magma itself? Are the abrupt boundaries in

▶ Fig. 3.16. Lower (LLST), Middle (MLST) and Upper (ULST) Laacher See Tephra deposits at Wingertsberg (Laacher See area, Germany). The MLST ignimbrite is separated from the more mafic, very crystal-rich cross-bedded gray phreatomagmatic deposits by three layers, about 50 cm thick, consisting of well-sorted pumice, separated from each other by darker ash layers representing overbank deposits laid down by pyroclastic flows (MLST C). (See also Figs. 10.13, 11.38, 11.40)

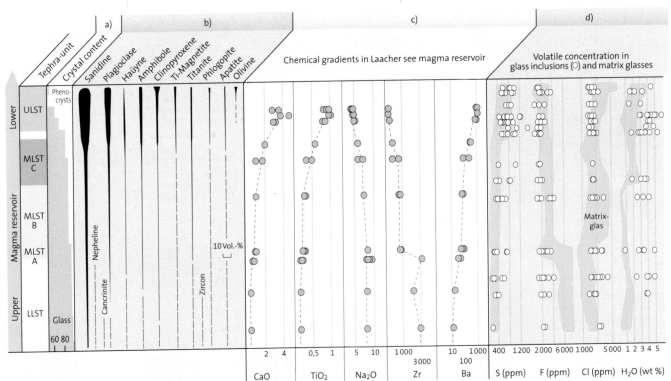

▲ Fig. 3.17. Change in the chemical and mineralogical composition of the Laacher See Tephra with stratigraphy resp. depth in the magma reservoir. The LLST represents the upper part (cupola) of a strongly compositionally zoned and layered magma column, which was partially emptied during the Plinian eruption. Lower, very crystal-rich portions (ULST) of the magma column were erupted during the late phreatomagmatic phase. The diagram shows a) the vol% of phenocrysts and glass matrix, b) the modal percentage of different crystal species, c) selected average major element (wt %) and trace element concentrations (ppm) of pumice lapilli (*gray dots*) and d) the volatile concentrations of non-degassed glass inclusions (white dots) and partially degassed pumice matrix glasses (*turquoise fields*) (33, 142, 315, 428 and unpubl). Note that volatile species S and H_2O are impoverished in the matrix glasses compared to glass inclusions in phenocrysts while F became enriched in the melt at a late stage after phenocrysts had crystallized. Moreover fluorine is not strongly partitioned into the gas phase and much remains in the glassy pumice lapilli

the deposit the result of episodic eruption of finite, compositionally distinct magma batches, subsequent reorganization and mixing of the remaining magma? If so, the sharp compositional break suggested by the field evidence may only be apparent and an artifact of episodic rather than continuous eruption. Abrupt changes, used for stratigraphic subdivision of a complex deposit, may have been caused, or become accentuated, by drastic changes in the dynamic evolution of the eruption, chiefly conduit / magma chamber wall collapse, vent migration and influx of groundwater. Another ever-present question concerns the processes that terminated the eruption. How much of the magma in the reservoir was erupted and how much remained beneath the surface? Did the eruption terminate because the remaining magma was no longer eruptible (due to its excessive crystal content and high viscosity)? Or were the volatile concentrations too low? Can the striking contrast in phenocryst abundance be explained by an almost crystal-free volatile-rich, highly evolved upper part of the magma column (*cupola*) contrasting with a very crystal-rich lower part of the erupted section? Did all crystals grow in the lower part of the reservoir or were some dense species transported downward by slow sinking or abrupt, catastrophic overturn of a crystal-rich top layer? At what depth (pressure) and temperature did crystallization take place and was the magma volatile-saturated at the time of eruption? Was the magma stationary for long periods of time at a particular depth interval, or did it move upward continuously as buoyancy increased with generation of volatile-rich low-density highly evolved melts?

Some of these questions can be answered, while the answers to others remain elusive. The depth interval over which the erupted part of the Laacher See magma column crystallized is now fairly well established by several independent lines of evidence (Fig. 3.18). For example, the composition of lithic fragments throughout the deposit shows a systematic increase in depth, apart from reversals associated with phases of lateral crater-widening. Some of the Devonian slates and phyllites must have formed the walls of the magma reservoir because they sometimes occur as contact-metamorphic rocks. The boundary between the Devonian slates and greenschist facies rock was inferred to lie at about 5 km depth below the surface, hence, the magma reservoir evolved at a high level, between approximately 4 and 7 km below the surface. Virtually identical results are derived by using the pressure-dependant composition of coexisting minerals, so-called *geobarometers*, especially two coexisting feldspars and the alumina concentration in amphibole. Finally,

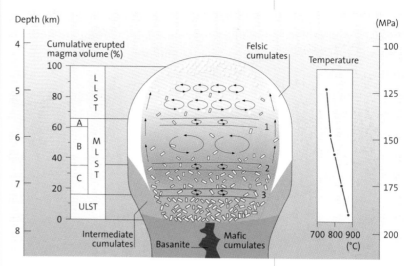

▲ Fig. 3.18. Model of the pre-eruptive Laacher See magma reservoir stratigraphy (33, 142, 374, 375, 428, 429, modified after 143) partially emptied during the eruption (not to scale). The tephra stratigraphy is approximately normalized to the partial volumes of erupted magma. The erupted part of the magma column shows three distinct breaks in composition, crystal content and abundance of crystal cumulates and contact-metamorphic country rock xenoliths. The uppermost zone (LLST), the cupola, is almost crystal-free, extremely enriched in H_2O, F and incompatible trace elements such as Zr, U, Pb and Th. This and the near-absence of crystal cumulates and contact-metamorphic country rock fragments indicate that this portion developed and moved upward late in the evolution of the magma reservoir. The most distinct compositional break occurs between LLST and MLST. The largest erupted volume is slightly more crystal-rich but with significantly lower concentration in incompatible trace elements. A significant increase in volume percentage of crystals and slightly more mafic composition characterizes layer MLST C. Some mixed pumice lithologies characterize the break between this and the basal batch of more mafic phonolite magma (ULST), erupted in phreatomagmatic explosions. These lapilli are extremely crystal-rich, contain abundant contact metamorphic rock fragments from the collapsed walls of the chamber and felsic to mafic crystal cumulates and, near the end of the eruption, mafic olivine-bearing basanite lapilli mixed-in with the phonolite. Temperature and pressure ranges based on experimental evidence, mineral barometry/thermometry and stratigraphy of country rock xenoliths

experimental studies of phonolite melts at various temperatures and pressures also suggest similar depth intervals (141).

Analysis of glass inclusions in phenocrystic mineral phases of the type shown in Fig. 3.19 has shown that the composition of the melt captured by the growing crystals also changes from more evolved compositions in crystals from lower stratigraphic levels (upper part of the reservoir) to more mafic in the crystal-rich uppermost ULST deposits (Fig. 3.16). This important evidence means that we cannot assume that the huge mass of crystals erupted during the terminal phase had

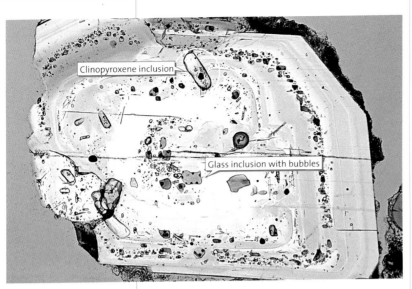

Clinopyroxene inclusion

Glass inclusion with bubbles

▲ Fig. 3.19. Photomicrograph (plain light) of a 1-cm-long plagioclase crystal of andesitic ash erupted 1976 in Augustine volcano (Alaska). The roundish, light-brown areas are glass inclusions with small gas bubbles. The elongate crystals are dominantly clinopyroxene

originally formed in the cupola of the magma column and then settled to the bottom. Instead, melt inclusion evidence indicates that the crystals formed near the level of the melt in which they were suspended when erupted. The uppermost magma did not pause long enough to cool and crystallize significantly, but instead was erupted "prematurely", perhaps because the solubility of volatiles, especially water, had been exceeded, leading to the generation of bubbles and increased buoyancy, a self-perpetuating process. Another factor that may have contributed to, or triggered, the eruption is the influx of basaltic magma into the magma column, a process that has been inferred in many magma systems (Chap. 4). A final factor contributing to the eruption was the heating of groundwater by rising hot gases and/or magma. The Laacher See eruption, like many other Plinian eruptions, started with a phreatomagmatic phase that produced almost exclusively lithic fragments, due to the rapid heating and expansion of groundwater in fractures in the otherwise dry Devonian basement very near the surface (Fig. 12.22). The enormous volume expansion as liquid water flashes to vapor is the physical reason for this thorough fragmentation (Chap. 12). Such a process would suddenly decrease the pressure on top of a rising magma column that might never have erupted without this environmentally-controlled process.

Summary

The observed broad spectrum of volcanic and plutonic rocks results from different types of source rocks, from which magmas are generated by partial melting, different mechanisms and degrees of differentiation during rise and cooling and variable degrees of contamination. Source rocks comprise many different types of peridotite, subducted ocean crust plus overlying sediments and lower continental crust. Three melting mechanisms are distinguished from each other: addition of heat, decrease in pressure and addition of fluid phases to the source rock.

Primary magma is defined as melt that is in equilibrium with its source rock. The composition of a primary magma is governed by the composition of the source rock and the degree of partial melting. Basalt magmas are generated in the Earth's mantle mainly by decompression but also by lowering of the peridotite melting curve by addition of fluids. Granitic magmas, on the other hand, are formed mainly by partial melting of lower crustal rocks.

Melts once generated along grain boundaries are mobile because of their low viscosity and are positively buoyant because of their low density. Once enough melt has accumulated, magma can rise relatively quickly along fissures that are generated at the top of the rising melt. Melt pathways either close again after a large batch of melt has risen, or survive as congealed dikes. The buoyancy of low density magma is only rarely sufficient to transport a magma directly to the Earth's surface. Exceptions are kimberlite magmas beneath old shields (Chap. 12), which rise from a depth of more than 150 km. Magmas that rise into reservoirs in the upper crust or to the Earth's surface are commonly not primary. Instead they have differentiated and mixed to varying degrees in transit from the source area to the final site of emplacement. A highly evolved rhyolitic magma can thus be generated from a basaltic one by differentiation. About ten parts of highly evolved rhyolitic, trachytic or phonolitic magmas are generated from about hundred parts of basaltic magmas, depending on their composition (tholeiitic, alkali basalt, etc.). The most common but least understood magma chambers are found below mid-ocean ridges. The formation of many ore deposits is also connected with strong differentiation of several types of magmas.

The explosivity of many volcanoes depends critically on an enrichment in volatiles, especially H_2O, during differentiation in near-surface magma reservoirs where H_2O (and SO_2) can form free gases at low pressures, CO_2 already exsolving at much greater depth. Before I discuss volcanoes in different tectonic settings, I will first deal with the rheology of magmas, magmatic gases and an overview of mechanisms by which an eruption is triggered.

Rheology, Magmatic Gases, Bubbles and Triggering of Eruptions

Eruptions in progress that may last from a few hours to many weeks often change their behavior drastically in mode of eruption and intensity. Lava may well out quietly from a vent (Fig. 4.1), breaking-up into glowing slightly vesicular fragments while descending the slope of a volcano (Fig. 4.2). At other stages, lava may rise dramatically to form impressive fire fountains up to hundreds of meters high (Fig. 4.3), the hot fluid blebs falling to the ground, coalescing thereby forming lava flows. In others, lava oozes out slowly at a vent (Figs. 4.4, 4.5) forming thick lava flows or dome-like lava bodies that episodically collapse, generating spectacular und often destructive pyroclastic flows. In still others a mixture of particles of fragmented magma and gases is shot out of a crater with close to supersonic velocity, the eruption column rising perhaps as high as 40 km (Fig. 4.6).

These are merely end members in a bewildering spectrum of eruptive phenomena. Most of the different types of eruptive behavior depend strongly on the composition and thus the temperature, fluidity, or viscosity, and gas content of a magma and their replenishment rate as well as the geometry of the vent, type of volcanic structure and eruptive environment. It is also obvious that hot, fluid, dense, gas-poor basaltic magmas are at one end of the spectrum, highly silicic, cooler, viscous, less dense and gas-rich evolved (rhyolitic, etc.) magmas at the other. But this is a great oversimplification and there is a

I found the reddest-faced set of men I almost ever saw. In the strong light every countenance glowed like red-hot iron, every shoulder was suffused with crimson and shaded rearward into dingy, shapeless obscurity! The place below looked like the infernal regions and these men like half-cooked devils just come up on a furlough. The smell of sulphur is strong, but not unpleasant to a sinner.

Mark Twain, Letters from Hawaii, San Francisco, 1866

◄ Fig. 4.1. Quiet extrusion of lava below a partially collapsed spatter cone on top of Mauna Ulu (Kilauea volcano)

▲ Fig. 4.2. Andesite lava flow on the lower slope of Arenal volcano (Costa Rica) inching slowly forward. Fragments of lava generated during flow of the viscous lava roll down the steep flow front and form a basal bed of boulders beneath the advancing flow

▶▲ Fig. 4.3. Lava fountain in Mauna Ulu volcano. East Rift Zone (Kilauea volcano, Hawaii)

▶ Fig. 4.4. Holocene rhyolitic lava with glassy (dark obsidian) and light-colored, more vesicular and locally more brittle zones with intricate folding. Mono Craters (California, USA)

▶ Fig. 4.5. Steep front, made of blocky talus of a viscous, rhyolitic lava flow. Big obsidian flow, ca. 1 300 years old. Newberry Crater (Oregon, USA) (See also Figs. 9.12, 9.13, 15.18)

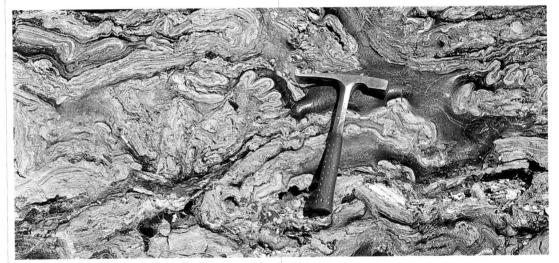

wide range of gas-poor to gas-rich magmas at the basaltic and similarly at the highly evolved end. Moreover, when external factors influence the eruptive behavior significantly as during magma-water interaction (Chap. 12), the importance of intrinsic factors (magma composition) diminishes significantly.

In Chapter 3, we followed the pathway of a magma from the mantle into the upper crust. Now we will take a closer look at the upper part of a magmatic system. For this purpose, we have to consider the fluid dynamic (i.e., rheological) properties of both magma and magmatic gases. A discussion of some of the dynamic processes is deferred to later in the book. In Chapter 10, for example, I will discuss the nucleation, growth and finally bursting of bubbles; processes that occur primarily beneath the Earth's surface. The processes in the most visible parts of the eruption that can be viewed from great distance (i.e. the eruptive column) are discussed in more detail in Chapters 10 and 11. These chapters will focus on two basic categories: basaltic (*Strombolian/Hawaiian*) and highly explosive *Plinian eruption* columns, illustrated by the case histories of Hawaii, Mt. St. Helens and Laacher See volcano.

Rheology

Many processes during the evolution of a magma are governed by its fluid dynamic or rheological properties and surface energies, such as:
- Aggregation of melt droplets following partial melting
- Ascent velocity
- Diffusional velocity of elements in the melt and their incorporation into crystal lattices
- Movement of crystals and bubbles in the melt
- Eruptive and flow processes on the Earth's surface

The study of rheology of silicate melts is a relatively young research field. One of the most important rheological properties of a magma is its *viscosity* or internal friction. The viscosity depends, as discussed in Chapter 3, on pressure (P), temperature (T) and especially on the chemical composition of a magma (denoted as X). For

▲ Fig. 4.6. Eruption column of the 1822 eruption of Vesuvius (Italy) (281)

example, lavas of basaltic composition are very fluid and can form extensive lava fields. In extreme cases, basaltic lavas can cover huge areas (several thousands to tens of thousands of square kilometers) within a few weeks or months (Fig. 4.7). Rhyolitic, sometimes called "acid" lavas, on the other hand, are highly viscous and generally form small, short and thick lava flows that break up into blocks during flow (Figs. 4.4, 4.5). The viscosity of a magma is thus significantly governed by its composition which determines its melt structure.

◄ Fig. 4.7. Spectacular columnar structures in a canyon-filling flood basalt lava flow. The lower *thicker columns* formed during slow conductive heat loss to the underlying ground, while the *thin columns* in the main upper part formed at a faster rate by convective and radiative heat transfer through the air. Horizontal to irregular orientation of upper columns is due to cooling of basalt lava against steep canyon walls. Columbia River Basalt (Washington, USA)

► Fig. 4.8. The basic structure of a silicate melt (magma) and that of the most common silicate minerals in the Earth's crust and Earth's mantle are the SiO_4 tetrahedron and the SiO_6 octahedron

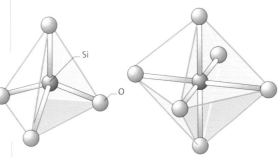

► Fig. 4.9. The viscosity of a silicate melt is largely determined by the contrast between the strong covalent Si-O bonds and the weaker ionic O–M bonds

Melt Structure

Contrary to some intuitive beliefs, melts are highly structured since their constituent elements are connected with each other into small units that become more interconnected when forming crystals. The basic building block of silicate minerals is a tetrahedron with oxygen atoms on each of the four corners and a silicon atom in its center (Fig. 4.8); this is termed *tetrahedral coordination of silicon*. Silicon can be substituted by aluminum or trivalent iron in the center of a tetrahedron. The elements Si and in part also Al and Fe^{3+} (ferric iron) are called *network formers* because many tetrahedra are connected along their corners via oxygen in strong covalent bonds to form one- to three-dimensional networks of highly polarized units (Fig. 4.8). Much activation energy has to be

expended to break these covalent bonds. Other cations, such as K^+, Na^+, Ca^{2+}, Mg^{2+}, Fe^{2+}, and in part Al^{3+} and Fe^{3+} form *ionic bonds* with oxygen. These are weaker than covalent bonds and are surrounded by more oxygen atoms, i.e. they are more highly coordinated. Such elements are called *network modifiers* (R), because the metallic or ionic bonds can be broken much more easily than the covalent bonds. These cations are generally surrounded by six oxygen atoms that form an *octahedron* (Fig. 4.8); hence we can say that the cations are *octahedrally coordinated*. When the tetrahedra of the network formers (T) are connected over all four corners, no non-bridge-forming oxygen-atoms (NBO) are free. The ratio NBO/T = 0 denotes a three-dimensional network (3-D unit). If tetrahedra are coordinated over only three corners, (NBO/T = 1), layer-like units form. Chains form by connection over two corners (NBO/T = 2). Even smaller units are called *dimers* (NBO/T = 3) and *monomers* (NBO/ T = 4) (316). *Complexes* form when the tetravalent Si^{4+} is substituted by Al^{3+} in a tetrahedron and the electric charge is compensated by a network modifier (R), e.g., by Na^+. If there are insufficient network formers in a magma, Al^{3+} and Fe^{3+} act as network formers and form $R + AlO_2$, $R + FeO_2$, $R_2 + Al_2O_4$ and a variety of other complexes. Only when the Na : Al ratio is greater than 1 does Al act as a network modifier.

Basaltic magmas have NBO/T = 0.6−0.9 (in very alkaline magmas NBO/T may be >1). Magmas of intermediate composition (andesites) have values of 0.3−0.5 and felsic magmas (rhyolitic) from 0.02−0.2. In other words: rhyolitic magmas are very viscous because much activation energy must be expended to break the strong covalent Si-O bonds of the 3-D units (Fig. 4.9). Basaltic magmas are more mobile, because they consist of small units, and because the R/O bonds are weaker than the T/O bonds and can be sheared during very low strain. Hence the viscosity of a magma is controlled by its chemical composition.

We know no natural magmas that are completely dry, i.e. do not contain any H_2O, CO_2, H_2S, F or Cl. Total volatile content and composition are additional important factors determining the viscosity of a melt. Highly evolved magmas contain up to ca. 7 wt % H_2O. For low to normal H_2O concentrations H_2O dissolves in the melt mainly as OH-groups, but at higher H_2O contents, it is also present in molecular form. The OH-groups act as network modifiers because they break oxygen bonds. A melt will thus become more depolymerized with increasing H_2O concentration, i.e. it becomes more "fluid".

On the other hand, CO_2 can form electrically neutral complexes with several network modifiers. In depolarized melts, this decrease in network modifiers increases the polymerization. Hence, a magma becomes more viscous. If aluminum complexes such as $NaAlSi_3O_8$ form in a melt, complexing with CO_2 causes the Na : Al ratio to decrease to less than 1 while building inner Na_2CO_3 complexes that leads to increasing polymerization and thus higher viscosity (321).

Viscosity

Because viscosity increases with decreasing temperature, a lava flow will become more viscous during cooling, which is intuitively easy to visualize. However, the fact that the viscosity of magmas slightly decreases with increasing pressure has only become known from experiments during the last few years. In other words, basaltic magmas are even more highly fluid at their place of origin in the upper mantle.

The viscosity of a magma (h) is defined as the ratio of shear stress to strain rate and has the unit *Pascal second*, formerly *Poise* (10 Poise = 1 Pa s). A fluid, in which shear stress and strain are proportional, such as clear water, is called a Newtonian fluid (Fig. 4.10). In *Newtonian fluids*, an infinitely small strain is sufficient to start flowage. In many fluids this ratio is nonlinear; a finite strain (*yield strength*) must be applied before they begin to flow, in other words, become deformed permanently. Such fluids are called *pseudoplastic* or *Bingham fluids*, their viscosity Bingham or plastic viscosity. Yield strength and Bingham and Newtonian viscosity increase with increasing crystallinity and polymerization of a magma. When one talks about the viscosity of magmas or lava flows or ash flows, one has also always to take in consideration that

- natural systems are dynamic, i.e., they constantly change, especially by cooling, degassing and crystal growth and
- only magmas of very low viscosity without bubbles and crystals can be considered as Newtonian fluids proper.

If we look at lavas in a simplified way as Newtonian fluids, the following viscosities are rough approximations: tholeiitic basaltic lavas of the type erupted in Kilauea volcano are around 10^1 –

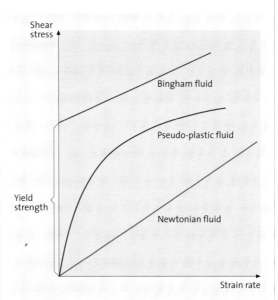

◀ Fig. 4.10. The viscosity of a magma can be shown in a flow diagram by the ratio of shear stress vs. strain rate. Most lava flows begin to flow when the shear stress is larger than the yield strength

10^2 Pa s at their liquidus temperature; rhyolitic magmas, however, between 10^6 and 10^8 Pa s. These viscosities depend, as shown above, also on the crystal, bubble and gas content and temperature of a magma at low pressures. For example, alkali-rich basaltic magmas have lower SiO_2 concentrations and contain more H_2O than tholeiitic ones and are therefore of lower viscosity. They are also richer in CO_2, which facilitates polymerization at higher pressure. On the other hand, CO_2 escapes from a magma early, because of its low melt solubility compared to other volatiles. Whether or not this leads to a significant decrease in viscosity in the upper few hundred meters or kilometers below the Earth's surface is still under debate. The influence of dissolved H_2O in lowering the viscosity is especially pronounced in granitic melts.

At low shear rates, bubbles in a melt increase the effective viscosity while the reverse is true for high shear rates. Magma viscosity increases with increasing amounts of crystals. For example, in tholeiitic basaltic magmas the viscosity at 25 vol % crystals is ten times higher than of the crystal-free magma at its liquidus temperature. Convection does not occur at such high viscosities. Hence, the magma can no longer erupt in an effusive manner, although can still explosively. Many dome lavas have > 25 % crystals and probably require shear localization along bubble-rich zones to flow.

The old problem of explaining the two most characteristic surface forms of basaltic lavas, namely *aa* and *pahoehoe* (Chap. 9; Figs. 4.11 – 4.16) is a good example of the effects of viscosity and lava yield strength. The lower the SiO_2 concentration in the lava (as a measure for the viscosity) and the higher the temperature, the lower the yield strength. In other words, the cooled surface of a lava flow becomes brittle and breaks up constantly during flow.

Yield strength is an important physical boundary for interpreting flow processes in both lava and ash flows. Flow only begins when the shear stress is larger than the yield strength. The initial formation of marginal levees, characteristic of many lava flows (Fig. 4.15), can be explained by

the high ratio of yield strength to shear stress at the cooler flow margins although most levees form at flow fronts where solid crust is pushed to the side.

Pahoehoe lavas in Hawaii generally form at higher temperature, lower viscosity and sometimes lower flow rates than aa lavas. They can change into aa lavas during flow, but not the other way around. The viscosity of flows increases during cooling, crystallization and bubble formation, as was described above. When temperature decreases, the pahoehoe (pahoehoe in Polynesian "on which one can walk") is replaced by aa because of increasing crystallinity (50). However, this is only one of the two main factors for explaining the formation of aa, either at the vent, or from transformation of pahoehoe during flow. The strain rate is the second factor (Fig. 4.16). In other words, when fluid lava stops before a critical viscosity or yield value has been reached, it is preserved as pahoehoe. If the critical value is reached during flow, because the viscosity, the strain rate or both are too high, the lava starts to form clots. These begin at its margins until, at high rates of deformation (e.g. when flowing over a steep slope), the clots separate from each other. This leads to the formation of breccias of irregular spinose clinker fragments that fall down at the front of a decelerating lava flow and are subsequently overridden by the advancing lava. This is how the characteristic zones of aa lava flows form: *top and basal breccias, marginal levees* and a massive center.

▼ Fig. 4.11. Aa lava flow on the east flank of Kilauea volcano, broken up into irregular blocks

◄ Fig. 4.12. Flowing pahoehoe lava. Pu'u 'O'o scoria cone, the eruptive center, in the background. Kilauea volcano (Hawaii, USA)

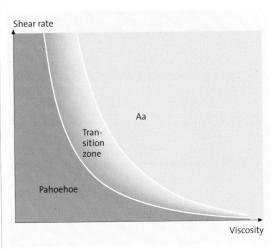

▼ Fig. 4.16. Fields of pahoehoe and aa lavas, depending on the rate of shear and viscosity (175, 261)

◄ Fig. 4.13. Alternation of pahoehoe and aa lava flows on the south flank of Kilauea volcano (Hilina Pali fault blocks)

▼◄ Fig. 4.14. Cross section through about 40 cm-thick aa lava flows with well-developed basal and top breccias. Kilauea volcano (Hawaii, USA)

▼ Fig. 4.15. Viscous phonolitic lava flow on the lower slope of Pico de Teide, Tenerife (Canary Islands), with well-developed lateral "moraines"

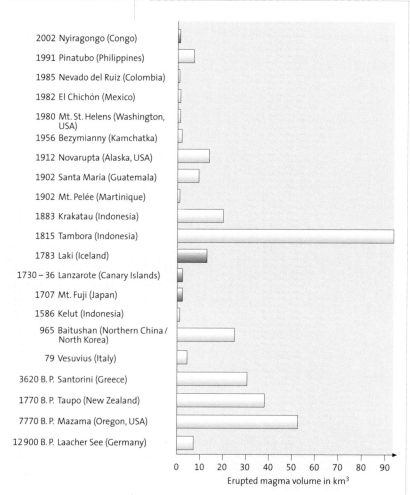

2002	Nyiragongo (Congo)
1991	Pinatubo (Philippines)
1985	Nevado del Ruiz (Colombia)
1982	El Chichón (Mexico)
1980	Mt. St. Helens (Washington, USA)
1956	Bezymianny (Kamchatka)
1912	Novarupta (Alaska, USA)
1902	Santa María (Guatemala)
1902	Mt. Pelée (Martinique)
1883	Krakatau (Indonesia)
1815	Tambora (Indonesia)
1783	Laki (Iceland)
1730–36	Lanzarote (Canary Islands)
1707	Mt. Fuji (Japan)
1586	Kelut (Indonesia)
965	Baitushan (Northern China / North Korea)
79	Vesuvius (Italy)
3620 B.P.	Santorini (Greece)
1770 B.P.	Taupo (New Zealand)
7770 B.P.	Mazama (Oregon, USA)
12900 B.P.	Laacher See (Germany)

Erupted magma volume in km³

▲ Fig. 4.17. Volumes (magma volume) of some historic, prehistoric, and Late Pleistocene evolved volcanic eruptions. Most eruptions were explosive, except for Laki, Lanzarote and Nyiragongo (modified after 387). The four basaltic eruptions are shown by dark shades

▼ Fig. 4.18. Volatile composition of basaltic (*left*) and rhyolitic (*right*) magmas

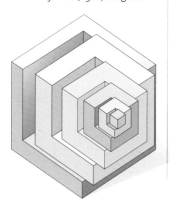

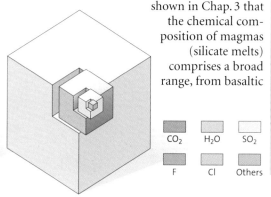

CO_2 H_2O SO_2

F   Cl   Others

Magmatic Gases

The importance of magmatic gases goes far beyond their role in volcanic eruptions most of which are explosive in nature (Fig. 4.17). The atmosphere, hydrosphere and biosphere, in other words, the entire organic life on our planet, owe their origin to degassing of the Earth during the past 4.6 billion years. The most important gases by mass are H_2O (35–90 mol%), CO_2 (5–50 mol%), SO_2 (2–30 mol%), HCl and HF. Some, such as sulfur (S), can occur in different oxidation states depending on temperature. It was shown in Chap. 3 that the chemical composition of magmas (silicate melts) comprises a broad range, from basaltic ($<$ca. 50% SiO_2 and about 25–35% $MgO + FeO + CaO$) to rhyolitic (70–75% SiO_2, 15–20% $Al_2O_3 + K_2O + Na_2O$ and $<$5% $MgO + CaO + FeO$). The three most common magmatic gases occur in these magmas not only in different absolute amounts, but also in greatly different relative concentrations (Fig. 4.18). These differences are based on the difference in type of magma source rocks, the degree of partial melting and the differentiation mechanism. Such mechanisms include, for example, the increasing concentration of magmatic gases during the course of differentiation of a mafic parent magma, and the liberation of volatile elements during subduction of oceanic sediments, oceanic crust and hydrated (serpentinized) mantle.

The amount of gases that can be dissolved in a magma depends, as discussed in the first part of this chapter, on the composition of a melt and hence on its structure and viscosity, as well as temperature and pressure. The solubility of most gases increases with increasing pressure, but decreases with rising temperature. Moreover, the volatile content of a magma rises with increasing differentiation, because the early crystallizing mineral phases (such as olivine, chrome spinel, clinopyroxene or plagioclase) do not contain volatile elements in their crystal lattices. However, some *volatiles* (H_2O, halogens) can become fractionated by mineral phases, such as amphibole, biotite and apatite. Sulfur can also be fractionated in sulfide melts and crystals or, in alkaline magmas, by the rare and beautiful mineral hauyne (Fig. 11.43). Broadly speaking, the theoretically possible, experimentally determined, and directly measured volatile contents of magmas (except CO_2) rise along with the increase of SiO_2 (Fig. 4.19). Alkali basaltic magmas are, however, much richer in volatiles than the SiO_2-rich tholeiitic basaltic magmas. Basaltic magmas characteristic for subduction zone settings are generally much richer in H_2O compared to the dry basaltic magmas erupted at mid-ocean ridges.

When the saturation point of a volatile constituent in a silicate melt is exceeded, a free gas phase forms. Volcanic gases are liberated in the following sequence in a magma: CO_2, S, Cl, H_2O and F. Some gases, such as N_2, CO_2 and the noble gases are not readily soluble in silicate melts and can therefore escape from a magma at relatively high pressures. The solubility for CO_2 is similar in most magmas from basalt to rhyolite. H_2O, on the other hand, as well as the more soluble gases S, Cl and F, are concentrated in the residual melts.

Relative proportions of different gas species emitted from a volcano are commonly constant, but this very much depends on temperature.

Deviations can herald a new eruption. The composition of volcanic gases depends on processes at depth, such as the separation of gases and melts during the formation and rise of a magma, as well as on processes that occur inside a volcanic edifice. Most remarkably, the relative proportions of gases in volcanic fumaroles agree with those that equilibrate at depth when gases separate from a magma. In other words: gases are not significantly fractionated under steady state conditions (118). The differences in the composition of gases released from volcanoes are thus determined mainly by near-surface processes as gases re-equilibrate during cooling and mixing with meteoric water, as well as during interaction with fluids of the hydrothermal system of a volcano (Chap. 15).

Basaltic magmas are so fluid that gases can escape relatively quickly and effectively from such melts. Thus, basaltic volcanoes can degas strongly during their non-eruptive phases. In more viscous magmas, such as rhyolites, however, a significant portion of volatiles remains in the melt even at moderate pressure. These can be released into the atmosphere in large amounts during explosive eruptions. Intermediate, especially andesitic magmas, are rich in volatiles that are largely derived from heated sediments, water-rich oceanic crust or even serpentinized mantle peridotite during subduction. They also have an intermediate viscosity and lower solidus temperature, so that fumaroles are most common in andesitic volcanoes. Such volcanoes are therefore the most suitable laboratory for the gas chemist.

H_2O

Water vapor in condensed form, i.e., water droplets emitted from craters, is commonly the visually most impressive aspect of an active, or at least of a still hot volcano when viewed from a distance. The vapor is a mixture of minor amounts of magmatic gases, together with larger proportions of heated ground or surface water, and atmospheric water, and commonly condenses on ash particles (Fig. 4.20). The role of magmatic H_2O is fundamental because it governs the acceleration and mass eruption rate of eruption columns. Water occurs in magmas in molecular form (H_2O) but also as hydroxyl ions ($OH-$). The presence of H_2O in magmas is reflected in crystallization of OH-containing mineral phases, such as amphibole or mica, which do not form in H_2O-poor, "dry" magmas. The concentration of water in magmas varies strongly. In general, H_2O is much more soluble in highly evolved magmas than in basaltic ones. When the chemical composition of very fresh volcanic glasses is analyzed by electron microprobe (a common instrument to

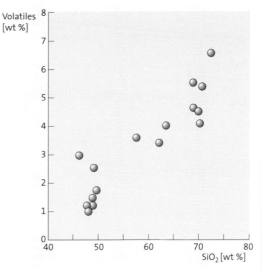

◄ Fig. 4.19. Volatiles (chiefly H_2O) increasing with the SiO_2 concentration of glass inclusions in minerals (84)

determine the chemical composition of glasses and mineral phases) the difference between the sum of oxide weight % and 100 % is a rough measure for the water content. New measurement techniques such as ion probe and FTIR allow more quantitative determinations. Fresh basaltic glasses from the deep sea (mid-ocean ridges) commonly show sums around 99 %, whereas fresh rhyolitic glasses sum to around 95 wt %. H_2O-concentrations measured in glasses or non-degassed melt inclusions (e.g., within quartz phenocrysts of rhyolitic magmas; Fig. 3.17) with the ion-probe are around 5–7 wt % (148). Magmas

▼ Fig. 4.20. Degassing of Bocca Nova crater in the summit area of Etna volcano. Hot H_2O gas is condensed to water at the boundary between the hot rising gas stream and the cold atmosphere at the summit of the volcano

erupted along subduction zones are very rich in H_2O and poor in CO_2, as shown by the analyses of fresh glasses, dredged from subduction zones and in melt inclusions of crystals (e.g., 243). Even basaltic magmas can have H_2O concentrations of up to 6 wt. percent in subduction zone settings. This H_2O is probably derived from marine sediments and altered basalts or serpentinite ("hydrated" mantle peridotite), becoming dehydrated during subduction. Water rises into the overlying mantle wedge and probably triggers its partial melting by lowering the melting point of peridotite (Chap. 8). Water and elements soluble in water, such as K and B, hence become part of the newly formed magmas. How much of the magmatic H_2O is generated during partial melting of the source rock and how much is added in the magma chamber, either by diffusion or water entering into partially emptied and refilled chambers, varies from case to case. Evidence is accumulating, however, that in basaltic as well as in highly evolved magmas a significant part of the H_2O is added only in crustal magma chambers (152, 187). Nevertheless, it is quite difficult to quantitatively determine the original water content of a magma, especially the amount ejected into the stratosphere. Approximate values, given in Table 4.1, have been derived from the analysis of quenched glasses and glass inclusions in minerals, as well as experimental studies on the stability of OH-containing phases and the maximum solubility of H_2O in natural magmas.

CO_2

The most important magmatic volatile component by mass in magmas erupted on Earth and the neighboring planets is CO_2. Carbon dioxide occurs in molecular form at high pressures (>10 kb, i.e., in the uppermost mantle, in the source area of most basaltic magmas). At low pressure, CO_2 forms CO_2- anions or metal carbonate complexes. In highly polymerized, very viscous, SiO_2-rich magmas, CO_2 cannot be dissolved as easily as in poorly polymerized melts.

Because of its high partial pressure, CO_2 can form a free gas phase even at high pressures, rise and be released into the atmosphere. CO_2 is the most common compound within fluid inclusions in crystals that form at high pressures, e.g., at the crust-mantle boundary or below. A magma with 1 wt % CO_2 will be CO_2-saturated at 2 kb, i.e., at a depth of about 8 km, where CO_2 can form a separate gas phase. It is therefore impossible to show a reliable list of CO_2-contents of natural magmas, as was done for H_2O (Table 4.1).

MORB-magmas are already saturated in CO_2 when they erupt along mid-ocean ridges at a water depth of about 2 500 meters, even though they are relatively poor in CO_2 and H_2O. The evidence for CO_2 degassing is that small bubbles found in the glass rims of submarine lavas contain CO_2, but no H_2O, which only forms bubbles at much lower pressure (228). The CO_2 released in submarine volcanoes is dissolved in seawater. In the relatively dry tholeiitic magmas, SO_2 forms a free gas phase at water depths of less than 150 m and H_2O only in very shallow water, at less than 50 m (233). The masses of CO_2 liberated along mid-ocean ridges and above subduction zones are very similar, but the source for about 80 % of CO_2 released along subduction zones may be exogenic, chiefly subducted sediments (111, 396). Intraplate volcanoes can also release large amounts of CO_2: the CO_2 emissions of Kilauea and Etna volcanoes together correspond to about 50 % of the CO_2 emission along the entire 70 000-km-long system of mid-ocean ridges (6, 111).

The CO_2 emissions of a volcano depend strongly on its activity. The estimated mean daily CO_2 output of Etna is about 35×10^6 kg/d (1975–1987) and varies, depending on its activity from $29–120 \times 10^6$ kg/d (Fig. 4.21). That of Kilauea is about 3.7×10^6 kg/d (1956–1983) (6).

Recent estimates of the CO_2 discharge at Kilauea demonstrate very rapid progress in the field and, at the same time, how incomplete our knowledge on magmatic gas budgets still is. The most current emission rates based on a number of experiments are about 8.5×10^6 kg day (t/d), several times larger than previous estimates (115). The CO_2-concentration of the primary magnesian basalt magma beneath Kilauea volcano is estimated to be about 0.7 wt %, saturation being reached at ca. 30 km below the summit. When the magma has risen to the base of the summit reservoir at about 7 km below the summit, most of the CO_2 has been exsolved, leaving some 0.09 wt % CO_2 in the tholeiitic reservoir magma.

The sudden non-eruptive release of CO_2-dominated gases is itself a significant direct or indirect volcanic hazard in some volcanoes.

▶ Table 4.1. Approximate H_2O-concentrations in natural magmas (modified from 96)

Magma composition	H_2O concentration wt %
MORB (tholeiites)	0.1 – 0.2
Island tholeiites	0.3 – 0.6
Alkali basalts	0.8 – 1.5
Basalts above subduction zones	2 – 4
Basanites and nephelinites	1.5 – 2
Andesites – dacites (island arcs)	1 – 3
Andesites – dacites (convergent continental margin)	2 – 5
Rhyolites	up to ca. 7

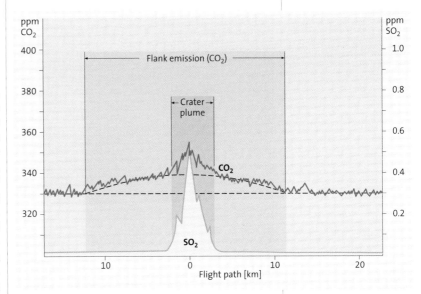

▶ Fig. 4.21. CO_2 and SO_2 degassing in Etna volcano. SO_2 degassing is confined to the actual volcanic edifice, while CO_2 is released over a much wider area, including the lower slopes and foreland of Etna volcano (6)

Around Lake Nyos in Cameroon, for example, about 1700 people died when a cloud of CO_2 was released from the crater lake (Chap. 13). In the young volcanoes of the Eifel in Germany, large masses of CO_2 are constantly released, as vividly shown along the east shore of Laacher See (Fig. 4.22). Several tens of thousands of tons of CO_2/day are recovered in the Eifel in many wells to be used in the mineral water and beer industry (119).

SiO_2-poor basaltic magmas (<45 or even <40 wt% SiO_2) are relatively CO_2-rich and of low viscosity. A famous example is the Nyiragongo lava lake in the Republic of Congo, whose gases contain up to about 50 mol% of CO_2. During an eruption in 1977, the lava lake was drained when a fissure opened along the volcano flanks. The nephelinitic lava flows were of extremely low viscosity, advanced at an incredible velocity of 100 km/h and killed about 300 people. This was one of the rare cases in which people perished during an effusive volcanic eruption. The viscosity of the lava was so low that banana leaves and elephants were covered with a thin skin of lava.

Not so long ago, scientists thought that H_2O was the main gas phase in Kilauea magma, but they are now convinced that CO_2 is much more abundant by mass.

SO_2, H_2S

The volatile compound that we notice most easily with our senses when visiting active volcanoes is sulfur. This is either by the acid stench of SO_2, or the smell of rotten eggs (H_2S), or its distinctive gray or yellow mineral deposits (Figs. 4.23, 4.24). The total amount of sulfur that can be dissolved in a magma depends strongly on its iron content. Basaltic magmas erupted along mid-ocean ridges (MORB) contain about 1000 ppm S (0.1 wt%), alkali basalts up to more than 5000 ppm. More highly evolved magmas contain much less sulfur. Exact concentrations are difficult to estimate, however, because most of the sulfur is released before and during an eruption – much supplied from basaltic parent magmas (Chap. 14) –, and there are no data available for felsic magmas at depth. The solubility of both S and HCl in magmas is complicated because of the formation of non-volatile compounds, such as sulfides, sulfates or metal chlorides.

A further complication in mass balancing of sulfur is its variable valency in magma. In general, sulfur is present in a magma in a reduced form as

S^{2-}, e. g., in the form of sulfide melt inclusions in crystals. The dominant S species in volcanic gases are SO_2 and H_2S. The solubility and relative proportion of SO_2 and H_2S largely depend on the composition and the degree of oxidation of a magma. At equal oxygen concentration, sulfur occurs as SO_2 at higher temperature and as H_2S at lower temperature. At the same temperature but decreasing oxygen concentration, SO_2 is reduced to H_2S. With increasing differentiation and therefore oxidation of a magma, the ratio of sulfate (S^{6+}) to sulfide (S^{2-}) increases, eventually even leading to

▲ Fig. 4.22. CO_2 bubbles rising along the eastern shore of late Quaternary Laacher See Volcano (12 900 years B.P.) (Eifel, Germany)

▶ Fig. 4.23. Crater of Vulcano (Aeolian Islands) that erupted in 1888/1890. Sulfur-rich gases rise along fractures (fumaroles). The islands of Lipari, Salina, Filicudi and Alicudi (from *right* to *left*) in the background

▼ Fig. 4.24. Bluish sulfur-rich gases rising from open central conduit of Masaya crater (Nicaragua). Complex evolution of alternating collapse and refilling is reflected in numerous intra-crater unconformities

crystallization of minerals such as beautiful hauyne (Fig. 11.42) or even anhydrite (202).

In fumaroles, H_2S and S_2 are the volumetrically most important sulfur components after SO_2. Even COS and CS_2, important trace gases, occur in fumaroles in equilibrium with CO_2, H_2, CH_4 and CO. In a magma column, the molecular species of sulfur may change drastically from reduced S_2^+ to highly oxidized ($SO_4^=$) in the more strongly evolved and gas-rich cupola (140).

The emission of SO_2 months or years prior to an eruption is one of the most robust types of evidence for the rise of magma from depth into high-level magma reservoirs. Likewise, massive emission of SO_2 following an eruption indicates the continuous presence or rise of new batches of largely undegassed magma. Most impressive is the large mass of SO_2 escaping from the top of Miyakejima Volcano following its spectacular caldera collapse and minor eruptions in July-August 2000, daily rates peaking at ca. 100 000 t (10^8 kg)/d in September 2000 decreasing to less than 10 000 t/d in December 2002 (Fig. 4.25; Chap. 9). Figure 9.44 shows a schematic cross-section through the volcano-magma system of Miyakejima volcano and a model of convective rise of gas-rich magma which entered the reservoir partially emptied by lateral underground drainage (Chap. 9), degassing and return flow of the denser degassed magma through the blocks generated during caldera collapse. The health hazards posed by the strong sulfur degassing has not allowed the return of the more than 5 000 inhabitants of the island evacuated in August 2000 following caldera collapse.

Emission rates of SO_2 at Kilauea volcano decreased from about 4×10^5 kg to about 1×105 kg/d between 1987 and 2000 (115). Because this drastic decrease is paralleled by steady summit deflation it most likely mirrors the decrease in magma in the summit reservoir, perhaps enhanced by increased scrubbing as SO_2 rises through the crust and traverses the groundwater/hydrothermal system above the magma chamber.

The effects of volcanic sulfur emissions on the atmosphere and the contrast between high SO_2-emission and much lower saturation concentrations in a given magma are discussed in more detail in Chapter 14 because of their importance for stratospheric chemistry and their impact on climate and ozone layer stability.

Halogens

The most important carriers of halogens during volcanic degassing are hydrogen halogenides, such as HCl, HF and HBr. Alkaline basaltic magmas contain significantly more halogens than sub-

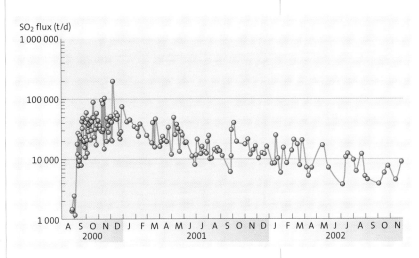

alkaline (tholeiitic) ones. Highly evolved alkaline magmas show the highest halogen contents, but reliable data are still sparse. The main Cl species in a magma are alkali chlorides and not HCl, the species released during degassing. Mass balances of halogen emissions during large eruptions (based on the composition of glasses and melt inclusions in phenocrysts) indicate HCl-emissions between 1×10^6 kg and 1×10^9 kg (mid-ocean ridge basalts and alkaline basalts, Iceland) and 10^9–10^{11} kg in andesites-rhyolites (253). Because of their relatively high solubility in magmas, only about 20–50% of halogens in a magma are released during an eruption as gas, with Cl being significantly less soluble than F. In Plinian eruptions, up to 5×10^9 kg HCl can be released, but most of it is probably adsorbed on tephra particles (249), or is removed from the eruptive column by super-cooled water and ice before the halogens can reach the stratosphere.

Fluorine (F), because of its high solubility in magma, does not readily enter the gas phase during an eruption, but remains largely in the melt. Fluorine is therefore often rapidly adsorbed onto solids, such as ash particles or vegetation, and can be washed out of an eruption column relatively quickly, HF being easily soluble in water. This is a severe problem for agriculture in areas that have been affected by volcanic gases for a long time. Fluorine-rich deposits are infamous during some eruptions because they can trigger fluorosis, a lethal disease for grazing animals. The most dramatic incidence occurred in 1783, following the eruption of Laki volcano in Iceland. In the resulting famine, more than 20% of the population of Iceland perished.

Halogen radicals that form in the stratosphere by photolysis can destroy ozone catalytically by various types of reactions. However, most eruption columns do not rise much above 20 km,

▲ Fig. 4.25. SO_2-emission rates of Miyakejima volcano from August 2000 (caldera collapse) to December 2002 (K Kazahayasan, pers. comm.)

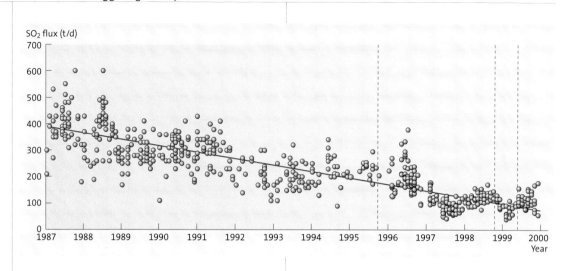

▲ Fig. 4.26. SO$_2$-emission rates at the summit of Kilauea volcano from 1987–2000 (115)

◀ Fig. 4.27. Contrasting gas compositions of Kilauea volcano. Gas of type I is released in the central conduit area (Halemaumau) (Fig. 4.28). Type II gas, released during flank eruptions, has a depleted composition compared to those in the magma rising from depth. Residual volatiles are those that remain in a lava after an eruption (112)

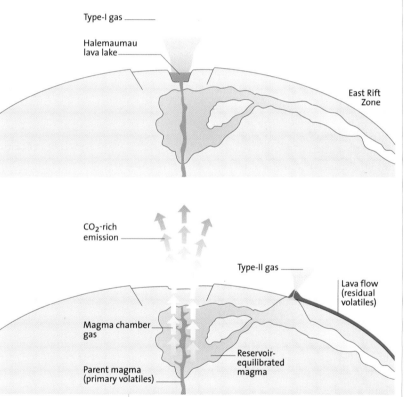

below the level of the highest ozone concentration. Destruction of ozone by volcanic gases, while significant for some very large eruptions, is thus of minor importance for most.

Volatile Budget of Kilauea Volcano

The difficulty in determining the original composition of the volatile components of a magma becomes evident in one of the first detailed reconstructions of the total gas budget of a volcano for five different volatile components (Figs. 4.26, 4.27).

The primary volatiles of tholeiitic magma in Kilauea volcano were determined by analyzing the composition of fluid inclusions in olivine crystals and the degassing of a magma rising from depth (ca. 40 km) in Halemaumau Crater (type I gas, Fig. 4.25). These are much richer in CO_2 than formerly assumed. At these high CO_2 concentrations, the gas could have been released from the magma as deep as 40 kilometers and risen as a separate fluid phase with the ascending magma (112). Once the magma has arrived in the main reservoir, the total gas composition is adjusted to the new low pressure. Here, the concentration of CO_2 decreases to about 5% and that of sulfur to about 50% of their original concentration in the magma at depth. The volatile components that are more easily dissolved in a silicate magma, H_2O, Cl and F, do not change their absolute concentration very much in the magma reservoir. The strong CO_2-degassing in Kilauea caldera (rarely active since 1925) is thus the result of degassing of magma that resides in the chamber (Fig. 4.26). We can similarly interpret the significant CO_2-degassing in many parts of the world (such as central Europe, for example in the Quaternary Eifel volcanic fields; Fig. 4.22), as the result of degassing of deep-seated magmas that have accumulated at the base of, or within, the crust.

When magma makes its way from the main reservoir beneath Kilauea caldera into the East Rift Zone, the composition of the volatiles adjusts to the new low pressure (type II gas, Fig. 4.27). At this stage, not only CO_2 and sulfur, but also H_2O occur as a free gas phase. The composition of gases (type II) that are released from magmas erupted along the East Rift Zone corresponds to the difference between the equilibrium concentrations in the magma and those on the Earth's surface. Agglutinates (magma blebs thrown out onto the crater rims of scoria cones) along the East Rift Zone contain only 30% of the original H_2O, 2% of CO_2, 12% of S while the concentration of the halogens Cl and F in the magma has changed very little. The remaining gases are called *residual volatiles* because they are easily dissolvable in these magmas. Lava flowing out of the crater will continue to degas with the gas bubbles becoming "frozen" or preserved in the lava during cooling.

Miyakejima volcano (Japan) erupted in July-August 2000. A detailed study of its gases and melt inclusions in phenocrysts allowed quantification of changes in CO_2, H_2O and SO_2-concentrations with decreasing pressure showing a similar correlation between solubility and pressure for the three main volatile species as in the Kilauea system (Figs. 4.28, 4.29). CO_2-concentrations decreased drastically from 0.1 wt% at 10 km b.s.l. (250 MPa) to 0 wt% at 20 MPa (0.5 km b.s.l.) (A) while H_2O (1.9 wt%) and SO_2 (0.12 wt%)-concentrations remained constant during magma rise from 10 to ca. 2 km b.s.l. (60 MPa), but then decreased strongly to 1 wt% (H_2O) and 0.05 wt% (SO_2) at 20 MPa ca. 0.5 km b.s.l. (B) (327). The evolution of the 2000 eruption is discussed in Chapter 9.

Formation of Bubbles

Scientists trying to explain volcanic explosions to laypeople commonly shake a bottle filled with champagne, sparkling mineral water, soft drink or beer and suddenly open the lid. What happens? The gas, in these beverages CO_2, is not visible when the lid is on. It was squeezed into the liquid at higher pressure and dissolved within it. During opening (decompression), the molecules of the gas collect and form a *gas phase* that separates from the liquid and forms bubbles which expand. The resulting foam shoots out of the bottle more or less explosively depending on the degree of shaking and dissolved gas content and gas composition. But what would happen if a thicker, more viscous liquid were in the bottle,

▲ Fig. 4.28. Change in the composition of gases in Kilauea volcano by extreme decrease in the CO_2-concentration, compared to the primary composition, above the central magma reservoir (type I) to the flank magma reservoir (type II). See also Fig. 4.27 (110)

▼ Fig. 4.29. Schematic cross section of Miyakejima volcano, upper part of conduit system, magma reservoir and changes in magma density, vesicularity and concentration of CO_2, H_2O and SO_2, with pressure (290)

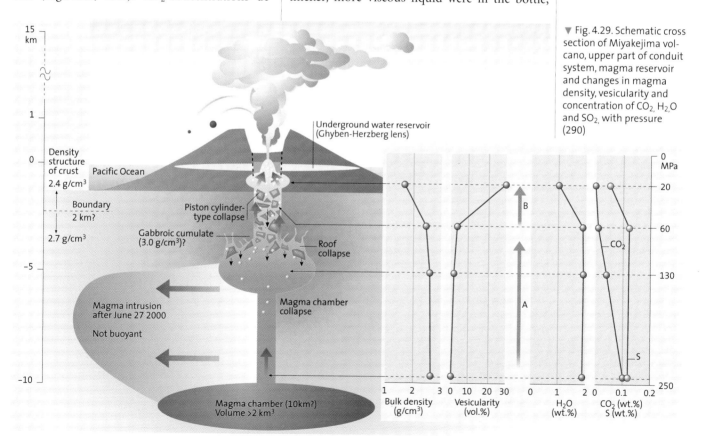

such as glycerin or honey? The simulated eruptions would look quite different.

Beginning with Scrope (319), expansion of magmatic gas during the rise of magma in the crust has been known to play a central role during the fragmentation of lava into single particles. But how do bubbles in a magma form? How do they grow? Do they rise in a passive magma or are the rise velocities of a magma so large compared to the growth velocities of bubbles that the bubbles are passively transported upward together with small crystals? What are the most important static and dynamic magma properties that determine formation, growth and rise of bubbles? Is it the composition of a magma, its viscosity, temperature, type and amount of volatiles, surface tension, or diffusion coefficients? Which of these variables depend on each other? Which are the most important processes during the actual fragmentation of a magma and what is the nature of the eruptive processes of the mixture of lava particles and gas?

Degassing of magma is mainly controlled by the dependence of the solubility of a volatile element on pressure, temperature and magma composition, as well as on the nucleation of bubbles and their growth. Magmas in the Earth's crust are normally undersaturated in volatile components. In other words, volatiles are dissolved in the melt as part of, or in between, molecules. The higher the lithostatic pressure at constant concentration of volatiles, the larger the undersaturation. With decreasing pressure, such as during the ascent of a magma, the partial pressure of volatiles increases until the magma becomes supersaturated with respect to one or more gas species. Bubbles start to grow in a melt – in other words, a free gas phase forms already at low degrees of oversaturation (<10 bar). Another process to decrease the solubility and generate bubbles is advanced crystallization of a melt to form volatile-free crystals such as feldspar and pyroxene, so that the volatile compounds become concentrated in the decreasing melt volume.

Bubbles are commonly thought to nucleate at the surface of extremely small heterogeneities, especially oxides, in a melt, just as crystals are thought to begin their growth at some impurities, a process called *heterogeneous crystal nucleation.* At surface tensions of $50-400$ dyne cm^{-1} (realistic values for natural magmas) a minimum bubble size of between $1-10\,\mu m$ forms at these low oversaturations corresponding to observations that bubbles less than $5\,\mu m$ in diameter are rare.

Bubbles grow by two processes: diffusion of volatiles into the bubbles and expansion of a bubble when the pressure is lowered. Which of the two processes responsible for the formation of bubbles (diffusion and decompression) is more effective, depends on the composition of a magma and the dynamics of a particular system, especially the rate at which a magma is decompressed or at which the bubbles rise. When magma rises to the surface rapidly, bubbles probably grow mainly by decompression. Bubble growth by diffusion is comparably slow and thus occurs only at low ascent rates. Bubbles rising in a relatively static low viscosity magma (basalt) probably expand largely by diffusion (Fig. 4.30).

The ascent rate of magma (decompression) is an important factor controlling the efficiency of degassing (271, 322). The presence of bubbles will reduce the density of a melt. Because the density difference between magma and country rock and volume expansion are the main reasons for magma buoyancy, the ascent velocity of magma will increase once bubbles have formed. However, the increase in ascent velocity is at the same time also counteracted by the viscosity that increases due to the loss of volatiles from the melt as well as by the formation of bubbles. At high shear rates, however, the presence of bubbles will decrease viscosity.

The expansion of bubbles and the rise of a magma into colder regions of the crust both lead to cooling of the magma and therefore a corresponding increase in its viscosity. With increasing tension of the bubble walls, their resistance against further expansion of the gases increases. Whether or not fragmentation begins when bubble growth stops or a short time later, when a minimum overpressure has developed, is still under debate. Perhaps it is not the overpressure in the bubbles that is responsible for fragmentation, but the high shear rate that occurs during their rapid growth in a melt, a topic discussed in more detail in Chapter 10.

Explosive Eruptions

Many people associate volcanic activity with explosive eruptions. The three most disastrous volcanic eruptions during the past 188 years were explosive: Tambora (Indonesia, 1815), more than 100 000; Krakatau (Indonesia, 1883), 36 000 and Montagne Pelée (Martinique, 1902), 29 000 fatalities (Chap. 13).

The importance of explosive compared to effusive eruptions (except for the submarine eruptions along mid-ocean ridges) is also obvious when we look at the volume of magma erupted in volcanoes. More than 90 vol% of the material erupted in historic times from volcanoes on land consists of fragments collectively called *tephra* (254) (Chap. 10). With the exception of the basaltic eruptions of Lanzarote in 1730–1736,

and Laki in 1783 all eruptions shown in Fig. 4.17 were dominantly explosive. To analyze explosive volcanic eruptions from nearby is obviously impossible. Some occur after a long time of dormancy with little recognized warning, sometimes in remote parts of the Earth. Survivors usually have made observations only from great distance, and death by volcano is a real occupational hazard for volcanologists.

The eruptive behavior of volcanoes is reflected in their products. Analysis of geologically older deposits shows that some volcanic eruptions in the geological past were much more energetic than many of the famous historic eruptions (Fig. 4.17). The deposits of the youngest, highly explosive volcanic eruption in Central Europe, e.g. that of Laacher See volcano 12 900 years ago, represent a magma volume of about 6.3 km³, more than the volumes of the Mount St. Helens' eruption (1980) and that of El Chichón (Mexico) of 1982 combined. The eruption of Taupo in New Zealand in 1 800 B.P. was almost an order of magnitude higher with a magma volume of some 35 km³.

Despite its relatively small magma volume, the great eruption of Mount St. Helens on 18 May 1980 (Chap. 10) was of fundamental importance for the science of volcanology. It occurred in an area that was relatively easily accessible and in a country with many well-trained and experienced volcanologists and sufficient financial resources. The eruption was well-monitored from the beginning by highly qualified scientists. The results of their studies have significantly advanced our understanding of the behavior of explosive volcanoes. This eruption is the best-studied example of an explosive volcanic eruption and papers on the deposits and processes are still being published (195).

External and Internal Forcing Mechanisms or Why Do Volcanoes Erupt?

Most magmas generated by partial melting in the mantle and ascending because of their lower density are either arrested in the mantle or pool at the base of the crust. Those that rise further remain temporarily or forever in one or more holding reservoirs at one or more physical boundaries within the crust. During stagnation in crustal reservoirs magmas cool, differentiate (fractionate), mix with older magma or subsequent fresh magma batches, become contaminated with wall rock or at least heat the groundwater in the pores and fractures of the surrounding country rocks. Only a fraction of magmas generated at depth ever reaches the surface.

The question of why magmas erupt at the surface of the Earth in the first place is not trivial. Indeed, triggering mechanisms vary widely and

▲ Fig. 4.30. Bursting gas bubbles in a stationary lava lake Mauna Ulu (1969). Kilauea volcano (Hawaii)

▶ Table 4.2. External and internal forcing mechanisms and triggering of volcanic eruptions

Intrinsic

Buoyancy

Differentiation and vesiculation

Second boiling

Injection of hot mafic magma into a cooler evolved resident magma

Extrinsic / Intrinsic

Decompression by sector collapse

Magma-water interaction

Extrinsic Far Field: Lithosphere

Earthquakes

Extrinsic Far Field Atmosphere and Climate

Ice sheet loading/unloading

Sea level changes

Long-term wind directions

many processes are likely interacting to provide the final push out. Below I will list several end-members of triggering mechanisms some being well-documented while others are still highly speculative. The latter are mentioned to emphasize that volcano-magma systems are an integral part of the lithosphere-hydrosphere-atmosphere-cryosphere system providing for infinite feedback scenarios.

Triggering of Volcanic Eruptions

Why does a volcano erupt?

Based on the current state of knowledge – always a safe phrase for a scientist – and attempting to view triggering mechanisms systematically we may distinguish three groups of processes, simultaneous interplay of several mechanisms being the rule rather than the exception (Fig. 4.31).

Internal Forcing Mechanisms

Buoyancy

Basaltic lavas may quietly pour-out of a conduit for hours, days or weeks or, as along the East Rift Zone of Kilauea volcano, where the current East Rift eruption has been going on for 20 years, for many decades (Fig. 4.1). The main cause of the magma rise to the surface may be the arrival of a new magma batch at the base of the high-level magma reservoir. This causes the magmatic pressure to rise until a crack is opened in the crust above the reservoir. Eruption will continue until the overpressure has dissipated. This triggering mechanism represents the most simple case and it may actually play a role in many other scenarios discussed below.

Differentiation and Vesiculation

Geologists asked how an explosive volcanic eruption is triggered would probably answer by exsolution of volatiles in a rising magma column and

resulting gas overpressure, having in mind the spectacular image of eruptive columns above a volcano (Fig. 4.6). That a gas phase plays a pivotal role during eruptions is obvious, but the answer would be in many ways incomplete.

Water and other volatiles, especially SO_2 and halogens, accumulate in the upper cupola of a magma reservoir in the most highly evolved lower density melts. These concentrations can dramatically increase during advanced differentiation and may be partly buffered by the crystallization of OH^--containing minerals. Overpressures may develop through bubbles that may force the expanding magma to eventually exceed the tensile strength of the overlying roof thus initiating explosive eruptions (26). This model may apply particularly to large explosive eruptions of highly evolved magmas, many of which are characterized by systematic chemical and mineralogical zonation as reflected in their deposits. Even the fact that highly explosive eruptive phases in multicycle volcanoes are interrupted by non-eruptive intervals (which may last for tens of thousands of years) argues for a causal connection between accumulation of a critical amount of highly evolved magma rich in volatiles as the major triggering mechanism – or at least boundary condition – for explosive eruptions. The actual fragmentation mechanism of the magma may be a combination of several processes and comprise rupture of bubble walls along a fragmentation front, shearing of the melt during ascent and by granulation, due to various degrees of thermal shock when encountering external water in the conduit system. These processes are all additional to the effects of expansion of a magmatic gas phase (Chap. 10).

Second Boiling

Another classic view holds that explosive eruptions occur because volatiles accumulate in a closed system contained magma body when many OH-free crystals – especially feldspars – have grown during cooling, leading to a concentration of the volatiles in a small volume of the more differentiated remaining liquid magma. When the saturation of this cooler rest magma with respect to these volatiles is exceeded due simply to the increasingly smaller volume of the melt fraction, bubbles form that lead to an overpressure, a process called *retrograde or second boiling*. Sudden rupture of the overpressured bubbles in turn can generate an explosive eruption. This mechanism may be responsible for the partial fragmentation of the rock roof above some evolved high level e. g., andesitic-dacitic intrusions (porphyry copper ore deposits). However, it unlikely represents a

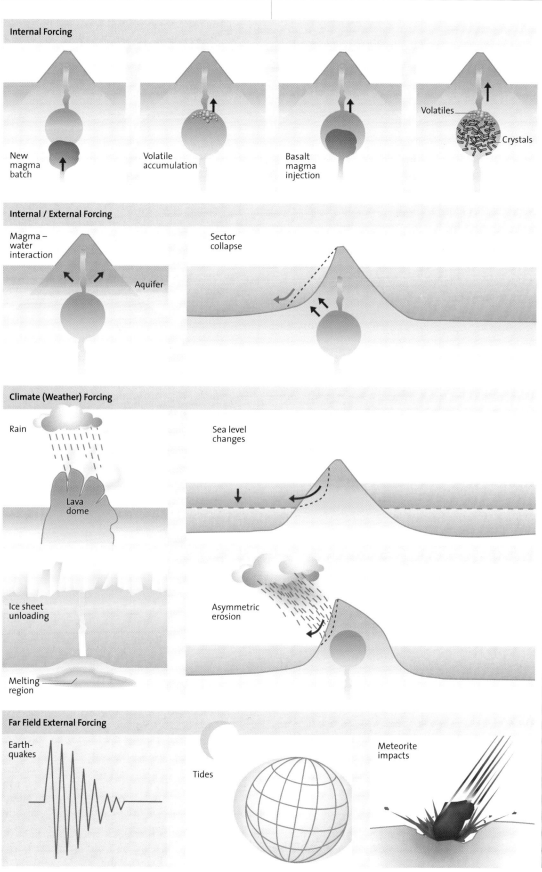

Internal Forcing

New magma batch

Volatile accumulation

Basalt magma injection

Volatiles

Crystals

Internal / External Forcing

Magma – water interaction

Aquifer

Sector collapse

Climate (Weather) Forcing

Rain

Lava dome

Sea level changes

Ice sheet unloading

Melting region

Asymmetric erosion

Far Field External Forcing

Earth-quakes

Tides

Meteorite impacts

general mechanism for explosive volcanic eruptions, because practically all well-studied deposits of explosive eruptions of highly evolved magmas begin with crystal-poor to crystal-free tephra. It should also be noted that magmas generally crystallize from the cool outer boundary of a magma reservoir inward. In other words, the explosive energy for these events developed in the still liquid part of a cooling magma body.

Injection of Hot Mafic Magma into a Cooler Evolved Resident Magma

An entirely new but physically plausible mechanism for triggering an eruption has been recognized since the mid 1970s. Geologists noted in products of eruptions the occurrence of crystals entirely out of thermal and chemical equilibrium with their host matrix. The most drastic case is magnesian olivine – a silica-undersaturated mineral phase stable in basaltic magmas – surrounded by rhyolitic, extremely silica-rich glass without a reaction rim. This is a very unstable situation and can only be explained by mixing of two or more melts of highly contrasting compositions, mixing having occurred immediately prior to eruption to prevent a reaction by quenching of the mafic melt during the eruption. The encounter may even have induced the eruption.

There are now tens if not hundreds of accounts in the literature documenting mixing of magmas and postulating the injection of hot mafic melt into a cooler resident magma as an actual eruption trigger. Processes invoked include heating of the cooler resident magma which in turn triggers convection and especially vesiculation, thus increasing buoyancy. The reported examples are so numerous that it is becoming hard to find eruptions where magma mixing did not appear to have triggered an eruption. Of course, mixing could be coincident and the actual trigger could be a different mechanism.

Internal / External Forcing Mechanisms
Triggering of Eruptions by Magma-Water Contact

Volcanic eruptions caused, or at least significantly influenced, by magma-water interaction are now recognized to be exceedingly common and occur in almost all types of volcanoes. The source of water may be seawater, lake water, groundwater and even rain. The many different scenarios for magma-water interaction are detailed in Chapter 12. In the face of overwhelming evidence from actual observations of eruptions and from the initial deposits for the ubiquitous encounter of rising magma and external water, the main problem boils down to the question of whether the explosive energy caused by sudden expansion of steam is coincident with an eruption that was triggered by another mechanism or whether it provided the critical pressure to open a crack for the final ascent of a magma to the surface.

One additional point. Why do magmas of identical chemical composition erupt in the same volcanic system explosively in some instances and effuse quietly at other times? Answers to these questions are not only important to better understand the mechanisms by which magmas erupt, but also fundamental when developing specific scenarios in preparing volcanic hazard and risk maps. There is some evidence that the availability and amount of external water may be decisive in whether or not a magma approaching the surface may stop its rise and become a subvolcanic intrusion, or become rapidly decompressed locally because the critical conditions for explosive magma-water interaction have been reached.

Decompression by Sector Collapse

The collapse of parts of the flank of a volcano by whatever cause (Chap. 9) will decompress the magma system and may allow the magma to erupt in a different style, at different rates and perhaps with a different composition than if a collapse had not occurred. An excellent illustration of this mechanism is given by the events at Stromboli volcano as this manuscript went to press. On December 26, 2002, part of the Sciara di Fuoco, the northern flank of the volcanic edifice, slid into the sea and generated a tsunamis (Smithsonian Institution). This collapse released the load on the conduit system to such a degree that lava was pouring out from the flank at 500 and 600 rather 926 m a.s.l. at much higher effusive rates than previously with the drastic side effect that the famous repetitive explosive eruptions (Strombolian activity) terminated for the time being. The first large and unusually energetic explosive eruption following the abrupt end of the Strombolian activity occurred on April 5, 2003 – following major rain storms in the area, the penetration of abundant water into the conduit system most likely triggering this eruption. Whether or not Stromboli will ever return to its famous intermittent activity remains to be seen.

Extrinsic Far Field Lithosphere

The idea that tides can exert stress on a high-level magma reservoir and trigger an eruption is old and physically plausible. Nevertheless, the number of well-documented examples is small. The general significance of this trigger mechanism is thus difficult to judge. Other far field mechanisms such as changes in spreading rate can be associated with increasing magma production and output. The main far field external triggering mechanisms concerning processes generated within the lithosphere are earthquakes.

Earthquakes

A glance at a map showing the location of most earthquake hypocenters and more than 90 % of all historic eruptions on land (see book cover) illustrates the fact that both types of major dynamic Earth events are concentrated along convergent plate boundaries, especially the Circumpacific Ring of Fire. Obviously, both types of events are in some way basically governed by the dynamics of processes above and within subducting slabs.

Nevertheless, when volcanologists are asked by the media at the occasion of a current volcanic eruption whether the eruption could have been triggered by an earthquake that occurred shortly before in the same region of the world, the standard answer is probably no. Or, with a more balanced answer: *based on the present state of knowledge*, reflecting the cautious attitude of scientists to stay clear of quick answers where temporal, areal and causal relationships between different types of Earth events are unknown or at least poorly understood. Several case histories including the 1991 eruption of Pinatubo (Chap. 13) provide tantalizing bits of evidence that there are more connections than apparent at first sight. Such answers may thus have to be modified in the future to: *in general yes, in this particular instance maybe – but we do not know exactly how.*

This change in paradigm is due to the increasing numbers of documented cases of close temporal and spatial correlation – beyond the often-cited classic example of volcano Cordon Caulle in Chile erupting two days after the great earthquake of 22 May 1960. The interactions between large tectonic earthquakes, periods of volcanic unrest and, in some cases, actual volcanic eruptions now emerges as a most interesting field of inquiry (156, 35, 192).

The arguments discussed in the literature concerning the nature of the critical interaction are based on several different mechanisms and scenarios. One view holds that tectonic earthquakes in subduction zone settings may potentially cause *stress release* around a magma reservoir and trigger the rise of the magma and surface eruptions. Much interest focusses on *earthquake-fluid interaction*, the fluid envisioned being magma or the hydrothermal system above a magma reservoir. In one scenario, bubbles in a basaltic chamber are dislodged from the surface of crystals or the walls of the reservoir by distant seismic waves, rise and thus increase the pressure above the reservoir. In another model pressure oscillations by seismic waves lead to alternating expansion and compression of bubbles thereby pumping more gas into the bubbles which in turn rise and increase pressure in the magmatic or hydrothermal fluid. It has also been postulated that *bubble nucleation* in resident magmas may be triggered by preceding seismic shocks and that dislodging crystal mushes accumulated at the top of a magma chamber may trigger or intensify convection and rise of undegassed magma to the top.

The reverse case is also possible. The rise of magma along a fracture or expansion of a magma reservoir may push the stress field beyond the critical point for failure to occur. A variant of this theme concerns the impact of volcanic earthquakes in one eruptive system on a neighboring volcano a few km or tens of km away that erupted a few hours or days later. Obviously, the second system was in a near-critical state to erupt but needed an external trigger to go. Whether or not this second volcano would ever have erupted without such a trigger or not remains an interesting but tantalizing question.

In any case, the perhaps not-so-remote possibility that major earthquakes may trigger an eruption hundreds of kilometers away – or be triggered by magma rise – has major consequences for understanding magmatic-hydrothermal systems and better recognition of the types of precursors that may herald an eruption. This is obviously very relevant for the detailed monitoring of volcanoes and planning of alert systems (Chap. 13). Finally, growing recognition that fluids may play a major role in earthquake source mechanisms suggests more causal relationships between magma generation processes beneath volcanic arcs in subduction zones and earthquakes.

Atmosphere and Climate

Analysis of direct or indirect climatic impacts on volcano-magma systems is understandably difficult and clear proofs are generally elusive. Nevertheless, there can be little doubt that volcano-magma systems can be strongly influenced by changes in sea level or ice sheet thickness. The processes invoked focus on the unloading of the lithosphere as a mechanism to decompress a fertile melting region in the mantle that is close to its melting temperature for a given pressure. A mechanism proposed from time to time is loading of the lithosphere by thick inland ice sheets and unloading during melting and subsequent rise of the lithosphere (Chap. 12). Iceland has been the focus of several studies discussing apparent increase in volcanic activity following melting of the last ice sheet some 10 000 years ago. However, the mechanism of lithosphere/asthenosphere decompression is mostly applicable to the generation of magmas but not necessarily an eruption trigger.

The effects of rises and falls in sea level and the concomitant change in pore pressure within the edifice are likely a significant factor influencing flank stability and therefore high level magma reservoirs and eruption dynamics in large ocean island systems.

Finally there is tantalizing evidence that persistent wind patterns operating for millions of years can lead to strong unilateral erosion on the windward side of large oceanic volcanic edifices (Chap. 1, Fig. 1.15). The asymmetric destabilization of volcano flanks can thus lead to increased rates of flank collapse and concomitant complex impacts on the magma chamber system.

The 864 A.D. and 1707 A.D. Eruptions of Mt. Fuji

A case in point for the contrasting eruptive modes of basaltic magma and the possible triggering effects of a preceding earthquake is Mt. Fuji, the highest and largest volcano in Japan, posing a major potential threat for a large and densely populated area, including the megacity of Tokyo. The last historic eruption occurred in 1707, 50 days after a very strong magnitude 8.2 earthquake, centered about 50 km south of the volcano. More than 90% of the two week long very explosive eruption in December 1707 A.D. was dominantly basaltic, but began with more evolved dacite (Chap. 9). Perhaps the stress release on the magma chamber beneath Mt. Fuji caused by the earthquake promoted vesiculation of a thin cap of highly evolved dacitic and underlying basaltic magmas resulting in the highly explosive eruption

(295). Experimental data show that the underlying basaltic magma, which contained only minor phenocrysts, was extremely water-rich (about 3 wt%), possibly the reason for the high explosivity of the eruption. There may also have been some magma-water interaction during the early stages of the eruption (Chap. 9).

In 864 A.D., 0.5 km^3 of basaltic magma was erupted from Mt. Fuji, volume and composition being similar to that of the 1707 eruption. Yet, both eruptions differed drastically from each other in the mode of eruption. The lavas of the A.D. 864 eruption are very rich in crystals, contrasting with the almost aphyric 1707 products. Experimental data (295) show that the 864 magma had evolved at much lower pressure (about 2 km below the surface of the volcano compared to 5 km for the 1707 eruption), having lost much of its water in the shallow storage chamber. This contrasting behavior of erupting basaltic magma necessitates two different types of hazard maps and risk management programs, one for local inundation of relatively small areas by lava flows and minor scoria cones, the other for large explosive eruptions that could affect the larger Tokyo area.

Classification of Pyroclastic Eruptions

How can the bewildering multitude of different pyroclastic eruptions be classified and named? Following Mercalli (219) and later modifications, different types of eruptions are named after the place where they were first observed or most commonly occurred: Strombolian, Vulcanian, Vesuvian, Katmaian, Pelean, Merapian, Surtseyan, etc. However, apart from the basic problem that qualitative observations are not very suitable for precise classifications, there are also insufficient systematic observations to precisely define the processes responsible for contrasting eruption styles. Moreover, this form of subdivision suffers from the fact that many volcanoes during their entire evolution, or even within hours, days or weeks during one eruptive phase, can experience several eruptive intervals with completely different eruptive styles. A further difficulty of this type of classification is the fact that only a few historic eruptions have been studied in enough detail, particularly those close to the working area of the volcanologists.

Walker (408) proposed to classify pyroclastic eruptions according to two quantifiable properties of their deposits (Fig. 4.32):
(1) The area (D), inside of that isopach (line of equal tephra thickness), which amounts to 1% of the maximum thickness: D = 0.01 Tmax. isopach.
(2) The degree of fragmentation of tephra (F), defined as the fragmentation index; the percen-

▼ Fig. 4.32. Classification of explosive volcanic eruptions, based on the areal distribution of their tephra deposits D (inside that isopach, which represents 0.01% of the maximum thickness) and the degree of tephra fragmentation F (= % < 1 mm) (408)

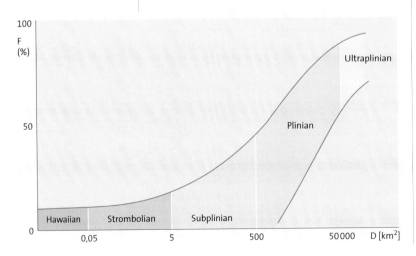

tage of tephra <1 mm at 1% of the maximum thickness, measured along the axis of the depositional fan. For this type of quantification, one has to map the thickness and distribution of the tephra fan very well, and has to carry out many grain size analyses, both very time-consuming jobs. Tephra resulting from Plinian eruptions commonly covers hundreds to thousands of square kilometers and is moderately coarse-grained. Tephra deposits of Strombolian eruptions cover areas generally <5 km^2 and are even more coarse-grained while deposits of phreatomagmatic eruptions (Chap. 12) are less widely distributed and are fine-grained. The classification of volcanic eruptions using another method, the so-called explosivity index (VEI), is presented in Chapter 13.

Summary

Temperature, viscosity and gas content of a magma can be roughly specified today based on the chemical composition of a rock and, more specifically, by examining its mineral components (which may or may not be in equilibrium with each other) and glass inclusions in the mineral phases, aided by experimental data based on whole rocks or simple systems. This data framework allows us to deduce the dynamic behavior of a magma during a pyroclastic eruption or during the flow of a lava. For example, basaltic lavas issuing from Kilauea volcano have liquidus temperatures around 1 200, while the very crystal-rich dacitic magma erupted on 15 June 1991 from Pinatubo volcano had a temperature of about 800. Most magmas have temperatures between these values, very few are hotter or cooler. Viscosities range from 10 Pa s in Hawaiian tholeiites to about 10^7 Pa s in rhyolitic lava flows, a difference of seven orders of magnitude. While H_2O contents of MORB basalt magmas are around 0.1 wt% and of alkaline basalt magmas around 0.5 wt% they can exceed 4 wt% in basaltic magmas generated in subduction zone environments. Felsic, SiO_2-rich magmas have primary H_2O-concentrations ranging from about 4–7 wt%, but very low CO_2-concentrations, generally much less than 0.25 wt%.

The most important processes that determine the composition of volcanic gases take place at depth and are intimately interwoven with the generation of different magma types and their ascent history. Basaltic magmas have already lost much of their CO_2 when they collect in near-surface magma reservoirs. Crystals that have formed at high pressures commonly contain fluid inclusions rich in CO_2, testimony to the low solubility of CO_2 in these magmas.

CO_2 and H_2O are the most important magmatic volatiles. In magmas with bubble content <1 vol%, CO_2 will be the main gas phase. At a ratio H_2O/CO_2 >1, H_2O will be the main phase that contributes to the internal pressure of a degassing magma. Pyroclastic explosive eruptions are mainly governed by expansion of magmatic H_2O.

Gases such as SO_2, H_2S and the halogens are less important for triggering explosive eruptions because they are not as abundant in a magma. The noble gases and N_2 (which are not very soluble in silicate magmas) occur in such small amounts that they also do not play a major role in the pressure build-up in a magma via gas exsolution. Some such as SO_2 are important precursors to eruptions or at least evidence that new magma has risen into the upper crust. The amount of gases that leave a vent in many sporadically active volcanoes is much higher than would be expected from the erupted magma masses. This means that gases already exsolve from a magma at greater depth and, of course, can leave a vent much more easily than the melts. This also implies that magmas in subduction zones are basically transitory storages for volatile elements, or a temporary transport medium for the volatiles generated in subduction zones and released into the atmosphere through open volcanic vents.

We know very little about the detailed processes that occur during formation of bubbles in magmas, including bubble growth and magma fragmentation, and the relative importance of dynamic parameters such as temperature, viscosity, differing gas species, internal bubble pressure etc. This is not only due to the paucity of focused experiments and theoretical models. Even the empirical analysis of eruptions and tephra particles is still in its infancy. Since there is little doubt that the overwhelming fraction of magma generated at depth never reaches the surface, the question of what actually triggers an eruption remains one of the fundamental problems in volcanology.

A systematic subdivision of triggering mechanisms is proposed. The overview lists major classic intrinsic mechanisms and several more speculative external forcings that are likely to impact volcano-magma systems. In reality, however, two or more mechanisms are likely to operate simultaneously in providing the final conditions for magma to appear at the surface of the Earth.

A volcano erupts when a critical strain is exceeded. To reach such a critical state is the result of the interplay of many different parameters and processes. Whether or not a magma is erupted or breaks through the surface depends on (1) Pre-existing conditions such as crustal structure, faults and pre-heated pathways; (2) Changes in the magma batch as it rises into cooler regions of the crust, cools, crystallizes, mixes with older resident magmas and exsolves volatiles; (3) Specific environmental conditions especially hydrology; (4) Far field lithospheric processes such as Earth tides and especially earthquakes; (5) Far field climatic conditions (atmosphere, hydrosphere, cryosphere).

Three basic types of explosive volcanic eruptions will be discussed in Chapters 10–12 using characteristic properties of their deposits. I will also discuss how scientists are attempting to ana-lyze physical processes that determine the path of a magma and its volatiles from the reservoirs into the stratosphere. Prior to that, the main types of volcanoes will be viewed in the context of their particular tectonic environment. I begin with by far the most voluminous and least understood volcanism on Earth, that which occurs at great depth beneath the oceans.

Mid-Ocean Ridges

Two thirds of our planet are covered by ocean, beneath which stretches basaltic oceanic crust. The eruptions that produce this crust are hidden from direct observation. Remote sensing instruments with which to record the countless volcanic eruptions that occur under the ocean waves each year are few and far between. Even the fact that *mid-ocean ridges* (MOR) are also belts of frequent, albeit low magnitude, earthquakes cannot be explained by magmatic activity alone. If we were able to remove the water layer of more than 3 km, we could stand in the center of the 20–30 km wide rift of the *Mid-Atlantic Ridge* (MAR), with its walls towering as high as 3 km above the valley floor at 4–4.5 km depth. In the center of this up to 8 km wide youngest and volcanically most active zone, we might never see a volcanic eruption in our lifetime. Moreover, our view along the axis of the ridge would be obstructed at less than 50 km distance by an offset of the ridge itself. Nevertheless, we can be certain that, if there were a sharply defined seam in the middle of the rift, a fracture about 3 m wide would open during a human lifetime of 70 years. This cleft would form between the Eurasian and African plates, moving eastward, and the American plate moving west. Lava would fill the fracture sooner or later. The fact that the most important crust-forming processes on Earth occur along the mid-ocean ridges and that these are magmatic in nature and associated with frequent volcanic eruptions, is based on both direct sampling and indirect geophysical measurement. Geologically older, uplifted ocean crust exposed on land, *ophiolite complexes*, provide another fundamental tool for our understanding of the structure and origin of present-day igneous ocean crust.

The Revolution in the Earth Sciences

The existence of the *Mid-Atlantic Ridge* (MAR) has been known for about 150 years. But not until the 1950s was it recognized to be part of a 70 000-km-long globe-encircling ridge system. These ridges, the most impressive continuous morphological element of the Earth, rise up to 3 km above the older adjacent deep ocean basins. Although Alfred Wegener in 1912 had recognized the fundamental difference between oceanic and continental crust, the significance of mid-ocean ridges in the creation of oceanic crust was not

As yet our attention has been confined to the phenomena of those volcanic vents which open at once in the atmosphere; but it must not be forgotten that such appertures are liable to be created on any point of the globe's surface, and therefore, as well on those which are covered by permanent bodies of water, as on the dry surfaces of our continents. Indeed, from the far greater extent of surface of the former character, which exceeds that of land in the proportion of three to one, we might expect the number of subaqueous eruptions infinitely to exceed those that take place in open air.
G Poulett Scrope, Considerations on Volcanos, London, 1825

understood until much later. These new insights were the result of systematic morphological mapping of the ocean floor by depth soundings, beginning with the Challenger Expedition in the late nineteenth century, and even earlier with the trans-Atlantic cable laying, culminating in the 1950s (Fig. 5.1, Table 5.1). The phase of morphological mapping was followed, and later accompanied, by detailed remote sensing employing a variety of geophysical techniques, most important among these being seismic, gravity, magnetic and heat flow methods.

▼ Fig. 5.1. Global distribution of mid-ocean ridges, representing different spreading rates (see Fig. 5.4)

Slow
Intermediate
Fast

The ridges in the Atlantic and Indian Ocean basins lie roughly halfway between the neighboring continents. By contrast, the *East Pacific Rise* (EPR) is located near the eastern margin of the Pacific Basin, with two offshoot branches pointed toward South and Central America. The EPR itself continues northward into the Gulf of California. It disappears beneath the continental crust of

▶ Table 5.1. Seismic subdivision of oceanic crust. Highly simplified (243)

Layer	Thickness (km)	P-wave velocities [km s⁻¹]
2	1.71 ± 0.75	5.1 ± 0.63
3	4.68 ± 1.42	6.69 ± 0.26
4	(Mantle)	8.13 ± 0.24

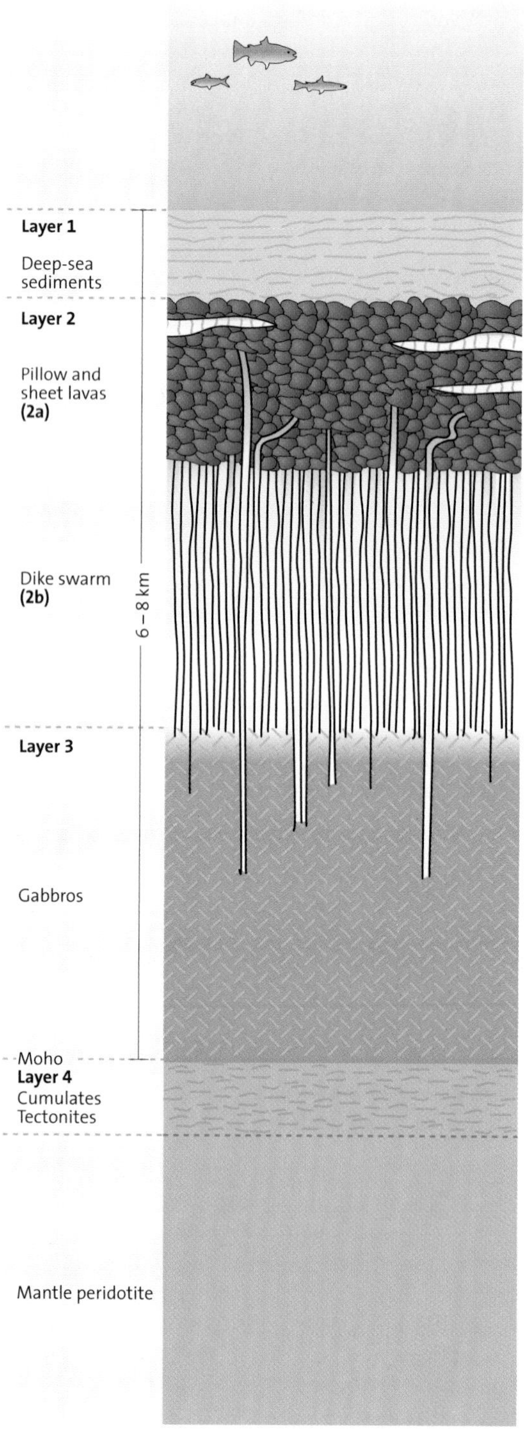

▶ Fig. 5.2. Cartoon of the structure and composition of uplifted oceanic crust (ophiolite). *Layer 4* Mantle, overlain by ultramafic cumulates at the base of the plutonic sequence of the magma reservoir, which are below the seismic Moho because of their high density. *Layer 3* consists in its lower part mainly of tectonically banded and in the upper part of massive and isotropic gabbros. *Layer 2* consists of basaltic feeder dikes in the lower part and an extrusive layer, about 1–1.5 km thick, in its upper part. The extrusives consist of pillow lavas and sheet lavas, more rarely brecciated lavas. This igneous crust is overlain by different types of deep-sea sediments, e.g., by umber (low temperature hydrothermal iron and manganese deposits), red deep-sea clays, chert and, depending on the environment and water depth, clastic sediments (turbidites) or limestones

western North America and the Basin-and-Range Province of Nevada (at whose northern end the birthplace of the Yellowstone plume is thought to be located; Chap. 7). The EPR re-appears off northwestern US, where it is called *Juan de Fuca Ridge* (Chap. 6) and disappears again to the north beneath Alaska. The ridges are offset laterally at intervals of between a few km to 50–100 km by many transform faults, especially in the Atlantic and Indian oceans. The lateral fracture zone continuations of present-day *transform faults* can be followed hundreds to thousands of km, extending even into the continents in some areas (Fig. 2.5).

Geophysical Studies of the Ocean Crust

Seismic studies turned out to be the most important tool to elucidate layering within the ocean crust. The ocean crust is on average 6 km thick and is subdivided into three main layers (Table 5.1): Layer 1 is composed of sediments that are very thin at or near the mid-ocean ridges, and up to several km thick next to continents. Sediments are especially thick along *passive continental margins*, where sediment has accumulated for more than 150 million years (such as in the older part of the Atlantic Basin).

Layer 2 is basalt. It is subdivided into two sublayers, contrasting strongly in structure and origin. The upper part, 2a, is the extrusive oceanic crust, which consists dominantly of *pillow lavas* and lesser volumes of *sheet lavas* (see below). Layer 2a is 1 to 2 km thick and is seismically characterized by a very high velocity gradient from about 2.5 km/s at the top to 5 km/s at the bottom (Table 5.1). Dikes that penetrate into layer 2a locally, make up practically all of layer 2b, forming a *sheeted dike swarm* with a thickness of roughly 1 to 2 km (Fig. 5.2). These dikes chart pathways for magma that rose from magma chambers (layer 3) to reach submarine volcanoes (layer 2a) (Fig. 5.3). This sheeted dike swarm is the most important structural argument in interpreting ophiolite complexes as having been formed by a rifting process. Layer 3 is distinguished from layers 2a and 2b by having relatively constant seismic velocities (6.5–7 km/s) and a thickness of approximately 5 km. This layer consists dominantly of gabbro, i.e., basaltic magmas that cooled slowly in the subsurface.

The lithological interpretation of the seismically defined layers within present-day ocean crust was achieved by comparing the relative thickness of these three layers and their seismic velocities with the stratigraphy of ophiolite complexes on land. The most important of these ophiolite belts extends roughly from the eastern Alps through the Balkans, eastern Mediterranean (notably

▶ Fig. 5.3. Pillow lavas, breccias and feeder dikes in the extrusive lava series of the Troodos Ophiolite (Cyprus). Geologist as scale. Klirou Bridge (Akaki River Canyon)

Turkey and the island of Cyprus) through Syria, Iran, Oman, Pakistan into China. These rock series may be several km thick and consist of basal peridotite overlain by gabbro, sheeted dike swarms and extrusive basalts (Fig. 5.2).

Ophiolites are interpreted as fragments of oceanic lithosphere that were once trapped and uplifted during plate collisions. The Troodos ophiolite on Cyprus and the Oman ophiolite in the northwestern part of the Arabian Peninsula are classic and well-studied examples. The analogy between the seismically subdivided ocean crust and ophiolites is still fundamentally accepted, even though detailed geochemical studies, beginning in the late 1970s, have shown that nearly all of these ophiolites did not form at mid-ocean ridges but rather at back-arc spreading ridges above subduction zones in relatively small ocean basins (257; Chap. 8).

The processes that form oceanic igneous crust were further elucidated by measurements of heat flow, showing that much more heat is emitted from the Earth along the mid-ocean ridges than along their flanks and in the basins. This has been explained plausibly by hot mantle rising beneath the ridges. Moreover, the famous magnetic stripes discussed in Chapter 2 run not only parallel to the ridges, but are also symmetrical, providing strong evidence for the decreasing age of the crust toward the center of the ridges. Most volcanic eruptions forming the ocean crust occur in the central accretion zone, where the oceanic lithosphere is split episodically and heals more or less continuously by emplacement of magma.

Further evidence for the youthfulness of the oceanic basins came forth when the famous "Glomar Challenger" drilling vessel began decades of fundamental research in 1969 by drilling deep holes in all major ocean basins. Unfortunately, the entire oceanic crust was never completely penetrated, although sufficient sections of basalts, less commonly of the dike swarms and more rarely of gabbros were recovered to provide ground truth to verify the seismic/ophiolite model of the ocean crust. From 1974 onward, diving expeditions of the American "Alvin" and the French "Cyana" manned submarines explored minute sections of very young volcanic terrain in the Atlantic and Pacific. These surveys dramatically increased our knowledge of the morphology and rock composition and processes along the central accretion zones in both the Pacific and Atlantic oceans. During such dives, the spectacular black smokers

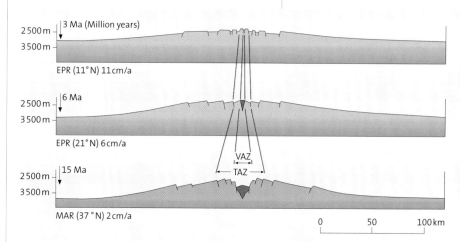

were also discovered in the Pacific in 1979, as discussed in Chapter 15.

Ridge Morphology and Tectonics

We can view the ocean crust as a gigantic jigsaw puzzle composed of millions of interleaved volcanic complexes, underlain both by basalt-filled fissures through which the magma rose (sheeted dikes), and by the plutonic rocks of the slowly cooled magma chambers beneath (Fig. 5.2).

The MAR (Figs. 5.4, 5.5a) consists of a central graben 3.5–4.5 km deep, bordered on both sides by very steep inner walls and more gently dipping outer flanks. Farther away, toward the continents, lie abyssal plains with an average depth of 5–6 km b.s.l. Within the central graben, flank mountains rise to about 1.5–2.5 km b.s.l. The axis of this ridge is marked by a 35 km-wide

▲ Fig. 5.4. Schematic cross sections through mid-ocean ridges representing different spreading rates. *VAZ* Volcanically active zone (dark gray); *TAZ* tectonically active zone; *EPR* East Pacific Rise; *MAR* Mid-Atlantic Ridge (53; 205)

▶ Fig. 5.5. Morphological comparison of the Mid-Atlantic Ridge and East Pacific Rise. The Mid-Atlantic Ridge a) is characterized by a central graben that is crossed in its upper part of the image by the Kane Fracture Zone. The East Pacific Rise b) shows an overlapping spreading center between the Siqueiros Fracture Zone in the south and the Clipperton Fracture Zone in the north. From Detrick (WHOI), RIDGE Initiative, 1992, p. 31

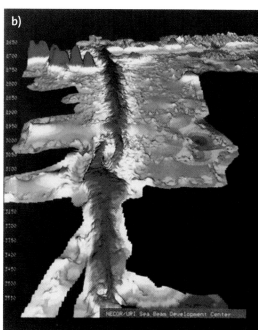

▶ Fig. 5.6. East Pacific Rise between 7°S – 9° 40°. Here, the deformation of the ridge occurs along a transform fault, with seamounts formed on both sides of the central ridge (60)

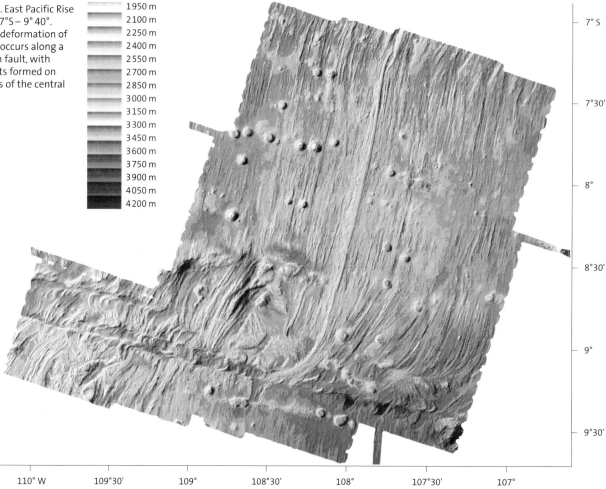

◀ Fig. 5.7. Tectonic graben with sheet lavas broken along several faults. Hydrothermally altered, intensely colored rocks on both sides of the faults. Width of photograph about 50 m. White chalk overlies the pillow volcanoes in the background. Pediaeos Canyon (Troodos Ophiolite, Cyprus)

and 1.5–3 km-deep tectonic and morphological rift zone along the axis (MAR, Fig. 5.5a). In some places, such as the Reykjanes Ridge, south of Iceland, the MAR lacks such a graben and changes into the characteristic oceanic rift morphology only as far south as 61°N, 700 km south of Iceland.

Volcanically active zones marked by volcanic edifices (VAZ, Fig. 5.4) are commonly less than 500 m wide, but can change their position frequently in the more than 8 km-wide inner graben zone, in a process called rift jumping (Fig. 5.5b). The main rift is formed by uplift of the lithosphere, which migrates away from the zone of accretion. The *tectonically active zone* (TAZ) is thus four to five times wider than the VAZ (Figs. 5.4–5.7). The irregular morphology of the volcanic crust caused by these faults is preserved as the crust migrates away from a mid-ocean ridge until it becomes completely buried by sediments closer to the continents.

There are significant differences between axial zones in the Atlantic and Pacific Ocean basins that are associated with contrasting spreading rates. At intermediate spreading rates (5–9 cm/a), the central rift is only 50–200 m deep, superimposed onto a gently rising central morphological high. A good example is the East Pacific Rise (EPR) at 21°N. At higher spreading rates (9–18 cm/a), no central rift zone is formed, instead, a 3 km wide and 500 m high axial swell is present, such as over a large part of the EPR (Fig. 5.6). At 12°50'N on the EPR, the central rift (VAZ) is only 10–50 m deeper than the flanks and about 600 m wide. The TAZ is characterized by horst-and-graben structures with minor offsets and is about 4 km wide.

The morphology in the central accretion zone of the MOR is largely governed by volcanic landforms. Lava flows have fresh glassy crusts and are free of sediment cover. Volcanic edifices are mostly elongate and parallel to the axis of the rift. Their height depends on the spreading rate. In the Atlantic, volcanoes are up to 250 m high, 1 km wide and up to 5 km long (Fig. 6.4) (342). Volcanic edifices are longer at intermediate spreading rates, but only up to 50 m high. At ridges characterized by fast spreading such as along the EPR where water depths are constant over large distances, volcanoes are 1 to 2 km wide, up to 100 km long and resemble the gentle shield volcanoes of the type found in Hawaii. Such dimensions cannot be reached by volcanoes in the Atlantic for the simple reason that axial zones are subdivided into segments by major transform faults, each segment being no longer than 50 km, whereas segments are about 200–300 km long along the EPR.

The magmatic, tectonic and volcanic crust-forming processes along MOR do not only depend on the spreading rate. The transform fractures which offset mid-ocean ridges are deep-reaching faults that allow water to penetrate far into the crust. This leads to drastic cooling with major consequences for the stability and evolution of magma reservoirs. Hence, long-lived magma chambers do not generally extend to the fault zones. The extrusive crust along these fracture zones is thus generally thinner and more alkaline basaltic magmas can rise from greater

▲ Fig. 5.8. Densely packed pillow lava along Akaki River Canyon (Troodos Ophiolite, Cyprus)

▲ Fig. 5.9. Pillow lavas in the uplifted submarine part of the island of La Palma (Canary Islands) with clear, tube-like lava structures (Barranco de las Angustias)

depths. Moreover, diapirically risen serpentinized mantle material (olivine, the major constituent mineral of peridotite, becomes serpentinized at temperatures less than 500° C) has been documented from a number of fracture zones and is also exposed directly on the sea-floor in several areas in the Atlantic.

Pillow Lavas and Pillow Volcanoes

Characteristic morphologies of lava flows interbedded with marine sediments have been recognized as early as the nineteenth century and taken as evidence for lava having erupted under water. Such lavas were believed to consist of round to elongate kidney-shaped basalt bodies, about 0.5– 1 m in diameter, and hence were called *pillow lavas* (Figs. 5.8, 5.9). Closer studies have shown, however, that such "pillows" are merely cross sections of lava tubes, although the idea that pillow lavas consist of isolated lava sacks still survives in many textbooks.

The morphology of volcanoes grown in the central rift zones, especially along the MAR (major central rift zones are rare along the EPR) have been mapped with present-day methods, such as swath bathymetry. But we must go to exposures of ophiolite complexes on land or uplifted ocean islands to study the cross section of submarine volcanoes and thus be able to reconstruct their evolution in more detail. Areas where such pillow volcano cross sections are particularly well-exposed are the Oman Ophiolite, the uplifted part of La Palma Island (Canary Islands) and especially the magnificent exposures along the Akaki and Pediaeos canyons of the Troodos ophiolite on Cyprus (Figs. 5.7–5.10).

Most submarine volcanoes consist chiefly of pillow lavas (Figs. 5.8–5.10). The tubes form at low to intermediate eruptive rates when lava is extruded under water, most likely above short fissures and, locally, central vents. The cross sections of the main feeding tubes at the base of such pillow lava volcanoes are up to 10 m wide and 2–4 m high and are called *megapillows*. These are the primary basal distribution systems, through which large volumes of lava are transported. As a pillow volcano grows, which can be compared to a complex root or branch system of a tree, the tube diameters decrease. During the very last growth stage of a pillow volcano, whenever decreasing amounts of lava can be transported through smaller and smaller diameter tubes, buds can form and become detached, especially if a quickly grown volcano has developed steep flanks. Such isolated pillows and collapsed pillow tubes may form a rubbly sheath, enveloping the flanks of a pillow volcano.

Sheet Lavas

Until the 1970s, opinion prevailed that basaltic lava extruded under water forms only tube (pillow) lavas. More massive basalts, several meters thick, found by deep sea drilling and also known from land, were commonly interpreted as intrusions (sills) and not as sheet-like lava flows. It is now recognized, however, that at least 20 to 30% of all submarine lava extrusions form sheets, generally 3–8 m thick and up to several km in length (Figs. 5.11, 5.12). Because pillow lavas and sheet

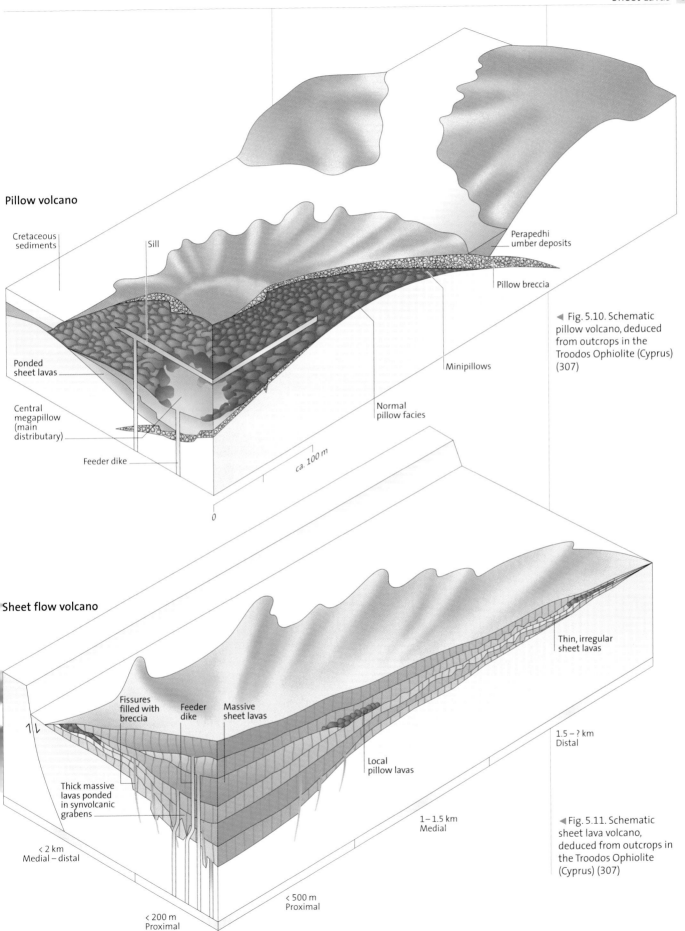

Pillow volcano

Cretaceous
sediments

Sill

Perapedhi
umber deposits

Pillow breccia

Ponded
sheet lavas

Minipillows

Central
megapillow
(main
distributary)

Normal
pillow facies

Feeder dike

ca. 100 m

0

◀ Fig. 5.10. Schematic
pillow volcano, deduced
from outcrops in the
Troodos Ophiolite (Cyprus)
(307)

Sheet flow volcano

Thin, irregular
sheet lavas

Fissures
filled with
breccia

Feeder
dike

Massive
sheet lavas

Local
pillow lavas

1.5 – ? km
Distal

Thick massive
lavas ponded
in synvolcanic
grabens

1 – 1.5 km
Medial

< 2 km
Medial – distal

< 500 m
Proximal

< 200 m
Proximal

◀ Fig. 5.11. Schematic
sheet lava volcano,
deduced from outcrops in
the Troodos Ophiolite
(Cyprus) (307)

▶ Fig. 5.12. Andesitic submarine sheet lavas with thick columns in the lower and irregular thin columns in the upper part. Graben in center of photograph shown by dashed lines. Width of photograph about 100 m. Pediaeos Canyon (Troodos Ophiolite, Cyprus)

▼ Fig. 5.13. Schematic of an eruptive cycle with lateral migration of the accretion axis based on evidence from Troodos ophiolite:
a) Sheet lava eruption at fast spreading episodes;
b) Young pillow central volcano at slow spreading (307) (based on 19)

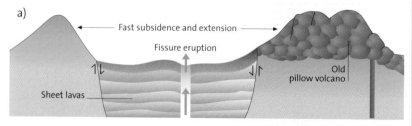

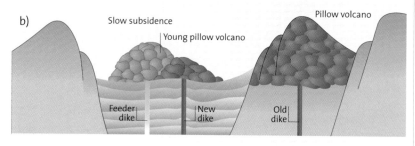

Based on observations on the EPR and the Troodos Ophiolite, sheet flows appear to commonly mark the beginning, and pillow lavas the end of individual eruptive phases or periods (Fig. 5.13). In many ophiolite complexes, more viscous lavas of andesitic to dacitic composition are extruded, which, because of their higher viscosity, form more irregular volcanoes, consisting of highly brecciated lavas (307). These types of lavas and volcanoes are rare on mid-ocean ridges.

Pyroclastic Eruptions in the Deep Sea?

The pressure at 2.5 km water depth amounts to about 0.25 kb. At these high pressures, gases dissolved in magma have little chance to form a free gas phase and therefore bubbles. The exception is CO_2, which is not very soluble in magma and which may form small vesicles even at great water depths. During the last few years, evidence has been mounting that basaltic magmas that erupt above subduction zones (Chap. 8) (and many ophiolites have basically formed above subduction zones) are much richer in water than basalts erupted along mid-ocean ridges or within plates. Vesicles formed by gas exsolution are thus common in such lavas. Similarly, even alkaline intraplate basaltic lavas erupted during the deep submarine early stage of oceanic islands can be surprisingly vesicular at water depths exceeding 1000 m. Some agglutinate-like deposits in the lava

lavas show identical chemical compositions and thus magma viscosities, the contrasting flow behavior must depend on factors other than chemical composition or physical properties. In analogy to certain eruptive styles in Hawaii (Chap. 6), submarine sheet lavas are thought to be the result of high eruptive rates, while pillow lavas form at lower rates (19). Such sheet flows generate morphologically subdued types of volcanoes.

series of the Troodos ophiolite on Cyprus having erupted at water depth of some 2 000 m are interpreted as resulting from submarine lava fountains either by degassing lava or by eruption of lavas at high pressure through narrow openings, analogous to spraying of water pressed through a hose.

How Common Are Submarine Eruptions?

Submarine eruptions, like eruptions on land, are episodic events. As mentioned earlier in this chapter, no submarine volcano has yet been observed to form directly along mid-ocean ridges, even though the entire upper crust consists of volcanic rocks. It appears that in the Atlantic, with its slow spreading rates, eruptive periods (lasting perhaps a few years) may be repeated at the same place in intervals of more than 10 000 years. Eruptive intervals along the EPR may be separated from each other by only hundreds to thousands of years. The probability of observing a volcanic eruption is thus much higher along the East Pacific Rise. Indeed, repeated surveys by research vessels or submarines of the same area along the Juan de Fuca Ridge (off the northwest coast of the United States) have detected lava flows less than a few months or years old (Chap. 6). Explosive eruptions of volcanoes, whose top is only a few tens of meters below the water surface, generate acoustic waves that can be recognized at distances of thousands of km. However, seismic signals of most deep-water eruptions (volcanic or harmonic tremors) are generally too weak to be detectable at larger distances with presently available instruments.

Because the extrusive ocean crust is 1–1.5 km thick and the visible volcanoes only 50–200 m high, the crust along the central zone of the rift must sink continuously or, more precisely, episodically, either by vertical subsidence or by rotation toward the axial zone. The entire extrusive crust thus forms by placing about 5–15 volcanic complexes on top of each other. After the end of an eruption, the crust continues to become fragmented structurally for millions of years, probably mainly by curved *listric faults* that are steep near the surface and shallow toward the axial zone (see below). Such faults are also typical of continental rifts (Chap. 7). These tectonic processes are complex in detail and poorly understood. The ocean crust is cooled in the axial zone mainly by seawater penetrating several km downward along deep-reaching fractures (a scenario discussed in more detail in Chap. 15). If one wants to better understand volcanic eruptions on the seafloor, one has to look at the rock complexes below volcanoes that make up about 75 % of the oceanic crust.

The Roots of Mid-Ocean Ridge Magma Chambers

Hot mantle rises beneath mid-ocean ridges from the asthenosphere to shallow magma chambers. This mantle material has previously lost part of its easily melting components probably early in Earth history and is called *geochemically depleted*. It partially melts up to about 20 % at relatively shallow depths (15–70 km, in the presence of small amounts of H_2O, but also as deep as 150–200 km).

Detailed work has been carried out in an area of rapid spreading, about 1 500 km west of South America, where the Nazca and the Pacific plates migrate away from each other with velocities of 15 cm/a. Seismic velocities show that melt occurs as deep as 100 km, possibly as much as 150 km, i. e., far below depths of about 70 km where garnet is stable. This zone also appears to be several hundred km wide (98) (Fig. 5.14). Electromagnetic measurements also show an area of high conductivity, which reaches as much as 200 km below the ridge. Whether or not this high conductivity is due to partial melts or higher amounts of water in the mantle is the subject of current studies. The results of these large experiments have not yet been completely processed. The extremely wide zone of mantle material with low amounts of partial melt (perhaps 1 % or less) confirms earlier hypotheses that the mantle rises passively beneath mid-ocean ridges and that the plates are probably pulled by the subducting plates in subduction zones.

Basaltic magmas, directly risen from the Earth's mantle, have MgO concentrations of 12–18 wt % and are named *picrites*. Since MORB lavas generally have compositions with less than 9 wt % MgO, the picrite magmas must have differentiated chiefly by fractionating olivine during their ascent, as well as in crustal and subcrustal magma reservoirs. The existence of magma reservoirs containing "pure" basaltic melt, their size, longevity, spacing along ridges etc have been highly debated.

Gabbros, the plutonic rock that forms when basaltic magma cools slowly, are the prime candidate for making up layer 3 based not only on seismic evidence. The chemical and mineralogical composition of the surface lavas shows that magmas within reservoirs cool at low pressure (0.5–3 kb) by crystallizing mainly olivine, clinopyroxene and plagioclase. Gabbros have been drilled in several ocean basins and dredged repeatedly from submarine exposures of uplifted parts of the lower oceanic crust. MORB gabbros derived from old ocean crust also occur as fragments in some

5

▶ Fig. 5.14. Asymmetric spreading rates, temperature distribution and region of partial melting beneath the East Pacific Rise. The *gray arrows* show the direction of mantle flow, deduced from seismically inferred crystal orientations (98)

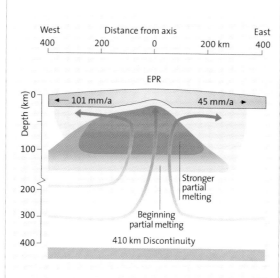

▼ Fig. 5.15. Models of magma chambers in cross section a) and in longitudinal section b), showing a ridge at high spreading rates and high magma intrusion rate, typical for the East Pacific Rise (EPR). The main elements of this model are thin, sill-like magma bodies, 1–2 km beneath the ridge axis, which grade downward into a partially solidified crystal mush zone (>50 vol% crystals). This is surrounded by a transitional zone, made up of solidified but still very hot gabbro (342)

oceanic islands such as the Canary Islands, brought to the surface by the much younger island-forming alkaline lavas. More than half of the igneous rocks in ophiolites are gabbros. But the problem as to the dimensions and structure of the reservoirs they crystallized in has been very elusive.

Geophysicists, using a variety of methods, have tried for decades to document active magma chambers below mid-ocean ridges. For example, since seismic waves become attenuated when passing through liquid reservoirs, magma chambers can be documented by a gap in earthquake epicenters. At the surface of such chambers, characteristic reflecting areas are observed, *bright spots* in a seismic imaging profile. In addition, measurements of electric conductivity and gravimetric characteristics are useful for detecting magma chambers or, at least, melts in some crustal areas. Earlier models, proposed in the 1970s to 1990s, envisioned large magma chambers, e. g., below the Mid-Atlantic Ridge that were commonly expounded in textbooks. Such models are, however, unrealistic.

Magma chamber dimensions differ significantly from each other, depending chiefly on the spreading rates (342). At fast spreading rates along the EPR, there is probably a relatively thin and narrow lens of melt along the ridge (tens to hundreds of m thick, 1–2 km wide). This overlies a crystal mush, itself surrounded by a transitional zone of largely solidified crust with isolated smaller magma pockets (Fig. 5.15). Magmas can differentiate in the melt lenses in the upper part of such complex magma reservoirs along fast spreading ridges. Lavas emitted from such melt lenses are often more strongly fractionated and enriched in iron, forming *ferrobasalts*. Episodic rifting characterizes plates at ridges with high magma supply rates, underlain by relatively large volumes of low viscosity melt. However, eruptions are probably not directly linked with the episodic rise of magma batches from the mantle into the crust.

Magma reservoirs at intermediate spreading rates appear to consist of small, discontinuous melt lenses that often form at the tip of rift zones as well as overlapping rifts. These are the optimum conditions for prolonged magma differentiation. The critical magma supply rate necessary to stabilize melt lenses in the upper part of such complex magma reservoirs corresponds to full spreading rates as low as 70 mm/a, a rate also correlated with an abrupt change in the morphology of the ridge axes.

No stationary reservoirs of largely liquid magma can develop beneath ridges with low spreading rates (and thus low magma supply) and also near transform faults (Fig. 5.16). Surface eruptions appear to be related to periods of injection of new magma rising from the mantle. Magma mixing reflected in disequilibria between the melt and different types of phenocrysts is especially pronounced in lavas from ridges with

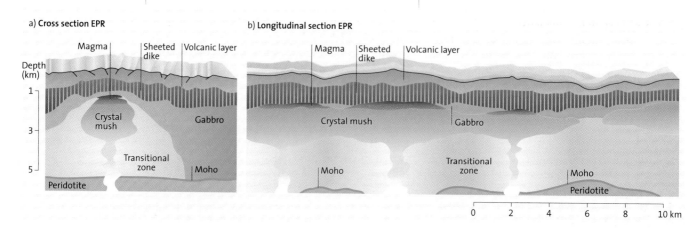

low magma supply. This may be due to reaction between crystals in the crystal mush and newly injected primitive mantle magmas. When areas with low magma supply rate are tectonically extended, eruptible magma is commonly not available. Such rifting episodes are called *amagmatic* and rift zones and deep rift canyons develop along such constructive plate margins. It is in these zones that serpentinized mantle peridotites appear at the surface, such as in many places in the North Atlantic.

Summary

The fundamental element of the revolution in the earth sciences in the 1970s was the robust evidence that ocean basins are geologically young features (the oldest ocean crust being around 200 million years old). Seismic, magnetic and heat flow studies coupled with the lithistratigraphy of ophiolite complexes (uplifted oceanic crust mostly formed in small basins) identify the mid-ocean ridges as sites where plates split apart and are healed by magma rising from decompressed mantle.

At fast spreading rates, episodically rising magma batches accumulate in the volcanically active zone and near-surface magma reservoirs, where they cool and start to crystallize. At the bottom of such magma reservoirs, i. e., in the boundary zone between the crust and the mantle, layers of early crystallizing and relatively dense mineral phases, such as olivine, pyroxene and spinel, accumulate followed at lower temperature by plagioclase. Most of the magma crystallizes to the relatively homogeneous rock, gabbro.

Magmatic differentiation beneath mid-ocean ridges occurs in two depth zones. If the rising melt flows through the transitional and crystal mush zone, chemical trends can be generated that correspond to the in-situ fractionation in the overlying melt lenses. Because magma in the melt lens is near its liquidus and is of low viscosity (ca. 10^2 Pa s), it can easily convect. The crystal mush consists of about 25 % crystals and can no longer erupt. Steady state magma lenses develop above a crystal mush (the main part of a magma reservoir), only at fast spreading rates, where they form connected melt lenses. They can also form at intermediate spreading rates, where the melt lenses are more isolated from each other, allowing especially pronounced differentiation. Magma chambers are not stable at spreading rates of less than 2 cm/a, in other words, the magma supply rate is lower than the cooling rate. Hence, one can only expect magma chambers with very small diameters and very short lifetime in the Atlantic.

Magma rising episodically in the central accretion zone through the 1–2 km thick roof

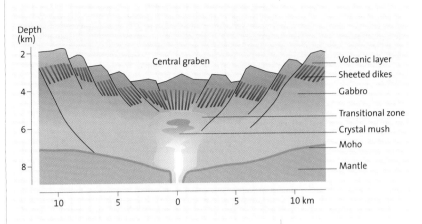

cools to form basaltic dikes above the magma chamber and is erupted on the seafloor either in large sheets or through tubes to form elongated volcanoes 50–200 m high. In the following eruptive period (which may be decades or thousands of years later), the crust splits either next to a volcanic edifice (perhaps at low spreading rates) or right through a volcano and a new volcano grows on top of the older one. Occasionally, the central accretion zone migrates sideways and thus generates asymmetric profiles in segments of a mid-ocean ridge.

After the primary volcanic processes have come to an end, the lithosphere continues to consolidate by cooling, a process lasting many millions of years. At low spreading rates, the central areas of crust formation are cut by more or less perpendicularly-striking transform faults, some of which reach giant dimensions, such as the Romanche fracture zone at 2°S which is about 1000 km long. At fast spreading rates, such as along the EPR, continuous rigid transform systems cannot develop easily in the thinner and mechanically weaker lithosphere. Instead, segments of about 100–200 km length generally develop, whose ends overlap and which can propagate along their tip when the spreading rate changes.

When magma production in the mantle is especially high at low spreading rates, a mid-ocean ridge can grow upward and downward, leading to the formation of volcanic islands as surface expressions of a significantly thickened crust. The classical example for such a process is Iceland, a volcanic island astride the MAR. Connected with Iceland is the Reykjanes Ridge which extends over a length of 700 km. Iceland is traversed by the so-called neo-volcanic zone, which is comparable to a mid-ocean ridge, since the crust is torn apart episodically and is healed again by volcanic intrusions and extrusions. Hence, the eastern and western parts of Iceland constantly

▲ Fig. 5.16. Model of a magma chamber beneath a mid-ocean ridge with low spreading rate and low magma transfer rates. There is probably a sill-like crystal mush zone below the central graben, out of which smaller sill-like intrusive bodies develop, which slowly crystallize. Eruptions correspond to magma injections rising from the mantle. Faults on both sides of the central graben zone may end in the transitional zone between ductile and brittle behavior inside the partially fluid magma chamber (342) migrate away from each other. The cause for the very high magma production beneath Iceland is thought to be due to an anomalous mantle *plume* extending at least 400 km below the island. For this reason, Iceland is an intermediate structure between normal ocean crust and the isolated intraplate volcanoes thought to grow above mantle plumes. The ocean crust is covered by many hundred thousand seamount volcanoes (Fig. 5.6). Only few of these volcanoes, such as Hawaii and the Canary Islands, have grown above sea level to enlighten geologists, provide enjoyment to tourists and benefit travel agencies. Ocean islands are the subject of the next chapter.

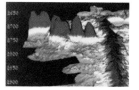

Seamounts and Volcanic Islands

During the years of European exploration, volcanic islands were portrayed as the image of the garden of Eden. The French explorer Bougainville even thought he had discovered the fundamentally good individual of Rousseau on Tahiti. The active volcanic Galapagos Islands were the birthplace of one of the most fundamental ideas of the nineteenth century: Darwin's theory of evolution. Indeed, the fauna of these islands represents the laboratory of life in a nutshell.

Nowadays, our dreams are not as idealistic as they were in Gauguin's day, our expectations of paradise have perhaps been reduced to slogans in travel brochures… Ocean islands have also lost much of their nostalgic or exotic exclusivity in these times of mass tourism. Madeira, the Canary Islands, Hawaii, Tahiti, Samoa, the Easter Islands and Réunion – all well-known and popular vacation places, are all oceanic islands. This group also includes more distant islands, such as St. Helena, Ascension, Tristan da Cunha, Gough or stormy Bouvet, of interest merely to seabirds, geologists or to the Royal Air Force to serve as natural aircraft carriers when defending the motherland on the Malvinas. The Malvinas (Falkland Islands) in the Atlantic and the Seychelles in the Indian Ocean are the exception to the rule; they are examples of the rare oceanic islands that are not of volcanic origin.

Oceanic islands have assumed a fundamental importance not only in the history of biology but also in Earth science, a role in no way corresponding to their small number and tiny surface area. Leopold von Buch, Charles Darwin and Robert Bunsen all formulated basic concepts in volcanology and petrology based on observations on oceanic islands. The age progression of linear volcanic island chains is the fundamental cornerstone to the hot spot and mantle plume models. These concepts were also crucial in developing the theory of plate tectonics (as discussed more fully at the end of this chapter). The most active and best studied volcano on Earth, Kilauea on Hawaii, has allowed more insights into the generation, rise, differentiation and eruptive mechanisms of basaltic magmas and evolution of submarine and subaerial volcanic edifices than any other competitor.

▶ Fig. 6.2. Cross section through a tholeiitic Hawaiian shield volcano with a cover of younger alkali basalts. Its size is compared to the classical shield volcano, Skjalbreidhur (Iceland) and the giant Olympus Mons on planet Mars (245)

Oceanic islands appear most impressive when approached by ship. Tenerife rises sharply to 3 718 m above the sea surface and La Palma at ca. 2 400 m a. s. l. and Gran Canaria at almost 2 000 m a. s. l. are also not easily missed. Mauna Loa and Mauna Kea on the island of Hawaii, both over 4 000 m high, are giants among volcanoes on Earth. Their huge dimensions are not immediately apparent, however, because their flanks rise only very gently (Figs. 6.1, 6.2). Ocean islands may be compared to icebergs in so far as most of their volume, generally more than 90 %, is beneath water. When measured from the sea floor, Mauna Loa and Mauna Kea are the highest mountains on Earth. The 80 000 km³ lava of the >10 000 m high, active volcano Mauna Loa on Hawaii would be sufficient to cover all of New Jersey with a basalt layer 4 km thick.

> I cannot help thinking, that upon a close examination, many islands at a great distance from continents, would be found to have been raised by explosions from subterraneous fires.
>
> *Sir William Hamilton, Campi Phlegraei, Naples, 1776*

▼ Fig. 6.1. Silhouette of the classic shield volcano Mauna Kea (4 214 m a.s.l.) on Hawaii. Diameter of the photograph about 60 km. The humps making up the slightly irregular surface on both flanks are alkali basaltic scoria cones, representing the late alkalic phase. A close-up of such a scoria cone is shown in Fig. 9.2

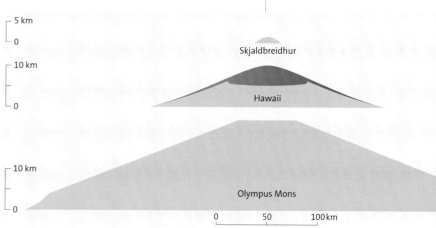

▶ Fig. 6.3. King Ferdinand of Sicily visiting the ephemeral new tuff cone Graham Island, also called Ferdinanda, grown upward from an active seamount. The short-lived island that appeared above sea level in June 1831 between Sicily and Africa became completely eroded and disappeared beneath the sea surface early in 1832 (author's collection)

Seamounts

As a child, I was particularly fascinated by those adventure stories in which a sailor wanting to reach a newly charted island never found the place. Were those all cases of the inexact science of early navigation and dead reckoning? In some cases, the original reports and charts were reliable but not the islands. This is because for the very small number of successful ocean islands, there is an immense number of unsuccessful underwater volcanoes. In these cases, their eruptive rates were just sufficient to reach the surface and create an ephemeral patch of land. This could not withstand the power of the waves and disappeared beneath the ocean surface within weeks or months.

A classical example is the fate of a volcano that appeared above sea level with powerful explosive eruptions on 18 July 1831 between Sicily and Africa (Fig. 6.3). At its peak, this island was 60 m high and 50 m wide. Its name Graham Island reflects the political dominance of the British in the Mediterranean at the time. The island was also called Ferdinanda in Italy, in reference to King Ferdinand II, who climbed it. This island never became a tourist Mecca because it disappeared beneath the waves only six months later in early 1832. Today, the top of the eroded volcano lies at about 5 m below sea level.

Ocean islands start to grow at great water depths, perhaps 2 000, 3 000 or 4 000 m b.s.l. These submarine volcanoes are called *seamounts* (21, 300). Seamounts are the most common volcanoes on Earth. In the Pacific alone, there are probably more than 1 million of them, but we know little about their structure and growth history. In the Atlantic, young seamounts are common in the center of the Mid-Atlantic Ridge (Fig. 6.4). Tertiary to Recent seamounts are abundant on 140–180 million year old ocean crusts off Northwest Africa (Fig. 6.5).

Seamounts grow preferentially close to mid-ocean ridges. For example, along the East Pacific Rise they occur mostly about 5–15 and more rarely 50 km away from the ridge axis. In other words, although magmatic productivity decreases with distance from a mid-ocean ridge, the number of seamounts per unit area actually increases because a few new ones are added to the old ones from time to time. With increasing distance, the apparent volcano density decreases again, because sediments increasingly cover the smaller seamounts. Along ridges with low spreading rate, seamounts grow almost exclusively in the central graben (Fig. 6.4), except along the boundary oceanic/continental lithosphere (Fig. 6.5). In the North Atlantic, the spacing amounts to about 200 per 1 000 km², more than an order of magnitude

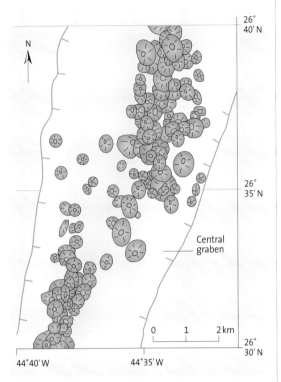

▲ Fig. 6.4. Seamount field in the central graben of the Mid-Atlantic Ridge (344)

higher than in the southern Atlantic (7 per $1\,000\,km^2$) and Pacific (9 per $1\,000\,km^2$). It is especially high (310 per $1\,000\,km^2$) along the Reykjanes Ridge, the southern submarine continuation of Iceland.

Lavas erupted on seamounts formed close to mid-ocean ridges are tholeiitic in composition, but lavas at some distance from a ridge become increasingly more alkaline. Intrusive and extrusive rocks occur in a roughly equal proportion in seamounts.

The vesicularity of pillows increases with decreasing depth of eruption, i.e., decreasing hydrostatic pressure. Seamounts also grow in height because abundant sills are intruded into their substructure (359). During their growth, the flanks of seamounts become steeper and therefore more unstable. Flank collapse and explosive processes that become more vigorous as sea level is approached, lead to the generation of clastic mass flows (debris flows and turbidity currents), whose deposits greatly enlarge the diameter of seamount bases.

Submarine volcano growth can be monitored today faster and more precisely than even a few years ago (Fig. 6.6). For example, on 25 January 1998, strong earthquakes were registered in the northeastern Pacific by the monitoring system of the American agency NOAA. The seismic swarm continued for two and a half days and resembled earthquakes observed in 1993 along a segment of the Juan de Fuca Ridge and 1996 on the north end of Gorda Ridge, off the coast of the state of Washington (USA). Marine geological surveys carried out immediately following the recorded signals in 1993 and 1996 found fresh lavas in the summit areas and along the flanks of a seamount. Typical for such types of eruptions is the high recurrence rate of relatively weak seismic signals and the slow migration of epicenters along the rift zone, probably reflecting the injection of magma and subsequent flank eruptions.

Large seamounts are significant morphological hindrances during subduction because they resist subduction and can trigger slides along the slope as well as earthquakes, both of which may also generate tsunamis (Fig. 6.7).

▼ Fig. 6.5. Magmatic belt parallel to the coast of Northwest Africa, north and south of the Canary Islands (304). The age of the oceanic crust east of the Canaries is about 180 Ma. Numerous seamounts are still being discovered during research cruises in between the three major archipelagos (Madeira, Canaries and Cape Verde islands) and larger seamount groups (Sahara and Senegal seamount clusters). Image based on Globe Project (NOAA)

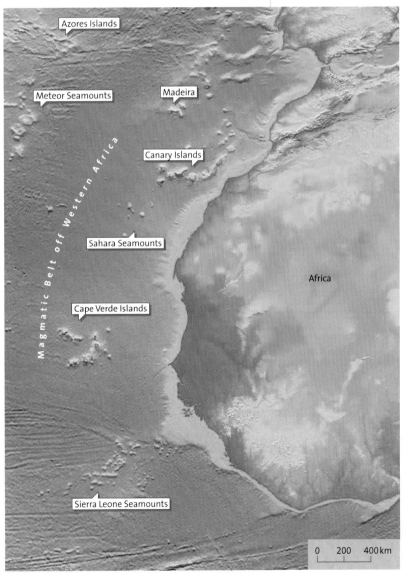

▶ Fig. 6.6. The active underwater volcano Axial Seamount on the Juan de Fuca Ridge off the coast of the state of Washington (USA) has been studied in detail since 1997, using several types of instruments. These include rumbleometers that measure deformation of the seafloor, volcanogenic earthquakes and compositional changes in seawater following magma injections. In late January 1998, eruption of basaltic lava in the summit caldera in the southeastern part of the volcano was accompanied by many seismic events, some of which were also registered on land. The details of the several instruments (e.g., OBH = hydrophones) and the latest results of the seismic and volcanic activity are described in http://newport.pmel.noaa.gov/axial193.html (Smithsonian Institution Vol. 23, No. 1, 1998)

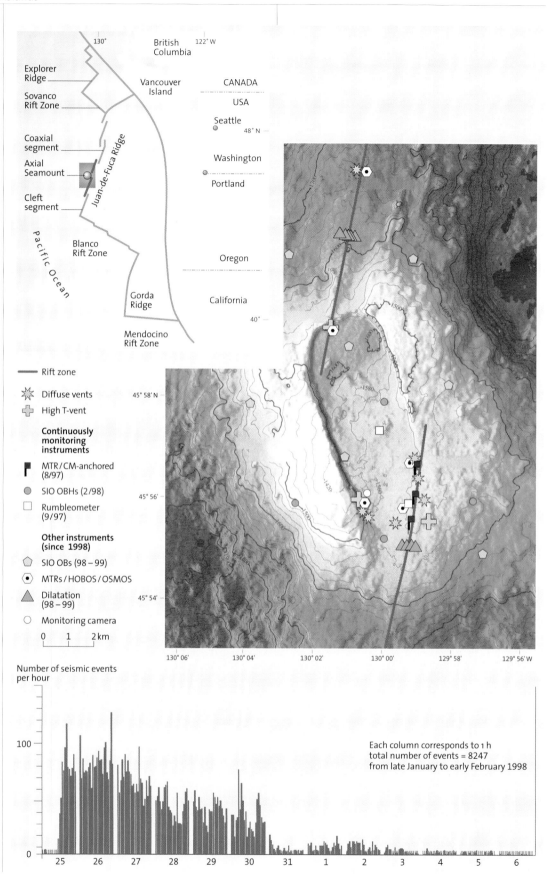

— Rift zone

✳ Diffuse vents

✚ High T-vent

Continuously monitoring instruments

▮ MTR/CM-anchored (8/97)

● SIO OBHs (2/98)

□ Rumbleometer (9/97)

Other instruments (since 1998)

⬠ SIO OBs (98–99)

⊙ MTRs/HOBOS/OSMOS

▲ Dilatation (98–99)

○ Monitoring camera

0 1 2 km

Number of seismic events per hour

Each column corresponds to 1 h
total number of events = 8247
from late January to early February 1998

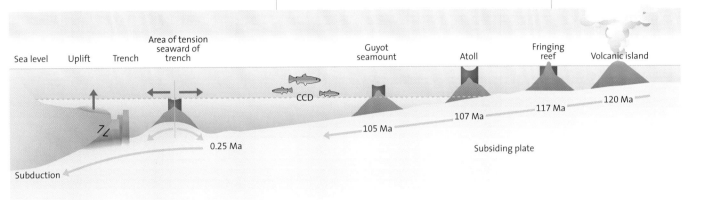

6

Guyots

Flat-topped seamounts are called guyots. Along the same line as the famous Atoll formation theory of Charles Darwin (75), guyots are formed when a volcanic island has been flattened by erosion and sinks below sea level. The subsidence is evidenced by fossil coral reefs, originally formed on top of an eroded volcanic island (230) and killed when subsidence of an island into greater water depth (>ca. 100 m) outpaces coral growth rates. Circular "dream" atolls, such as Bikini or Muroroa, form an intermediate stage (Fig. 6.7). Although the flat top of some seamounts can be explained by caldera subsidence (Chap. 9), Darwin's explanation is still the most convincing scenario for the origin of many guyots.

Today, we can understand why ocean islands sink. As shown in Chapter 4, the deep abyssal plains (on average 4 500 m b.s.l.) subside because of contraction of the lithosphere through cooling when mantle material, after ascent and partial melting beneath mid-ocean ridges, migrates away from the ridges. When 100 km thick lithosphere cools by 600 °C, its density increases so much that the plate shrinks by some 3 %. In other words, the surface of the sea floor subsides by 3 km.

Today, the origin of true oceanic intraplate volcanoes is best understood as dynamic uplift of the lithosphere above a rising stream of mantle material within a plume (see below). Sinking of the islands is explained by isostatic subsidence (due to the weight of the newly formed island) and also cooling of the heated lithosphere and thermal erosion at the base of the lithosphere by the lateral spreading of hot mantle material (279).

Volcanic Islands

Apart from the microcontinent Iceland (see below) and the colossal island volcano Mauna Loa, volcanoes of well-studied island groups, such as the Hawaiian or the Canary archipelagoes, have volumes in the order of about 20 000–60 000 km³.

These are huge in comparison to many andesitic-dacitic stratovolcanoes grown above subduction zones as around the Pacific, whose volume rarely exceeds 500 km³ (Chap. 8). In fully developed ocean islands, less than 10 % of the total volume of a volcanic edifice rises above sea level, depending on how its base is defined and how well it can be documented.

Volcanic island complexes grow vertically and laterally by several fundamentally different processes and evolve in distinct stages or phases defined by contrasting magma compositions, eruptive, intrusive, erosional, partial collapse and subsidence rates and hence dynamic changes in the magma source regions modulated by several external factors. This complex evolutionary pattern has been deduced principally from the sub-aerial edifices. Study of uplifted island cores, deep sea drilling, swath bathymetry mapping of the sea floor and submersible observations increasingly help to document the submarine evolution as well. It must not be forgotten, however, that some 90 % by volume of an island complex has accumulated under water and is hence extremely poorly known.

The Submarine Stage

The little we know about the bulk of oceanic island volcanoes hidden beneath the ocean waves, is based on deep sea photographs and samples taken from the submarine flanks in Hawaii (231) or the Canary Islands (308, 312), holes drilled in the sediment aprons around Gran Canaria (Fig. 6.8) or on the island of Hawaii or analysis of the uplifted and tilted submarine part of the western Canary Island of La Palma (359). These data provide us with a mere glimpse of the submarine structure of ocean islands (Figs. 5.9, 6.9).

When magma rises above an active melting anomaly through the generally much older ocean crust and its cover of soft young sediments, it is likely to intrude laterally into the low density

▲ Fig. 6.7. Evolution of volcanic islands and guyots. During subduction of an oceanic plate large seamounts form effective asperities that can trigger earthquakes and larger slumps (300). *CCD* (Carbonate Compensation Depth) is the depth below which the calcareous shells of organisms sinking to the seafloor begin dissolving

▶ Fig. 6.8. Schematic cross section of a seamount/ocean island and the clastic fan surrounding an island, based on drilling into the submarine flanks of Gran Canaria (308)

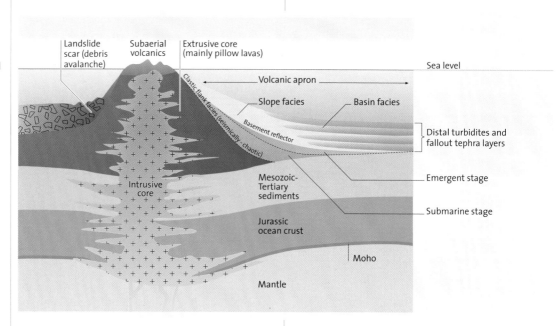

muds. This creates a complex of intrusive lavas mixed with sediments, above which grows the submarine edifice. The initial deep-sea seamount will consist mainly of pillow lavas, with intrusives increasing toward the core of the edifice. These coarser-grained plutonic rocks have formed in central magma chambers that grow upward with time. The plutonic core is separated from the volcano surface by a 2–3 km thick roof of extrusive rocks and dikes (Figs. 6.9–6.11), an igneous stratigraphy much like that of the ocean crust (Chap. 5).

With increasing height and therefore decreasing pressure of the overlying water column, gases dissolved in the magma at higher pressure can start to exsolve, forming discrete bubbles. Once the melt consists of about 65 % of such bubbles, it is torn apart by the gas expansion and shearing of the liquid (Chap. 10). This explosive underwater phase begins, depending on the amount of gas and viscosity of the melt, at water depths of 1 000 m or more, or, in the case of very gas-poor magmas at only a few hundred meters. Glass shards, pumice or scoria-type particles (pyroclasts) pile

▶ Fig. 6.9. Pillow breccias cut by basaltic dikes in an uplifted submarine seamount complex. Barranco de las Angustias (La Palma, Canary Islands). Hammer for scale

◀ Fig. 6.10. Basaltic scoria cones cut by dike swarm on the eastern end of Madeira (Ponta de São Lourenço)

up as unstable sediments that repeatedly slide down the flanks of a volcano. Widespread sediment aprons form and may extend for more than 100 km away from the steep-sided volcanic edifice (Figs. 6.8, 6.11). A seamount thus grows above the sea surface when the eruptive rate is significantly higher than the erosional rate, even during the emergent phase when wave erosion, steam explosions and lava debauching into the sea produce abundant clastic material. Most clastic material produced is reworked and deposited in adjacent sedimentary basins. Indeed, there are few winners in the battle between constructional processes and destructive erosion, and most aspiring seamounts have to capitulate. As the successful island grows, its core will become increasingly isolated from the water and is buttressed by lava flows and dikes. This was demonstrated on a small scale in 1957–1958 with the newly formed island volcanoes Capelinhos (Fayal, Azores) and in 1963–1964 in Surtsey (Iceland) (Fig. 12.2; Chap. 12). Lava compositions during the early submarine stage may vary widely as discussed below.

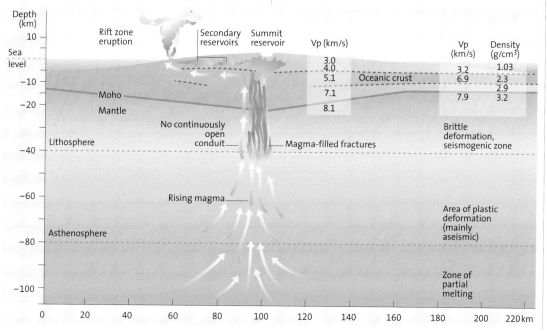

◀ Fig. 6.11. Cross section through the volcanic island of Hawaii, beneath which the oceanic crust has thickened to 16 km (386)

▶ Fig. 6.12. Map of the island of Hawaii with prominent shield volcanoes younging from north to south: Kohala, Hualalai, Mauna Kea, Mauna Loa and Kilauea. Most eruptions occur along the rift zones, in historic times especially along the East Rift Zone of Kilauea volcano

The Shield Stage

Most oceanic islands are dominated by large basaltic shield volcanoes, comprising more than 90 vol% of the subaerial edifice (Figs. 6.1, 6.12–6.14). The growth of a classical shield volcano has been studied particularly well on Kilauea on the Big Island of Hawaii, the most active volcano on Earth (Figs. 6.15–6.19). The steady growth of volcanoes can nowhere be followed in more detail as in Hawaii.

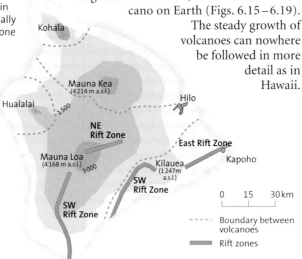

Mauna Ulu (Polynesian for *growing mountain*) is the official name for a shield volcano that started to grow in 1969 on the East Rift Zone of the island of Hawaii, about 10 km east of Kilauea Volcano. The growth of Mauna Ulu ended in 1974 after it had erupted about 0.5 km³ of lava (369). The lava flows of shield volcanoes are commonly only a few meters thick (Fig. 6.14) and erupt from central craters, or, more commonly, from rift zones (Figs. 6.12, 6.18). Lava may issue at the surface in almost uninterrupted streams for years, the longest period in history having started in January 1983 along the East Rift Zone of Kilauea Volcano, with no end in sight (Fig. 6.16). With time, activity along the initially 8 km fissure concentrated in one spot, the still growing scoria cone Pu'u 'O'o (about 300 m high in 2002) and its adjacent lava lake (433). Such continuous outpourings must be facilitated by hot stable pathways at depth, enabling mantle melts to make their way quickly to the surface, commonly via the main reservoir beneath Kilauea.

Lava flows issue episodically from these eruptive centers and spread across the southern slopes of Kilauea volcano. Destruction of the settlements Kalapana and Royal Gardens in the pathways of the almost continuous lava streams has so far amounted to more than US$ 100 million of dollars. Some flows enter the Pacific Ocean after flowing underground in lava tubes for many kilometers and extend the island into the sea. By pushing the coastline outward during the past 18 years about 2–3 ha of new land has been generated. Lava deltas form seaward of the lava front and gradually form benches that episodically collapse to form thick breccia aprons mantling the submarine flanks (237).

Even a non-specialist observer will look in vain for evidence of prolonged pauses in eruptive activity such as erosional unconformities or soil horizons between lava flows of shield lava successions (at least on the southern side of Kilauea). Detailed age determinations of rocks from the Hawaiian and Canary Island shields have shown that the main phase during the subaerial evolution of a volcano is completed in less than 1 or even 0.5 million years (e.g., 214). At Kilauea, about 0.1 km³ of new magma is added each year to the volcano (366, 89). Obviously the lavas were piled on top of each other fairly rapidly, reflecting high melt production rates in the mantle melting anomaly. The chemical composition of the shield lavas in most ocean islands is alkali basaltic or transitional, but in some, notably Hawaii and the Galapagos Islands, tholeiitic. In the latter case, it is possibly because the ascent rate of mantle material and therefore degree of partial melting beneath these islands is significantly higher than below most other volcanic islands.

Magma Chambers and Rift Zones

One magma chamber on Earth whose size, depth, episodic filling and partial emptying has been well-documented by several methods is that beneath Kilauea volcano (Figs. 6.17, 6.18). The highly simplified three-dimensional model of the complex magma reservoir system shows a central area 2–6 km beneath the summit caldera of Kilauea. This column, about 4 km high, consisting of magma and crystals, has a roughly elliptical cross-section and a volume of about 5–10 km³.

This magmatic feeder system is not a smooth magma-filled pipe, but probably consists of a multitude of laterally branching sills, through which magmas flowed during recent years with a remarkably constant rate of about 3 m³/s (= 0.1 km³/a). During the past 40 years, basaltic magma was mainly channeled along the East Rift Zone and much more rarely along the Southwest Rift Zone, whose feeding channel runs from the upper or lower levels of the main reservoir. When eruptions are separated by longer non-eruptive periods, the magma in the main or lateral reservoirs can cool to such a degree that it also crystallizes phenocrysts of plagioclase and pyroxene, (i.e., not only olivine, the common phenocrysts formed in magma chambers when the newly arrived magma has crystallized just below its liq-

◄ Fig. 6.13. About 300-m-
▲ high lava fountain
above the vent of the
newly developed shield
volcano Mauna Ulu (1969–
1974) along the East Rift
Zone of Kilauea volcano
(Hawaii)

▲ Fig. 6.14. Thin Miocene
▲ shield basalt lava flows
and red oxidized lapilli layers representing edge of
small local scoria/aggluti-
nate cone cut by feeder
dikes on Gran Canaria
(Canary Islands)

▲ Fig. 6.15. Degassing vent
area between two small
spatter ramparts in the
center of the young shield
volcano Mauna Ulu (East
Rift Zone, Kilauea volcano).
The two spatter ramparts
are about 50 m apart

◄ Fig. 6.16. About 80-m-
wide lava lake east of the
scoria cone Pu'u 'O'o (East
Rift Zone, Kilauea volcano,
Hawaii)

uidus). Depending on its residence time, this magma is chemically evolved to some degree when it erupts after having been arrested near the surface for a few years. It is also cooler than the primitive magma from which it differentiated. Many larger eruptions on Kilauea finish with newly supplied, hotter and chemically more primitive magma.

The updoming of the roof of a volcano and its vibrations due to the ascending magma are of great importance for predicting an eruption (Chap. 13; Figs. 6.19, 13.18). It takes a few decades until a portion of the magma formed by partial melting at a depth of approximately 60–90 km is erupted on the Earth's surface. More than 50% (in the Canary Islands probably more than 75%) of basaltic magmas that rise into the crust, however, do not erupt at the surface, but cool slowly in the interior of a volcanic edifice to form dikes, sills and larger coarser-grained plutonic rock bodies.

The fundamental structural elements of basaltic shield volcanoes are morphologically prominent, straight to slightly curved ridges marked by axial grabens. These rifts dissect the huge lava piles and extend from the summit calderas to the sea floor. *Rift zones* on land may be more than 50 km long (Fig. 6.12) but may exceed 150 km in length including their submarine extension. In many volcanoes three rift zones form a star-shaped pattern, individual arms being separated by 120° from each other. Rift zones may be several km wide and are underlain by closely spaced dense dikes, some of them actually having fed surface lava flows. For this reason, rift zones are marked by large positive magnetic and gravity anomalies and high P-wave velocities.

Much has been written on the structure, physical parameters and origin of these impressive rift zones, but many questions on their evolution are still highly debated. One of the main problems concerns the processes by which rift forms. Are dikes forcibly injected pushing the flanks apart? Or do the flanks exert a gravity pull and thus generate a favorable stress field that facilitates dike injection? In any case, rift zones are used time and again by ascending magma because the subparallel planar elements and the preheating facilitate the rise of newly ascending magma. Rift zones thus appear to

form early in the evolution of a volcano and may persist until its late waning stage.

The Late Alkaline Phase

The main shield phase is followed by very much smaller volumes of alkaline, more mafic and, in many islands in the Atlantic, more highly evolved lavas, intrusions or pyroclastic flows. Local scoria cones forming during the declining stage may each correlate with a major lava burst but are volumetrically minor compared to the lava flows.

Posterosional Phases

The shield phase is followed in many volcanic islands by a late phase in which the composition of lavas is much more alkaline. This rejuvenated or posterosional phase is separated from the shield phase by a period of erosion that may last 3 to 5 million years in the case of the Hawaiian islands. An island can become strongly eroded, depending on climate during such non-volcanic phases. The very small volumes of highly alkaline, very mafic and strongly SiO_2-undersaturated

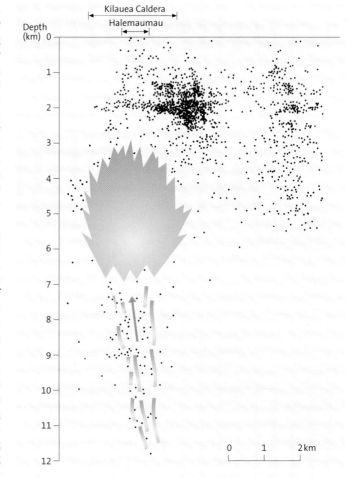

► Fig. 6.17. Distribution of earthquake hypocenters beneath Kilauea volcano. The hypocenter-free area is interpreted as more or less liquid magma reservoir (180)

lavas (basanites, nephelinites) commonly contain mantle-derived peridotite xenoliths such as lherzolites, harzburgites and pyroxenites. Derivative magmas (for example phonolites) may also be generated during these later phases. The alkaline lavas mostly form scoria cones and small lava flows (Fig. 9.2) and resemble the volcanoes of intracontinental volcanic fields, such as those of New Mexico or the Eifel with respect to eruptive mechanisms, chemical composition and volcanic evolution.

Magmatic Evolution

On many oceanic islands, the decrease in volume of erupted lava with time is accompanied by systematic changes in magma composition becoming more primitive and enriched in incompatible elements. The causes for these remarkably similar evolutionary trends (despite much intra- and inter-island and inter-archipelago differences) remain unclear, however, and have been the subject of numerous studies. The compositional change is most commonly explained by strongly

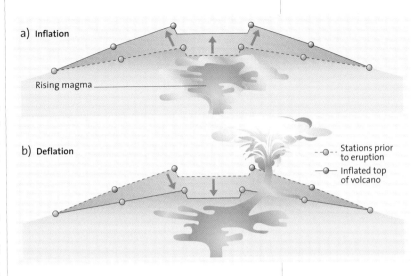

▲ Fig. 6.19. Schematic, not to scale. Cross section through the upper part of Kilauea volcano and the man rift zones. a) Magma rises from the upper mantle and elevates the volcano roof around the summit caldera. b) Following an eruption, the Earth's surface subsides again, but not quite to its previous elevation (89)

▼ Fig. 6.18. Geometry of the main magma reservoir of Kilauea volcano and the East Rift Zone (288, 386) (6.17 lower part)

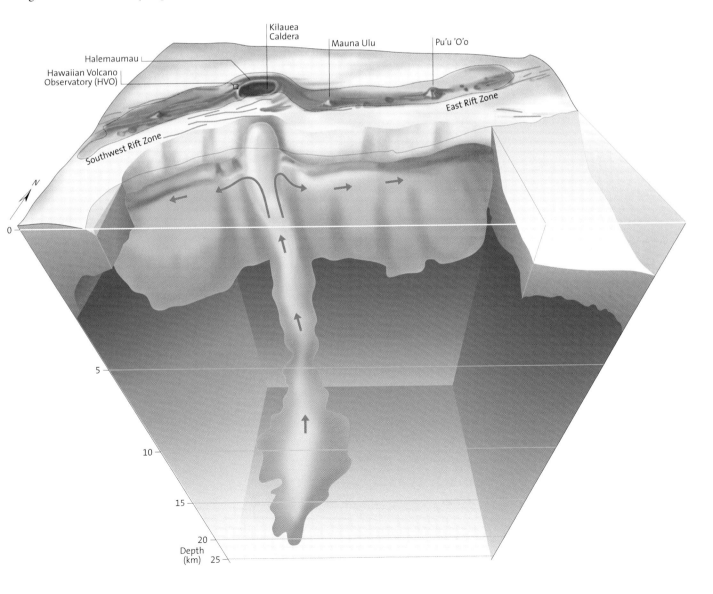

decreasing degrees of partial melting, only in some cases accompanied by changes in source rock composition, as reflected most clearly in the radiogenic isotope ratios (Sr, Nd and Pb) of the erupted magmas. In a presently popular model applied to Hawaiian volcanoes, the volcanoes on an island are fed successively by the margins of a plume during the compositionally more heterogeneous early and late stage, and from the central part of the plume during the main shield stage (144, 145). The alkali basalt magmas erupted on Loihi seamount south of Hawaii or the highly undersaturated lavas of the submarine rift zones of La Palma and El Hierro of the Canary Islands are interpreted to have been generated at an early stage. In other words, contrasting compositional zones of a stationary plume over which the plate migrates, are tapped at different stages. The less radiogenic post-erosional magmas formerly explained by a decreasing lithosphere component are nowadays also interpreted by remelting of the crustal (shield stage) and gabbro (post erosional stage) components of recycled subducted oceanic crust (349).

Destructive Stage and Lateral Apron Growth

Large compound oceanic edifices not only grow by addition of volcanic products to their surface and flanks and intrusive expansion from within, but change their shape and volume by slow erosion or sudden collapse processes. With time, a volcanic island also strongly grows laterally by accretion of the submarine volcanic apron. In particular, submarine deposits in the foreland of the main cone accumulate with time at the expense of the volcano height, until the subaerial island is finally completely eroded to the point where it may sink below wave base. The destructive processes have then largely terminated, except for some submarine collapse and isostatic subsidence.

As already mentioned, more than 90% of a volcanic island (and 100% of a seamount) are below sea level (Fig. 6.20). It is only in the last two decades that slumps and debris avalanches were recognized to be characteristic of the submarine and subaerial evolution of an island. When the offshore national boundaries were extended to 200 nautical miles about 15 years ago, many countries intensified their research efforts to better analyze the coastal zones. This was especially with respect to economically important deposits, such as oil and gas. Pioneering studies around Hawaii used the radar system GLORIA to systematically map the 200-mile zone around the Hawaiian Islands. This revolutionized our understanding of the evolution of ocean islands. Each island is surrounded by huge debris fans, some of which extend more than 100 km and contain blocks with volumes of hundreds to several thousand km³. These huge submarine landslide blocks are generated as commonly as those around many subaerial volcanoes (Chap. 9). Such submarine slides alternate with magmatic processes during the evolution of an island (239, 240). Around the Hawaiian Island chain, about 68 slumps with lengths up to 200 km and volumes up to several thousand km³ have been mapped (Fig. 6.21). Some debris-flow and turbidite deposits can extend for >1000 km. Two types of such clastic deposits have been distinguished off the flanks of Hawaiian islands (239). Slumps are relatively short but wide. They move episodically with ca. 10 m/100 a, such slump events being associated with large earthquakes. Slumps on the Hawaiian islands can trigger major tsunamis that may cause havoc as far east as the coast of California. Slumps have steep

▼ Fig. 6.20. Subaerial and submarine morphology of Tenerife and Gran Canaria. The Güimar sector collapse scar and the submarine debris avalanche fan are indicated on the southern subaerial and submarine slopes of Tenerife. For details see Fig. 9.39 (182)

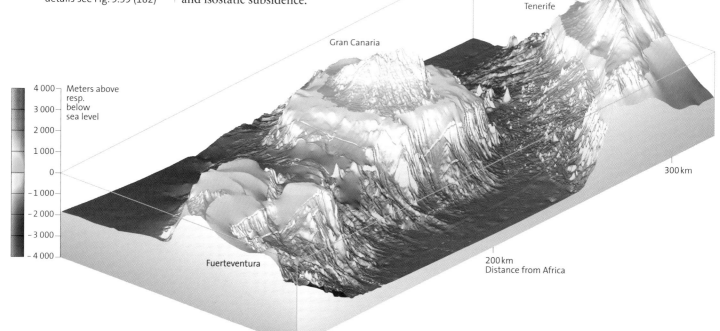

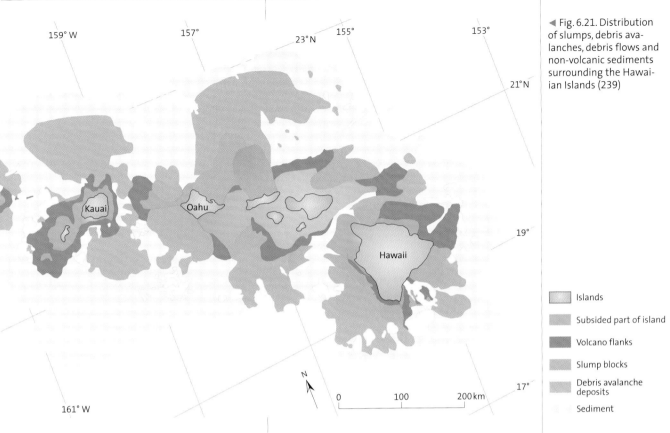

Islands

Subsided part of island

Volcano flanks

Slump blocks

Debris avalanche deposits

Sediment

fronts, show terraces, move on slopes greater than 3° and their thickness can be in excess of 1000 m!

The second type of mass transport, debris avalanches, can flow for several hundred meters upslope at gradients of less than 3° to 1.5° and leave long and hummocky deposits. They may be several hundred meters thick and can form the continuation of slumps. Some debris avalanche deposits can be traced through submarine canyons (which reach to water depths of 1500 m) to the huge bowl-shaped scarps where they originated, e. g., on Kohala volcano.

The cause for the destabilization of subaerial and submarine flanks are complex and vary between volcanoes. Slumps in Hawaii are most commonly interpreted as sliding of huge flank sections on a thin layer of oceanic sediments, such as on the south flank of Kilauea (e. g., 84). The force for pushing the flanks outward has been attributed to hydrostatic pressure exerted by repeated intrusions of dikes into the rift zones. In addition, the weight of the rapidly accumulated and therefore oversteepened unstable flanks has been considered responsible for flank collapse. Clague and Denlinger (58) hypothesize that large volumes of easily deformable dunite (which forms in magma chambers by accumulation of olivine) can push the southern flank of Kilauea volcano to the south and therefore trigger the destabilization of the

flanks. Several scientists argue that elevated groundwater pore pressures are a major cause of destabilizing island flanks (166). The morphological studies of the seafloor and the seismic fine structure of the clastic apron surrounding an island are important to aid the evaluation of the hazard potential of volcanoes for nearby densely populated islands.

How Representative Are the Evolutionary Stages of the Hawaiian Islands?

Kilauea is commonly regarded as a typical example of an ocean island volcano, understandably so because of its almost continuous volcanic activity and decades of pioneering research. However, many oceanic islands in the Atlantic and other ocean basins have developed differently. Despite some basic similarities, such as the dominance of the basaltic shields, there are significant differences in the evolution of volcanic island groups. Some of these differences will be highlighted by contrasting two well-studied oceanic archipelagoes, Hawaii and the Canary Islands, both serving as end members of a wide spectrum of oceanic volcanic island systems. The main characteristics of the Canaries compared to the Hawaiian Islands can be summarized as follows:

● Much longer lifetime of individual islands;
● Several posterosional phases;

▶ Fig. 6.22. Cross section through the Canary Islands and the underlying lithosphere (311)

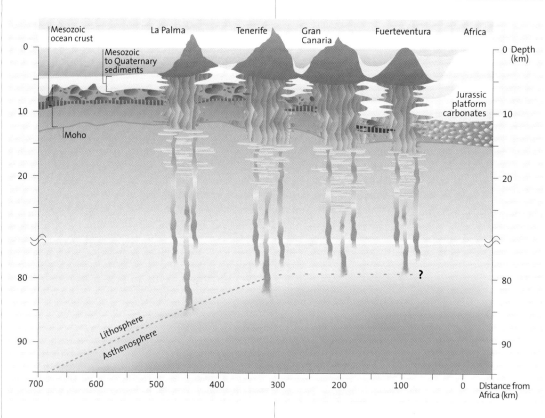

▼ Fig. 6.23. Map of the Canary Islands, ages of main shield (circles) and post-shield magmatic phases. Part of the islands of Fuerteventura and La Palma have been uplifted by several 1 000 m, probably by magma intruded from below

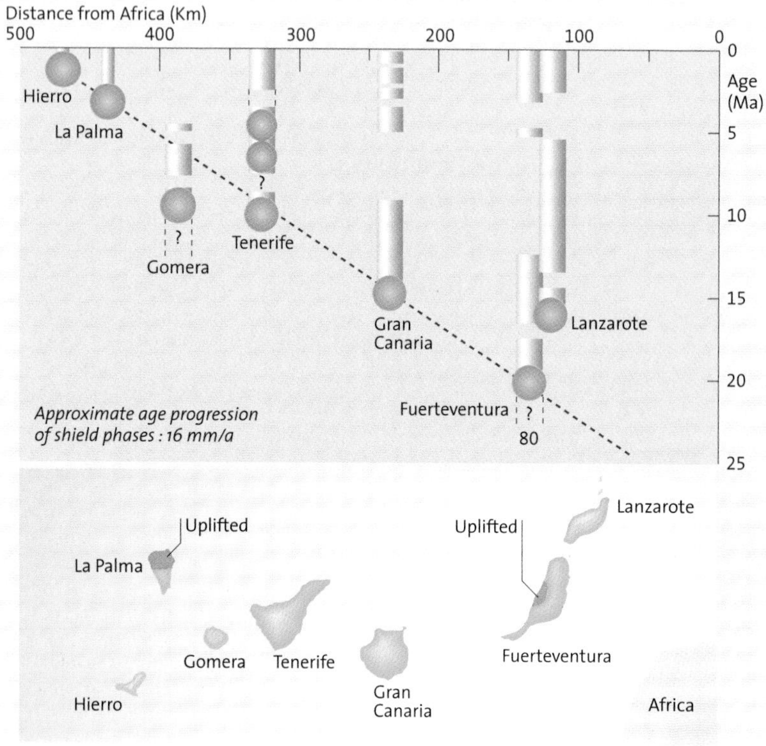

- More alkaline composition of the mafic shield magmas;
- Large volume of highly evolved volcanics in some;

- Abundance of pyroclastic deposits;
- Steep submarine and subaerial flanks;
- High rates of flank collapse;
- Growth on older (140–170 Ma) and thus thicker, cooler and more rigid oceanic lithosphere;
- Stability of the islands with respect to sea level;
- Proximity to continental lithosphere;
- High biogenic and volcaniclastic sedimentation rate in the volcanic apron;
- Significant compositional and evolutionary differences across the archipelago;
- Differences in magma production and eruptive rates.

Duration of Island Evolution

In the Hawaiian Islands, the three-stage volcanic evolution of each island terminates after about 3–5 million years. In contrast, volcanic activity following the initial shield phase lasts very much longer on the Canary Islands (Figs. 6.22–6.25); on Fuerteventura over 20 and on Lanzarote and Gran Canaria about at least 15 million years (all these islands are still active). Because the islands young to the west, the youngest islands La Palma and Hierro are still in the shield phase, while the oldest dated lavas on Tenerife are about 11 million years old. On Gran Canaria, the 14 million year old basaltic shield is overlain by thick sequences of rhyolitic, trachytic and phonolitic ignimbrites

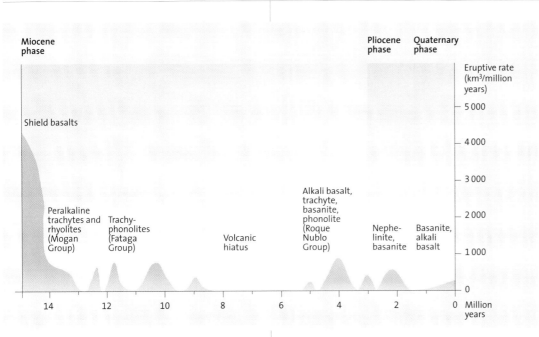

(Fig. 6.25). These were erupted from a spectacular 20-km-diameter caldera, unusual for oceanic islands. This huge volcano was largely eroded between approximately 9 and 5 Ma, the erosional intrusive rump becoming covered subsequently by thick younger volcanic deposits (Figs. 6.26, 9.35). All Canary Islands, except for La Gomera, are either still volcanically active or have erupted in the Holocene. In other words, the melting anomalies beneath single islands remained active, locally for more than 20 million years, although the composition of the mantle source and the degree of partial melting changed with time, possibly being fed from episodically rising blobs of fertile mantle peridotite, separating from a deeper mantle reservoir (157). The fact that volcanic activity in at least the older Canary Islands, once started, continues for tens of millions of years is most plausibly

◀ Fig. 6.25. The layers on the 700-m-high canyon walls (*right hand side*) in southern Gran Canaria are ignimbrite and minor lava cooling units in the upper part, overlying basaltic shield lavas (about 14 Ma old) (SB) at the canyon bottom (*left*). The ignimbrites represent the Miocene Mogán Group (lower reddish ignimbrites, 14–13.4 Ma) (M) overlain by ignimbrites, lavas and unconsolidated tephra and breccias of the Fataga Group (F), the uppermost ignimbrites being ca. 10 million years old. A major fault (white dashed line) parallel to the caldera wall (ca. 1 km upstream) with ca. 200 m displacement separates a graben (upstream) from a horst on the left. Barranco de Arguineguin (Gran Canaria, Canary Islands)

▶ Fig. 6.26. Trachytic to phonolitic dikes of a giant cone sheet swarm dipping left represent the eroded central part of a volcano about 10–12 million years old in the western center of Gran Canaria (Canary Islands). The old erosional surface is unconformably covered by about 4-Ma-old dark basaltic lavas and a thick breccia flow, about 100 m thick, in the upper part of Mesa de Junquillo. Tejeda Canyon (Gran Canaria, Canary Islands). A schematic cross section of the cone sheet swarm is shown in Fig. 9.50

explained by little differential motion between lithosphere and magma source region and maintenance of pathways (heated channels) between the melting domain and the upper crust.

Post-Erosional Phases

The volume of posterosional lavas in Hawaii is very small (56, 386). In contrast, all the older eastern and central Canary Islands have experienced three or more magmatic phases interrupted by longer periods of erosion and weathering. For example, the basaltic shield phase on Gran Canaria lasted about 1 Ma and was succeeded between 14 and 8 Ma by large volumes of highly evolved magmas, mafic parent magmas being rarely erupted. This shield phase was followed by a non-volcanic interval, lasting approximately 4 million years. During the post-erosional Roque Nublo volcanic phase, mostly active around 4 Ma, a huge composite volcano grew with flanks exceeding 500 m in thickness. It was built from lava flows, breccia flows and numerous smaller domes (Figs. 6.27, 6.28), an example of a particularly voluminous second magmatic phase on an oceanic island. The total volume of erupted lavas and pyroclastics of the younger volcanic phases is also very much smaller than that of the shield phases, as in Hawaii. However, the magma volumes of some younger phases such as the Pliocene Roque Nublo phase on Gran Canaria or the post-shield volcanics on Tenerife, amount to hundreds of cubic kilometers. This may reflect the pulsating activation of new melting domains, which are possibly blebs of distinct rising mantle bodies.

Composition of Shield Magmas

The shield lavas of the Canary Islands are enriched in certain incompatible elements. These are elements that, during partial melting of the source rock in the Earth's mantle, are concentrated in the melt. Hence the Canary magmas are more alkaline than the shield-building basalts in Hawaii. K_2O concentrations are rarely below 0.6 wt%. Most analyses fall into the field of alkali basalts of Macdonald and Katsura (204), only some in the field of tholeiites.

The magmas of the Canary Islands are unusually rich in iron and titanium. This is presumably due to formation of the magmas at higher pressures (>30 kb), under a much thicker lithosphere (ca. 100 km), compared to the 60 km thick lithosphere beneath Hawaii. In addition, the composition of the parent source material in the mantle may be significantly different (304). The lithosphere appears to be thicker beneath the western than beneath the central Canary Islands, a conclusion supported by the probable greater depth of magma generation in the west and occurrence of tholeiitic lavas only on Gran Canaria and the eastern Canary Islands (Fig. 6.22).

Abundance of Highly Evolved Magmas

The central Canary Islands Gran Canaria and Tenerife are instructive examples for the abun-

▶ Fig. 6.27. Risco Blanco, ca. 3.7-Ma-old phonolite intrusion, in the northern part of Barranco de Tirajana (Gran Canaria, Canary Islands). The lava flows to the left of the dome have been upturned in the vicinity of the intrusion. Width of the dome at the base about 500 m, height about 400 m

dance of highly differentiated rocks on the Canaries. This is shown immediately by contrasting colors of widespread pyroclastic rocks on the two islands. The highway from Santa Cruz to Los Cristianos on Tenerife, well known to hundreds of thousands of European tourists, crosses a white to cream-colored landscape for many kilometers. This is anything but basaltic. The bleak terrain mainly consists of light-colored ash flow deposits, thick layers of fallout pumice and lahar deposits. Similarly, the highly fragmented, bizarre obsidian lava flows in the Las Cañadas depression (Fig. 9.14) on Tenerife add to this non-basaltic volcanic scenery. Even more impressive is the south of Gran Canaria, whose scenic attraction stems largely from deep canyons (barrancos), showing dominantly green, white and brown to bluish, highly-welded ignimbrites. Nowadays, some are becoming increasingly overgrown by "concrete conglomerates" where the barrancos meet the sea, temporary shelter for tourists. Even in the interior of Gran Canaria, white to greenish phonolite intrusions, such as Risco Blanco (Fig. 6.27) or Pajonales are grand samples of such highly evolved rocks, as are the phonolite domes of La Gomera. When these felsic magmas erupt explosively, they form layers of hot ash flows (ignimbrites), or ash clouds at high eruption columns, whose deposits form wide-spread thin ash layers on the lee side of volcanoes in the marine sediments (Fig. 6.28). Further examples of ubiquitous highly differentiated rocks on ocean island volcanoes are the widespread pumice sheets and ignimbrites as well as the very scenic calderas Sete Cidades and Furnas on São Miguel (Azores), or the trachyte domes on St. Helena or on Porto Santo, near Madeira.

Evolved rocks also occur in the form of intrusive and extrusive domes, picturesque examples occurring on several islands, especially La Gomera and Gran Canaria. These domes are also one of the reasons for the steep slopes on the islands, both on their subaerial and submarine flanks. Gradients are 20° to 30° compared to gradients of a few degrees for the near-shore flanks on Hawaiian volcanoes although slopes may also approach 25° in Hawaii particularly where a thick cap of late-stage alkaline lavas has developed in the summit areas.

The enigma of the origin of felsic magmas on oceanic islands has been debated among scientists

for more than 150 years, the explanations falling into two major groups: are the trachytic, rhyolitic or phonolitic rocks products of partial melting of underlying continental crust or have they formed by differentiation from a basaltic parent magma? Earlier ideas that e. g. the rhyolites on Iceland represent partial melts of granitic continental crust lost credibility on geodynamic grounds and when seismic evidence in the 1970's did not reveal continental crust beneath the island. On the other hand, there is robust evidence that crystal frac-

▲ Fig. 6.28. Phonolitic fallout ash, deposited in shallow water, into which a basaltic lava flow was emplaced (pillow structures). Pillow lava grades upward into a subaerial lava flow. Roque Nublo Formation, ca. 4 Ma old, west of Las Palmas (Gran Canaria, Canary Islands)

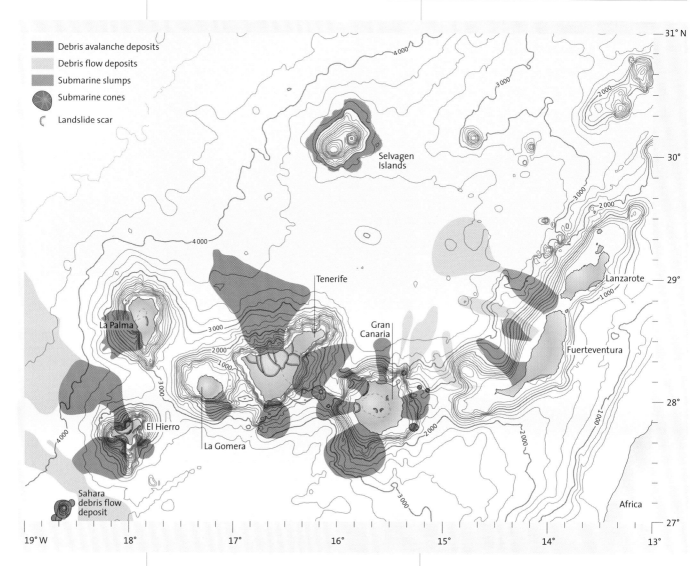

▲ Fig. 6.29. Distribution of deposits of slumps, debris avalanches and debris flows on the seafloor around the Canary Islands (182)

tionation and minor contamination can explain the generation of evolved magmas very well. Ocean drilling has even provided clear evidence that the parent magmas for different types of rhyolitic rocks on Gran Canaria erupted at the submarine flanks of the island but not at the surface because of their higher density.

Abundance of Pyroclastic Deposits

Even though pyroclastic and hydroclastic deposits are becoming increasingly recognized on the Hawaiian islands, it is fair to say that Hawaiian shields are made up almost exclusively of lavas. In the Canaries, in contrast, pyroclastic deposits form a significant fraction of the rocks exposed. They appear to be slightly more abundant in the shield successions compared to Hawaii, most likely due to their more alkaline and more volatile-rich composition. Much more significant, however, are the large volumes of pyroclastic deposits, especially

ignimbrites as on Gran Canaria and Tenerife. The Pliocene Roque Nublo stratocone, for example, produced many debris flow sheets, one reason why pore water pressures may have increased to such a degree that the flanks collapsed repeatedly, resulting in some of the most spectacular debris avalanche deposits known on Earth.

High Rates of Flank Collapse

Slumps and debris avalanches with magnitudes comparable to Hawaiian ones have been documented in the Canary Islands (Fig. 6.29). These lie off Hierro (211) northwest of Tenerife (413), as well as southwest of Tenerife and north of Gran Canaria (107, 182, 308, 412). Slumping occurred in the Canarian Archipelago throughout the entire development of the islands, however, not just in the shield stage as in Hawaii. This is shown especially well by the subaerial morphology of Tenerife (Fig. 9.40). The most likely reason for the

◄ Fig. 6.30. Recent fault with striated surface on the eastern scarp of the western Canary island of El Hierro. The seaward sector (*right*) has subsided

higher rates of flank collapses from the Canary Islands is the common development of near-surface magma chambers, around which the volcano flanks become more strongly destabilized by deformation and hydrothermal alteration. When one looks at a radar image of Tenerife, many flank scars are immediately obvious reflecting numerous sector collapses (Fig. 9.40). Fault scarps such those on eastern El Hierro testify to Recent major slumps although it is unclear whether or not they will develop into debris avalanches (Fig. 6.30). Slumps in the Hawaiian Islands begin on submarine and subaerial flanks, whereas on the Canaries, they also commonly start from the subaerial volcano flanks that are much steeper than in Hawaii. The number of well-documented sector collapses is still increasing because such processes had been almost completely overlooked until the 1980s.

Marine geological studies have made it possible to follow submarine volcaniclastic debris fans for more than 50 km (Fig. 6.29). Detailed studies of the Madeira abyssal plain about 1 000 km west-northwest of the Canary Islands and of the sea floor in between, shows that larger debris flow deposits can transform into more diluted, finer-grained turbidites, whose very fine-grained distal ends comprise a significant part of the abyssal plain sediments (414).

Island Stability

Volcanic islands form a major load on the underlying lithosphere and hence sink into it with velocities depending on: their mass, their growth (mass emplacement) rate and the age and elastic thickness of the lithosphere. In the Hawaiian archipelago, for example, the lowest terrace of Mahukona, the oldest volcano of the Big Island, is found in the northwestern continuation of Hawaii at 1 325 m b.s.l. (57, 230, 235). The large Hawaiian Islands sink isostatically with a velocity of 2 – 2.5 mm/a, supplemented by the subsidence of the cooling lithosphere, as well as the cooling partially molten mantle as it moves away from the hot spot (82, 263). Islands may rebound after volcanic activity has terminated and erosion removes much of the subaerial edifice such as on Oahu. Vertical isostatic movements of similar magnitudes have not yet been documented from the Canary Islands, probably because of the much higher age of the oceanic crust (311) and hence the greater rigidity of the older and thus thicker lithosphere. Some authors (413) postulated that the island of Tenerife subsided by about 2 500 m starting about 3 million years ago, but seismic data and the rocks dredged from the submarine flanks of Gran Canaria and Tenerife only show subsidence of at most 600 m, and possibly very much less (107).

Intra- and Inter-Island Differences

Given the infinite complexities of geologic systems it goes without saying that each island evolves somewhat differently. This is true even for the Hawaiian islands although the similarity in evolutionary stages between Hawaiian islands is impressive. Not so in many other archipelagoes such as the Azores or the Canaries. In the latter for example, the older eastern islands basically represent a very long rift zone. The Miocene shields grown over those rifts are almost entirely basaltic, intrusive rocks of highly evolved rocks being rare, in that respect resembling the Hawaiian islands. The western islands of La Palma and El Hierro likewise expose only minor phonolite but they are younger and may or may not evolve differently when matured. In contrast, the two central islands, Gran Canaria and Tenerife, expose very large volumes of evolved rocks. While these contrasts provide some symmetry to the archipelago, compositional differences can be recognized along an east-west traverse. In a very simplified way, magma becomes more alkaline westward and there are several lines of evidence indicating that magmas beneath the western islands were generated at greater depths and lower degrees of partial melting.

Difference in Magma Production and Eruptive Rates

The volumes and eruptive rates of lavas of the basaltic shield phase of the Canaries are significantly smaller than those on Hawaii. Calculations of the total volume of an island such as Gran Canaria are more difficult, however, because a much larger part of the volcanic apron of the seamount (representing the early submarine stages) is hidden beneath several hundred meters of biogenic and volcaniclastic sediments accumulated during the long history of these islands. Only a 50–80 m thick sediment blanket covers the lower flanks of the Hawaiian Islands, owing largely to the very low biogenic sedimentation rates in the center of the Pacific (Fig. 6.8). Present-day bulk eruptive rates of the Canary Islands are probably in the order of $0.001 \, km^3/a$ or roughly 1 % of the Hawaiian rate when shield-building stages are compared.

The shield phases of the Canary Islands are also much more heterogeneous in their temporal evolution than those on the Hawaiian Islands. Several overlapping shield volcanoes grew over 10 million years on the eastern Canary Islands, Fuerteventura and Lanzarote. They form part of the north-northeast elongated, 400-km-long East Canary Ridge that also includes the shallow Conception Bank 150 km north-northeast of Lanzarote (10). The shield phase of Gran Canaria consists of at least three coalesced basaltic shield volcanoes that, similar to Hawaii, each formed within about 1 million years or less. The young western Canary Islands El Hierro (ca. 1 Ma old) and the young southern ridge of La Palma may still be in their initial shield stage, the older inactive shield of northern La Palma being ca. 3 Ma old.

When the production rate of magma in the mantle decreases and the island lithosphere slowly drifts away from the center of a melting anomaly or if the positive buoyancy does not allow further growth of the island above ca. 3 500–4 000 m e.g., the supply rates of melts nourishing the magma chambers in the interior of volcanoes dwindle. This pattern has been inferred especially for the Canary Islands or the Azores, islands grown on a slow-spreading plate. Magmas stagnating longer in reservoirs are able to differentiate, thus generating significant volumes of highly evolved magmas. By contrast, in Hawaii, with its high magma production rates and very fast plate velocity, differentiation of basaltic magmas occurs only locally and produces very small amounts of trachyte magmas. The eruption of large volumes (exceeding $500 \, km^3$!) of highly evolved rhyolitic, trachytic and phonolitic magmas on Gran Canaria and Tenerife (Fig. 6.25) requires optimum filling rates of the magma reservoirs. These rates must be large enough to keep the system thermally and compositionally alive (on Gran Canaria between about 14–9 Ma), but must be slow enough to allow for extreme fractionation.

Speculations on the Cause of Contrasting Evolution of Oceanic Islands

Why do some island groups evolve differently than others? Several geodynamic factors may be responsible in causing the contrasting evolution of the Hawaiian compared to the Canary Island volcano-magma systems, although mantle processes beneath the islands remain difficult to specify.

Firstly, the African Plate is very slow moving, practically stationary in the global hot spot reference frame. Secondly, a lithosphere about 100 km thick in the western part of the island chain is likely responsible for the greater depth of magma generation beneath the Canary Islands. Hawaii is the foremost textbook example of a volcanic system above a highly productive, powerful and long-lived melting anomaly, the Hawaiian Plume. In contrast, island groups in the eastern central Atlantic Madeira, Canaries and Cape Verdes and probably hundreds of seamounts in between form a roughly 3 000-km-long magmatic belt approximately parallel to the boundary between the African continental and the Atlantic oceanic lithosphere, the rifting having started about 180 million years ago.

Thus, significant differences in growth and dynamics of the melting anomaly along the ocean-continent lithosphere boundary (158, 304) may be the most important reason for these differences. Smaller fertile mantle diapirs may be triggered by the "edge effect" between thick cold continental and thin hot oceanic lithospheric mantle. Moreover, perhaps the fertile mantle material migrates beneath the lithosphere from east to west, a model which would explain the migration of the main shield phases of the islands westward as well as the continuous, albeit episodic, eruptive activities on all of the islands (304).

Large Igneous Provinces and Oceanic Plateaus

Iceland, closely following Hawaii as the second most active volcanic island on Earth, is a small continent that in many respects represents a special case (Fig. 6.31). In principle, Iceland represents part of the Mid-Atlantic Ridge with a particularly high magma production rate. This had already started when the Eurasian and American plates separated about 60 million years ago. Evidence for the high magma production includes huge piles of flood basalts in eastern Greenland, Scotland and the Faeroe Islands. Iceland is geodynamically roughly symmetrical in structure because the age of volcanoes increases in both directions away from the neo-volcanic zone that separates Iceland into an eastern and western half. Rhyolitic magmas have formed repeatedly in Iceland and have erupted explosively several times during historic and pre-historic times mostly from Hekla volcano. Ash from Hekla eruptions

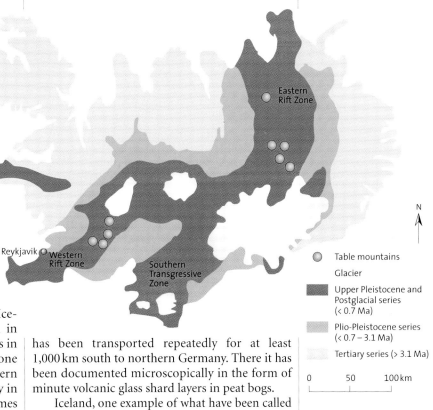

has been transported repeatedly for at least 1,000 km south to northern Germany. There it has been documented microscopically in the form of minute volcanic glass shard layers in peat bogs.

Iceland, one example of what have been called Large Igneous Provinces (LIPs) (61) (Fig. 6.32), is dwarfed, however, by still mysterious gigantic submarine basaltic plateaus, some more than 1 000 km in diameter and up to several kilometers in thickness (e.g., Ontong-Java (almost 2 million km^2 in size and more than 35 km thick), Hess-Rise, and

▲ Fig. 6.31. Simplified geological map of Iceland with neo-volcanic zone and selected subglacial table mountain volcanoes (417)

▼ Fig. 6.32. Geographic distribution of large LIPs (Large Igneous Provinces). *FB* Flood basalt provinces. The *black lines* parallel to passive continental margins, e.g., in the northern and southern Atlantic Ocean, represent thick basaltic packages (dipping reflectors) that formed when the continents started to separate (61). The *red areas* are flood basalt provinces and large oceanic plateaus. The *green areas* represent silicic LIPs (SE Bryan, pers. comm.)

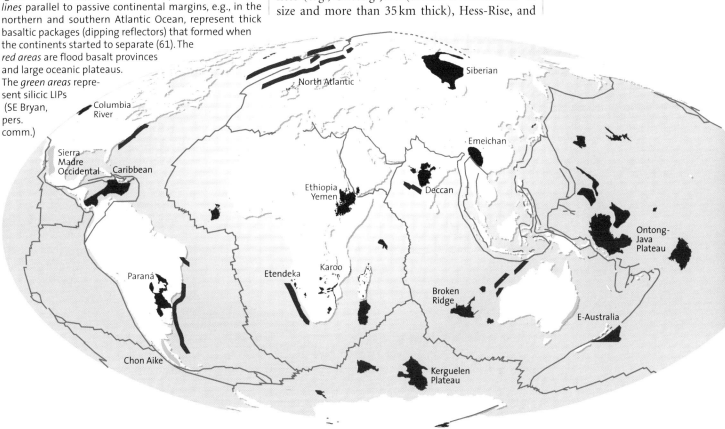

Manihiki-Plateaus in the Pacific and Kerguelen Plateau/Broken Ridge in the southern Indian Ocean). Deep sea drilling during the last three decades has helped to unravel the nature of these enormous piles of intraplate lava fields. Ontong-Java and Manihiki probably formed during very short periods around 122 Ma ago with a second peak around 90 Ma (190). The lavas are very uniform evolved tholeiites, in that resembling many flood basalts. These huge manifestations of intraplate igneous activity pose several problems, the two most hotly debated questions being the nature of their mantle roots and their environmental impact.

Two observations surround the mystery of the mantle roots of oceanic plateaus. One is the extremely high magma production rate. The magmas must have been generated by unusually large and/or very rapidly rising mantle plumes, possibly originating in the lower mantle, large magma volumes being produced in the plume heads by decompression. The intense igneous activity creating the very voluminous Cretaceous LIPs is perhaps the result of mantle convection being more vigorous at that time than at present and there are other periods in Earth history where magma production was particularly high. What is intriguing about the Cretaceous LIPs is the fundamental synchronicity of at least some of them. This calls for short-lived global episodes of particularly vigorous mantle convection.

Did such hypothetical *superplumes* also result in major global impacts as they temporarily increased the flux of mass and energy from the mantle to the crust? Examples of postulated impacts include the changing frequency of reversals of the Earth's magnetic field, changes in the spreading rate along mid-ocean ridges, as well as global changes in climate, hydrosphere, sea level, oceanic anoxia, seawater composition and biosphere (biological radiations and extinctions).

Despite the huge size of some LIPs and their potential role in contributing to our understanding of mantle circulation and environmental change, they are among the least understood features in the ocean basins and on continents (Fig. 6.32). Some authors also include within LIPs, larger oceanic island groups, such as the Canary Islands and the large flood basalt fields on the continents.

The fundamental geodynamic model that has revolutionized the interpretation of volcanic islands and LIPs and of magma production and surface volcanism in general since the late 1960s is that of mantle plumes.

Hot Spots and Mantle Plumes

Oceanic islands have played a pivotal role in the development of plate tectonics. Foremost was the more quantitative documentation of systematic age progressions along an island chain. In the nineteenth century, von Fritsch (1865) had recognized an age progression in the Canary Islands, where the islands apparently became increasingly older toward the east. Dana (1891) made similar observations when studying the Hawaiian islands, where the big island is the active end, and islands towards the west-northwest become progressively older. These conclusions were based purely on a qualitative evaluation of the degree of erosion, in part supported by a concentration of historic eruptions at the most active end. J. Tuzo Wilson (420), one of the fathers of modern plate tectonics, built on these observations when formulating his famous "hot spot" hypothesis. According to this, the oceanic lithosphere migrates above a melting area (hot spot) anchored in the mantle (Fig. 6.33). A submarine volcano and eventually an island actively grow as long as the moving lithosphere is located above such a hot spot. Volcanic activity terminates when the plate has moved away from this melting anomaly. Morgan (241) expanded this theory and defined hot spots, loci of high magma production and eruption, as points on the Earth's surface, below which columns, some 100 to 150 km in diameter, of mantle material rise. These columns he called mantle plumes. The plumes, according to Morgan, not only transport heat and magma but also can lead to a breakup of plates and represent a fundamental mechanism to drive plate motion. The plume concept is central to plate tectonics that itself is the governing paradigm of solid Earth science.

Voluminous Volcanism

Practically all areas of the oceanic and continental plates characterized by voluminous volcanism, some volcano groups showing a more or less linear age progression, are loosely called hot spots (believed to be the surface expression of rising mantle material). The total volume and the mass eruption rates are highly variable among such hot spots, however. On the one hand, there are the highly productive hot spots such as Hawaii or the Snake River-Yellowstone area, while the other and more common type of hot spot is exemplified by the Eifel in Central Europe where only small volumes of magma were erupted during the Quaternary. The erupted magma volume is only one criterion, however, to define hot spots/plumes.

Hot Spot Traces

The fundamental observation, on which the existence of relatively stationary melting anomalies in the sublithospheric mantle is based, is the more or less linear increase in age of volcanoes along a chain away from the point of youngest and most active volcanism (214). These linear age progressions have been extremely important to postulate relative and absolute plate motions and their directions (vectors), even though the model of plumes fixed in the mantle as a stable reference frame has recently been questioned. The most clear-cut age progressions are found in the Pacific, within the three major island and seamount chains: the Hawaii-Emperor-Chain (6 100 km long, 85 Ma old, 129 volcanoes); Tuamoto Islands and Gilbert Marshall Islands. These chains are most active or youngest at the southeastern ends, show similar age progression of volcanism (ca. 8–10 cm/a) and changes in direction about 43 million years ago (Figs. 6.33–6.35).

Age progressions along other chains of volcanoes are rarely as clear-cut as these examples from

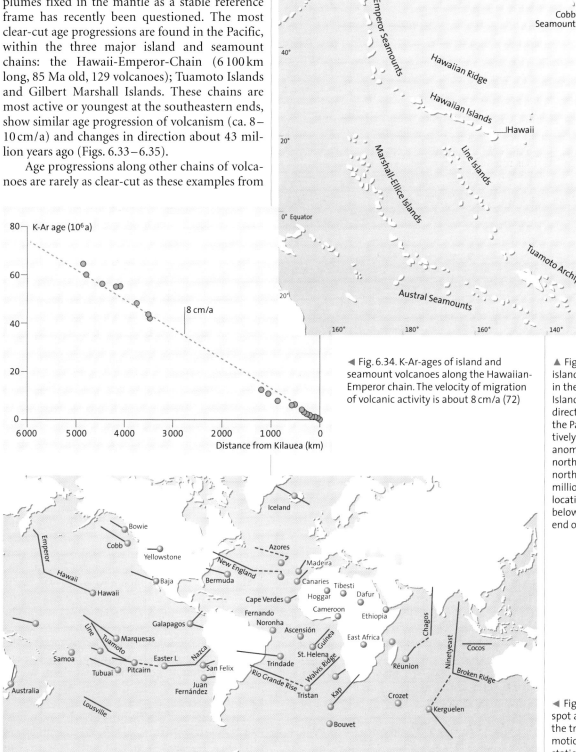

▲ Fig. 6.33. Map of larger islands in seamount chains in the Pacific ocean basin. Island chains indicate the direction and velocity of the Pacific Plate over relatively stationary melting anomalies (plumes) to the northwest, but to the north prior to about 42 million years. The present locations of plumes are below the southeastern end of an island chain (71)

◄ Fig. 6.34. K-Ar-ages of island and seamount volcanoes along the Hawaiian-Emperor chain. The velocity of migration of volcanic activity is about 8 cm/a (72)

◄ Fig. 6.35. Selected hot spot areas. The *lines* show the trace of the plate motion over relatively stationary melting anomalies (69)

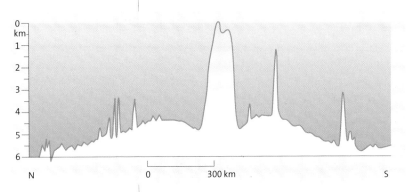

▲ Fig. 6.36. Anomalously elevated areas on both sides of the Hawaiian island chain (69)

▶ Fig. 6.37. Two models explaining the Hawaiian swell. a) Thinning of the lithosphere by partial melting of its base; b) dynamic push of ascending mantle material

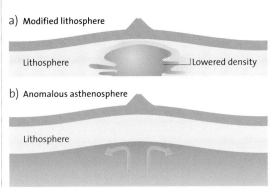

the Pacific and some of the hot spot traces are equivocal. The young Yellowstone volcano complex in the western USA marks the present end of a major progression in volcanism that started about 13 million years ago in northern Nevada and migrated through the Snake River Plateau to its present position, reflecting plate motions of about 2–3 cm/a. Volcanism is still active throughout this and many other chains, evidence that melting anomalies, once started, can be very long-lived and very broad, rather than extremely focused as in the three prior examples from the Pacific (Chap. 7). The arguments for the origin of the migration of the Yellowstone Plume are discussed in more detail in Chapter 7.

Positive Topographic and Gravity Anomalies

Hot spot areas are not only characterized by high rates of magma production, but also by positive morphological anomalies that can be very much broader than the volcanoes themselves. For example, around the Azores Islands in the center of the Atlantic, there is a residual depth anomaly, (i.e., an area of the sea floor that is significantly higher than would be expected, based on the cooling and thus subsidence curve of the lithosphere) that can be followed for more than 1 000 km to the southeast and is at least 60 million years old. Hence, there is not only an atmospheric Azores high, guarantee for good stable weather in Europe, but

also a topographic high of the lithosphere. Many continental intraplate volcanic fields are found on uplifted crustal blocks, some of which in Africa having been uplifted as much as 3 km and have diameters up to 3 000 km. The Hawaiian Islands are small dots on the huge Hawaiian swell with a diameter of about 2 000 km and a height of about 1.2 km. This swell is more than three times as wide as the zone of subsidence of Hawaii into the 30-km-thick elastic lithosphere and the marginal swell around the islands (Fig. 6.36).

The volcanic islands merely represent the tip of an iceberg, because the broad swell can be followed from the youngest island to the bend in the island/seamount chain over a distance of some 1 000 km. The former model to explain this swell by heating and therefore extension of the lithosphere (Fig. 6.37) is interpreted differently today. When the mushroom-like plume head (possibly as much as about 200 times hotter than the sur-round-ing mantle), impinges on the relatively rigid and melt-depleted lithosphere, it can expand laterally, analogous to heated air that rises in the summer and forms mushroom-like thunderstorm heads. This dynamic uplift can be several 100 km in diameter and amount to 1 km or more (343). Perhaps surface eruptions start above such an initial plume in the form of flood basalts, discussed in the next chapter. About 10–20% of the Earth's surface consists of similar hot spot "highs". Seismic proof for the existence of plume heads is still lacking, however.

Positive gravity anomalies also occur above oceanic uplifts, as shown by sea surface measurements by satellites. These probably form because hotter, less dense mantle material rises and can be compensated by the denser surrounding mantle material.

The Deep Roots of Hot Spots and Subduction Zones

During the last two decades, global seismic tomography has become the central method to map the depth extent of plumes and cold slabs in the mantle. Indeed, global tomography has provided remarkable three-dimensional views into the density distribution of the entire Earth's mantle (e.g., 395, 283). The correspondence of denser areas with cool subducting slabs on one hand (Fig. 6.38) and the less dense and therefore hotter major hot spot areas are most impressive results (Chap. 5; Fig. 6.39). For example, there is a zone of reduced density beneath the young rift of the Red Sea that can be recognized at least as deep as 550 km. Intraplate volcanic fields are thus not only located on topographically higher areas with positive gravity anomalies, but also show roughly cylindrical roots that can be seismically detected to a depth of at least 400 km. These low density "chimneys" probably consist of partially molten mantle material that is about 50–150 °C hotter than the surrounding mantle. Seismic waves, especially those from far away (called *teleseisms*) propagate in the hot mantle rock more slowly than in the surrounding cooler and denser peridotite. Thus they arrive with a slight delay in broad arrays of numerous seismometers employed around areas suspected to be underlain by plumes. Such anomalies have not only been found beneath Iceland, the Yellowstone volcanic area (Chap. 7) and Hawaii (areas of especially high magma production), but also beneath the smaller Neogene intraplate volcanic fields on the European continent (274).

The search for the deep roots of plumes recognized in the upper mantle has been on for some time. Yet, the only two plumes that have been traced to the core-mantle boundary (CMB) are two superplumes, one beneath the Pacific off-centered from Hawaii, and the other one beneath Africa (283). These superplumes appear to impinge at the base of the lithosphere, their material spreading laterally to feed the asthenosphere. Nearly all smaller hot spots may be fed from these two superplumes.

The Birth Place of Plumes

Plate tectonics and seafloor spreading have become firmly rooted in the theory of Earth science, at least for the Mesozoic and Cenozoic. But the models of hot spots and mantle plumes are still surrounded by many uncertainties, even though they are among the most important and challenging geodynamic hypotheses. Still unknown are their dimensions, their depth extent, their spatial and temporal stability and, of course, the place and mechanism of their origin. In the 1970s, it was thought that they originated in old so-called primordial mantle material that rises in the form of plumes from great depths. Today, most scientists believe that the chemically enriched nature of lavas erupted above presumed plumes can be better explained by recycling of subducted oceanic lithosphere (160). Nevertheless, there is no unequivocal proof that a major portion of the specific element inventory contained within a subducted slab does not disappear during its long and complex journey in the mantle. Most likely, there is a broad range of intraplate magma systems, preserving different degrees of mixing of older mantle and recycled lithosphere.

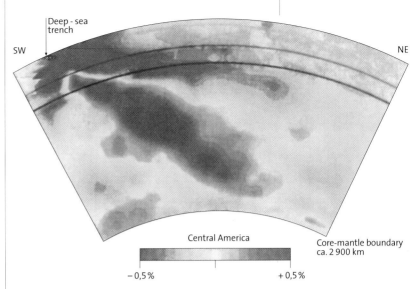

▲ Fig. 6.38. Cross section through the Earth's mantle from the Earth's surface to the core-mantle boundary in the area of the subduction zone in central America. The *Farallon Slab* can be documented, based on its seismically identified high density (*blue*), from the uppermost mantle to the core-mantle boundary (126)

▼ Fig. 6.39. Depth extent of partially molten mantle material beneath Iceland (*Iceland Plume*). The plume has been detected seismically by mantle tomography to a depth of about 400 km (432)

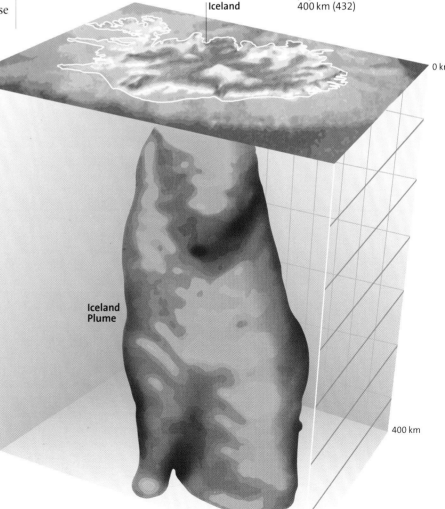

Several basic questions have not yet been answered by mantle topography. For example, do plumes start from the core-mantle boundary and rise to the lithosphere? Or does convection in the lower mantle (i.e., below or at about 660 km) produce secondary convection in the upper mantle, which leads to the rise of mantle domains, decompression, magma production and surface volcanism? Presently available tomographic data do show that cold and dense subducted slabs can reach much deeper than 700 km in subduction zones, in some areas as deep as 1 500 km, or even to the core-mantle boundary (126) (Fig. 6.38). A layer that has received much attention during the last few years is a shell, about 200 km thick, the socalled D''-layer (Chap. 2). This represents the very base of the lower mantle that according to some interpretations may consist of subducted lithosphere that has made it all this way. Perhaps plumes start from this D"-layer thus facilitating cooling of the surface of the outer liquid Earth core. Some subducted slabs may not penetrate as deep, but may remain in the upper mantle or may be transported to the 660 km discontinuity, the upper mantle convecting faster than the lower one.

Summary

The volume of many oceanic islands is very much higher than that of subduction zone volcanoes on active plate margins and island arcs. Their number is small, however, compared to that of the continental intraplate volcanoes. If one adds the seamounts (of which some, such as the Emperor Seamount Chain, were once islands, but of which the vast majority never grew to breach sea level), then one can assume probably much more than 1 million volcanoes on the seafloor. This is very much higher than the number of volcanoes on the continents. If the oceanic plateaus are added (a type of submarine flood basalt field), then the global contribution of intraplate compared to subduction zone volcanism, is significantly higher than commonly assumed.

Ocean islands and seamounts consist almost entirely of basaltic rocks. Alkali basalts dominate, at least in the subaerial part of ocean islands, the main exception being the Hawaiian and Galapagos Islands. Tholeiites seem to dominate on seamounts, but only a minute fraction of those have been studied and alkali basalts have also been found on some. Pu'u 'O'o, which started to develop about 20 km east of Kilauea caldera, has been an almost inexhaustible source of lava up to the present day. Several km³ of lava issued from the vent, and a nearby lava lake, with lava flows proceeding largely in sub-terranean channels to the sea. Early lavas from this center destroyed the picturesque Kalapana Blacksand Beach, the Royal Gardens residential development, as well as a famous worshipping place of the Polynesians. At the same time, however, about 5 km² of new land has been created by the lava flows by pushing the shoreline slowly day by day into the Pacific ocean.

Many ocean islands in the Atlantic also contain large volumes of more evolved volcanics, especially trachytes, phonolites, but also rhyolites. The origin of these highly evolved magmas can be explained largely by differentiation of basaltic magmas. Intraoceanic volcanic plateaus, such as the small continent of Iceland, cannot be classified very easily; the composition of Icelandic lavas resembling that of ocean floor basalts much more than that of the majority of other oceanic islands.

Based on the present state of knowledge, the basaltic magmas of ocean islands are probably generated in rising hot mantle diapirs, so-called mantle plumes. These may originate at the boundary between the Earth's core and mantle and/or along the boundary of lower and upper mantle, at a depth of approximately 660 km. The reason for the rise of these plumes is unknown and may be due to their specific composition or higher temperature or both. About 10% of the heat loss of the Earth may be due to rising plumes. The debate whether the entire mantle convects or whether the upper and lower mantle convect separately, and whether or not the plumes originate at the core-mantle boundary or higher is still in progress. The geochemical signature of plumes shows that they represent mixtures of mantle depleted to varying degrees by partial melting and variable volumes of subducted lithosphere, contrasting with a strongly depleted mantle beneath mid-ocean ridges. Age, volume and chemical composition of volcanic fields thus allow some conclusions as to the age and chemical evolution of even very deep root zones beneath highly productive volcanic areas.

Whether or not mantle plumes are also effective in pushing plates apart is uncertain. At present, the main mechanism for plate motion is thought to be pull by the descending dense plates in subduction zones. Before I discuss volcanism above subduction zones, let us take a look at intraplate volcano fields on the continents, for which there are many examples in central Europe and in the interior of continents, such as the western United States and Australia.

Continental Intraplate Volcanoes

ontinental intraplate volcanoes were the premier objects of scientific curiosity and inquiry when the field of volcanology came of age in the second half of the eighteenth century (Chap. 1). One of the pivotal localities where Abraham Gottlob Werner (416) in 1776 thought he had found evidence for the hypothesis that basalt columns had crystallized from cold water is Burg Stolpen in Saxony, about 20 km east of Dresden (Fig. 7.1). In the words of Werner, translated from the German: *In the summer of 1776 I visited the most famous basaltic hill in Saxony near Stolpen. I did not find a single trace for volcanic action, not a speck of evidence for a volcanic origin. The internal structure of the hill showed in fact the opposite. Now I dared to declare for the first time publicly that not all basalt is of volcanic origin which also includes the basalt from Stolpen. In short, basalts are not of volcanic but of wet origin.*

The actual Mid-Tertiary volcanic edifice at Stolpen has long been eroded. The rump volcano is a typical example of an intraplate volcano as shown by its alkaline basaltic composition and its occurrence in the interior of a continental plate far away from any plate boundary. Crucial proofs that basalts had actually crystallized from hot lava were discovered by Nicolas Desmarest in the Quaternary intraplate volcanic field of the Auvergne in France. He based his evidence for high temperature on baked soils beneath lava flows (Fig. 7.2) and the transition of columnar basalt to scoriaceous lava flow tops (Chap. 1). Intraplate volcanic fields such as those in Saxony and central France are characteristically located on top of uplifted crustal blocks that are traversed by morphologically prominent rift zones.

Rift Zones and Rift Shoulders

Rift zones are narrow graben structures that range from a few hundred kilometers in length such as the Rhine Rift, to broad rifted areas characterized by many horst and graben structures such as the Basin and Range Province in western North America. The characteristics of rifts are high heat flow, suggesting shallow upwelling of asthenospheric mantle and thinned crust and lithosphere,

contrasting with the adjacent uplifted lithosphere (254). The lithosphere has been rifted apart in these belts by up to many tens of kilometers, the ductile lower crust apparently having experienced the bulk of the stretching (Fig. 7.3). Fully developed rifts are 40–50 km wide and have subsided 2–3 km. Of interest here is the volcanism that character-

"This stretch of road is spooky, everything looks so old."
"Basalt lava," Christine said.
"Old volcanoes."
"Very old volcanoes."
"Like fifty million years."
"Even older."

David Guterson, East of the Mountains, London, 1999

▲ Fig. 7.1. Columnar alkaline basalt representing the conduit-filling of an eroded Tertiary volcano. Stolpen near Dresden (Germany)

▶ Fig. 7.2. Clay-rich red soil (weathered tuffs) heated at the contact by an overlying lava flow. Columns formed by volume contraction in the baked soil during cooling. Gran Canaria (Canary Islands)

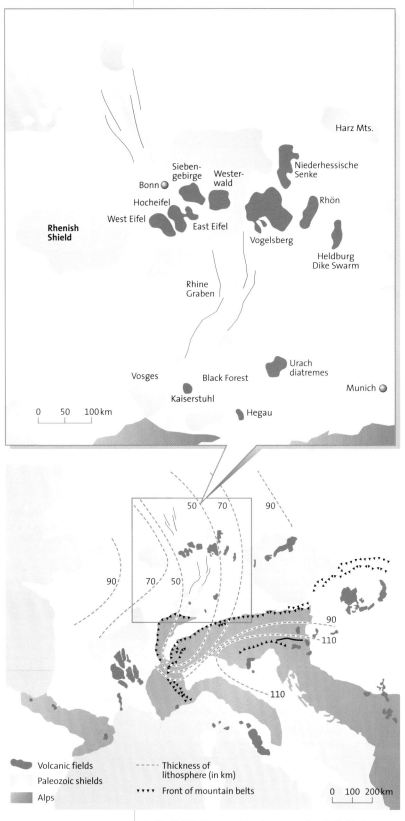

▲ Fig. 7.4. Tertiary and Quaternary volcanic fields in central Europe. The *dotted lines* in the lower map of Europe indicate the thickness of the lithosphere (254)

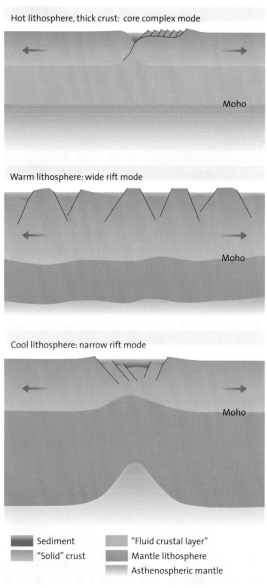

▲ Fig. 7.3. Three modes of lithosphere rifting and graben formation zones (38)

izes both the rifts as well as their shoulders, the uplifted blocks extending for hundreds of kilometers on either side of many rifts.

Most geologically young volcanic provinces in central Europe developed during the Tertiary between about 5 and 40 million years. Some of these such as near Cheb (northern Bohemia, Czech Republic), Clermont-Ferrand (France) and the West and East Eifel (Germany) also formed in the Quaternary, but they are of much smaller volume. These central European volcanic fields can be related to two tectonic domains: some volcanic areas (for example Westerwald, Siebengebirge, Eifel, Auvergne) grew on uplifted and in part still rising huge Paleozoic blocks (Fig. 7.4). Other volcanic fields are located in the grabens, in between the uplifted areas. Examples include the Eger graben, upper Rhine graben, which continues in the Niederhessische Senke northeast of Frankfurt, with the Kaiserstuhl in the south and, east of Frankfurt, the Vogelsberg, the most prominent volcanic structure in Central Europe, the rift in France being the famous Limagne graben.

Uplifted blocks and rift zones are highlighted by intraplate volcanic fields in many other continents. Major examples are the East African Rift with its uplifted shoulders (Fig. 7.5), Rio Grande Rift and the adjacent Colorado Plateau, Yellowstone Plateau, Tibesti, Baikal Rift, Tibet, the highland of eastern Australia with its numerous Cenozoic volcanic fields, or Auckland (New Zealand).

A few continental rift zones have developed into oceanic ridge systems. The classical example is the Red Sea Rift that opened about 20 million years ago and developed into an ocean divergent plate boundary producing oceanic tholeiitic basalts. The Afar Rift, the northernmost end of the East African Rift is transitional between continental and oceanic rift zones, becoming "oceanic" toward the Red Sea Rift judging from the very thin crust and the huge volume and "depleted" tholeiitic composition of its lavas. The central and southern part of the two East African rift zones that developed farther south, however, and other major continental rift zones have been active for millions to tens of millions of years. There is no evidence at present that the continental lithosphere might one day be completely rifted apart along these seams. There are other reasons for assuming that most continental rift zones form tectonic domains in their own right including:

- Geologically ancient rift zones, such as the Permian Oslo Graben, well known for its calderas and ignimbrite sheets, never develop into an oceanic rift zone.
- The magma composition and volcanic character, characteristic all over the world for the early stage of oceanic rift zones that later developed into ocean basins, are typically tholeiitic flood basalt, a classical example being the flood basalts of eastern Greenland.

The tectonic position of some active large volcanoes, such as Mt. Etna in Sicily, is more complex. The chemical nature of its basal tholeiitic lavas erupted about 0.5 million years ago closely resemble those of unequivocal intraplate volcanoes as in the famous Monti Iblei, south of Etna, where they form widespread subaerial and submarine volcanic deposits. But the younger Etna magmas also exhibit some chemical subduction-zone signatures, especially in the 2001 and 2002 lavas, reflecting the complex tectonic setting in a ten-

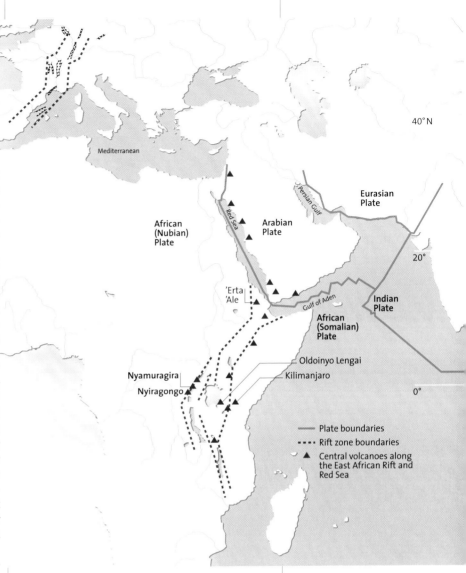

sional area at the contact of several colliding and subducting plates (Chap. 2).

Scoria Cones

The most typical volcanic landforms of continental intraplate volcanic fields are *scoria (cinder) cones* (Chap. 9), but they occur in some subduction zone settings as well. They result from locally confined eruptions of variable intensity but very small volume, many finishing their growth within a few weeks to months. Because of such short eruption periods, they are often called *monogenetic volcanoes.* This term should not be taken too literally, however, and does not apply everywhere. Many extremely well-exposed scoria cones in the Quaternary Eifel volcanic field, e. g., consist of several different volcanic phases of both, pyroclastic, phreatic and phreatomagmatic activity despite their simple external shape. These phases

▲ Fig. 7.5. Major rift zones in eastern Africa and central Europe. The *red triangles* show large central volcanoes along the East African Rift and the Red Sea (442)

▶ Fig. 7.6. Cross section through scoria cone (Wannenköpfe) about 200 000 years old in the Laacher See area (Germany). The main scoria cone overlies several initial tuff rings (ITR) formed by magma-water contact in the initial stages of the growth of the scoria cone (*dashed line*). The central crater (C) is shown by short dashed lines. The boundary between the black outer and the reddish central scoria deposits, oxidized by hot gases streaming through the volcano during its late stage, is shown by arrows

contrasting in eruptive mechanism and commonly also in composition may be separated from each other by one or more soil zones, evidence for vegetation and other signs of major interruption lasting thousands to tens of thousands of years (310) (Figs. 7.6, 9.24, 9.25).

A recent case in point is Cerro Negro volcano (Nicaragua), a scoria cone formed in a subduction zone setting which first appeared on the Earth's surface about 50 km north of Managua in October 1850 (Fig. 7.7). It erupted about 27 times in irregular intervals, the most recent eruption occurring in 1999. Lava flows issued from the scoria cone repeatedly and ash generated in lava-spray eruptions and transported westward has been a major hazard in the city of Léon, about 20 km west of the volcano (155). When approached from distance, the smooth and regular shape of the cone, at least when viewed from the west, belies its complex history and many geologists not being familiar with its evolution would call it monogenetic.

The pathways at depth beneath many scoria cones apparently remain hot long enough to provide convenient preheated routes to facilitate the repeated rise of new magma batches. A corollary is that many if not most attempts of very small-volume magma batches to erupt remain unsuccessful; the rising magma freezes in the crust on the way up before reaching the surface. The term monogenetic must thus be understood in terms of a construction period that is short compared to

the hundred thousands or even millions of years that are frozen in the lifetime of large stratocones or of huge oceanic island volcanoes.

Scoria cones are gregarious in the sense that they usually occur in clusters, forming *volcanic fields* (Fig. 7.8). Many such intraplate volcano fields on top of uplifted blocks have diameters between 30 and 80 km, covering areas between 100 and 5 000 km² and generally consist of several tens to at most several hundreds of scoria cones. Well-known examples in central Europe are the Quaternary volcanic fields East Eifel and West Eifel, Chaîne des Puy in the Auvergne and, in the Canary Islands, the famous scoria cones of the 1730–1736 eruption, on Lanzarote, the second most voluminous basaltic eruption in history. Almost 1 000 volcanoes occur in the unusually large Transmexican Volcanic Belt, covering an area of some 60 000 km² (64, 411).

The size of volcanic fields appears to be independent of greatly contrasting crustal and lithospheric thicknesses. The dimensions of many volcanic fields must thus reflect those of the melting anomalies, which must be similar. Apparently they reflect a minimum amount of magma, generated and transported into the crust from the melting anomalies in the mantle. Most likely, magmas that do not reach this minimum volume get stuck in the uppermost mantle and in the crust without ever erupting at the surface.

Primitive alkaline basaltic magmas are typical of scoria cones and their lava flows. Composition-l zonations are not uncommon, however. Magmas that show higher degrees of differentiation and that were probably erupted from larger magma reservoirs, occur toward the center of a field (Figs. 7.8–7.14). A typical example is the phonolitic Laacher See volcano in the East Eifel (Chap. 11), a crater surrounded by older tephritic and basanitic scoria cones. Less strongly pronounced compositional zonations also characterize individual scoria cones, the most evolved lavas having erupted first.

Volcanoes are commonly larger in the center of rift zones and at the boundary between the rifts and the uplifted blocks compared to the uplifted rift shoulders. Also, their lavas may only show mildly alkaline to tholeiitic composition (such as the Vogelsberg in central Germany), or may comprise thick trachytic and phonolitic lava flows and ignimbrites (such as in the East African Rift in Kenya). This contrasts with the small-volume more alkaline and primitive volcanic fields on top of the uplifted blocks. Most likely, hot mantle material rises higher beneath a rift, or partial segments of a rift, compared to the uplifted area, and can undergoe higher degrees of partial melting at lower pressures resulting in higher magma production rates.

The Quaternary Volcanic Fields of the Eifel

The Quaternary Eifel volcanic fields are in many ways typical intraplate volcanic fields (303, 310). Few intraplate volcanic areas on Earth are better exposed in numerous quarries, allowing detailed studies of the internal structure and evolution of scoria cones.

About 240 volcanic centers in the West Eifel and about 100 in the East Eifel have been recognized (Fig. 7.8). The West Eifel field is about 50 km long (East Eifel 35 km) and extends from Ormont, at the border to Belgium in the northwest, to Bad Bertrich at the Moselle River in the southeast. Both fields can be subdivided into an older northwestern and a younger southeastern field (Fig. 7.8), with a sharp temporal and compositional boundary between both subfields in the East Eifel. The primitive parent magma compositions in both fields differ significantly from each other, indicating that they have been generated in different mantle domains. Both fields show the regular orientation of the entire field and of most directions of dikes, volcanic axes and growth of volcanoes from northwest to southeast (Figs. 7.10, 7.11). Individual volcanoes in the fields are commonly not distributed statistically but increase in frequency per unit area toward the center of the

▼ Fig. 7.7. Active Cerro Negro scoria cone (500 m high, summit at 728 m a.s.l.) (Nicaragua) which erupted 27 times since it was born in 1850. It erupted last in 1999

field (Fig. 7.9). The degree of differentiation of the magmas also increases from the periphery, where only primitive lavas have erupted, to the center (Figs. 7.12, 7.13).

These interesting parameters, which have been noted in several other volcanic fields on Earth, can be interpreted as follows. The lithosphere north of the Alps is generally under compression, the direction of maximum tension being southwest-northeast west of Rhine River. Fissures can thus form most easily mechanically at right angles to this direction (22) (Fig. 7.10). The fissures, along which the magmas rose, and thus the feeder dikes are oriented northwest-southeast and are overlain by strings of volcanoes showing similar orientations. Secondly, the lithosphere beneath the Rhine Graben is thinner than in western and eastern Europe (Fig. 7.4), possibly due to rise of mantle material, a prerequisite for partial melting by decompression. Seismic waves, e.g. generated

at the other side of the globe and traveling through the Earth are attenuated beneath the Eifel volcanic field, at least as deep as 400 km (280), most easily explained by partially melted mantle material in the deep root zones of the volcanic fields. Moreover, the analysis of mantle xenoliths brought up from depth by the lavas, the *peridotite nodules* (Fig. 3.5), show that the minerals equilibrated at relatively steep geothermal gradients, a further hint for hot rising mantle rock (320). The chemical difference between the older and the younger subfields in the West and East Eifel is most plausibly explained by migration of the melting anomalies during the past 700 000 years from northwest to southeast. Moreover, two different mantle domains had become partially melted, which, at least in the West Eifel, must have been stacked on top of each other. The increase in volcano density and degree of differentiation toward a center of a volcanic field cannot be explained by crustal parameters, especially because many volcanic fields worldwide show similar structures. A zoned volcanic field thus most likely represents a melting anomaly 30–50 km in diameter in the Earth's mantle, in which the production of magmas has increased toward the center, where the flow velocity and therefore rate of decompression may be at a maximum.

The model of plate migration over a plume was once applied to the Eifel (88). It was thought that a hypothetical Eifel plume represented a

◄ Fig. 7.8. Areal distribution of Quaternary eruptive centers in the West Eifel and East Eifel volcanic fields (303). Freeways (A 61, A 48, A3) shown by double lines

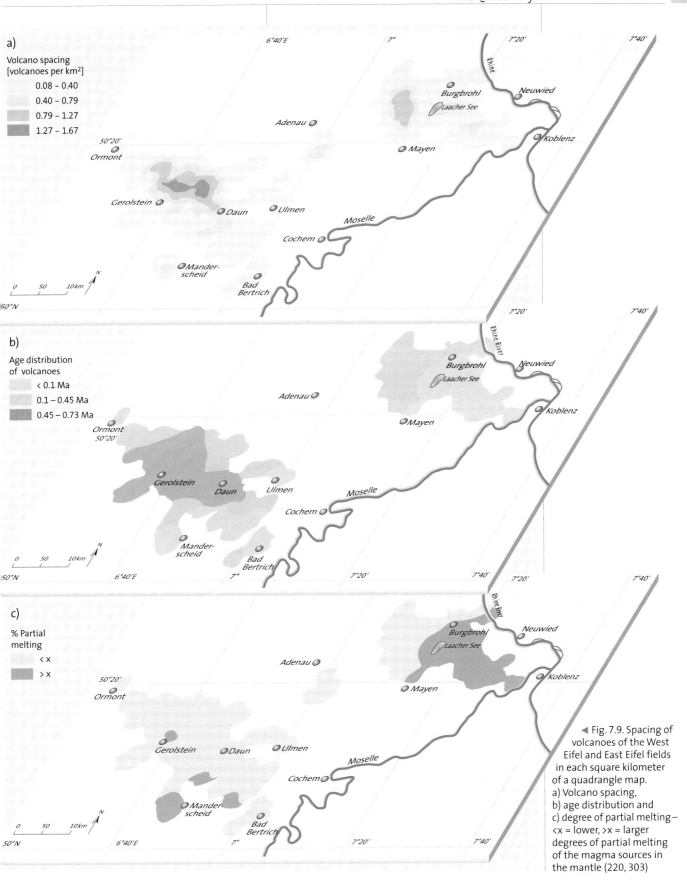

◄ Fig. 7.9. Spacing of volcanoes of the West Eifel and East Eifel fields in each square kilometer of a quadrangle map. a) Volcano spacing, b) age distribution and c) degree of partial melting – <x = lower, >x = larger degrees of partial melting of the magma sources in the mantle (220, 303)

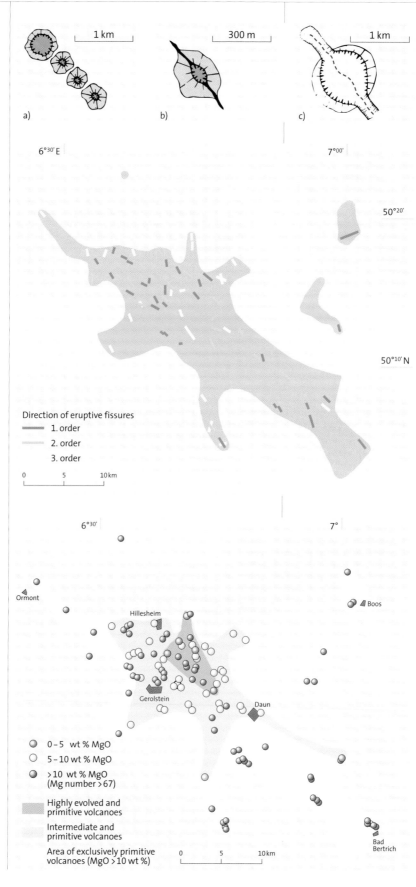

a) b) c) d)

▲ Fig. 7.10. Several criteria to determine the orientation of the feeding fissure beneath volcanoes:
a) linear rows of scoria cones and one maar;
b) dominant dike direction in a scoria cone;
c) zone of tectonic weakness, enlarged by erosion and a creek overprinted by a maar;
d) several overlapping scoria cones

6°30' E

7°00'

50°20'

50°10' N

Direction of eruptive fissures
— 1. order
— 2. order
3. order

0   5   10km

◄ Fig. 7.11. The orientation of the entire volcanic field mirrors the dominant dike orientations of volcanic centers. The orientations of the first order are most clearly shown (220)

6°30'

7°

Ormont

Boos

Hillesheim

Gerolstein

Daun

○ 0–5 wt % MgO
○ 5–10 wt % MgO
○ >10 wt % MgO
   (Mg number >67)

Highly evolved and primitive volcanoes

Intermediate and primitive volcanoes

Area of exclusively primitive volcanoes (MgO >10 wt %)

Bad Bertrich

0   5   10km

◄ Fig. 7.12. Spatial distribution of lava compositions of Quaternary West-Eifel volcanoes (220)

700-km-long belt of Tertiary and Quaternary volcanic fields in central Europe, showing an age progression from Silesia in the east to the Eifel in the west. This was interpreted as the result of migration of the European plate over a hot spot, whose present position was believed to lie beneath the Quaternary Eifel volcanic fields (Fig. 7.15). Available age dating, however, shows a dominant phase of volcanism between about 20 and 30 million years throughout Central Europe with much younger Pliocene and Quaternary phases of very much smaller volume, except for the slightly younger Vogelsberg, at least 700 m thick in its central part. The volcanic activity was thus fed from a string of melting anomalies that were simultaneously active over a wide area. The Eifel comprises not only relatively old (Eocene to Oligocene) but also very young Quaternary volcanic fields, as discussed above. Whether or not

▼ Fig. 7.14. Cross section through the Earth's crust beneath the Quaternary volcanic fields of the Eifel showing four different types of volcanoes and the hypothetical position of the magma reservoirs (not to scale). a) Volcanoes of primitive composition with magma reservoirs along the crust-mantle-boundary; b), c) scoria cones with intermediate to mafic zoned lavas = magma reservoirs in the middle crust; highly evolved, mostly phonolitic volcanoes such as d) Laacher See volcano with its main magma reservoir a few kilometers below the surface. The position of these former magma reservoirs have not been detected seismically but inferred, based on different types of xenoliths, thermobarometry of phenocrysts and other criteria (303)

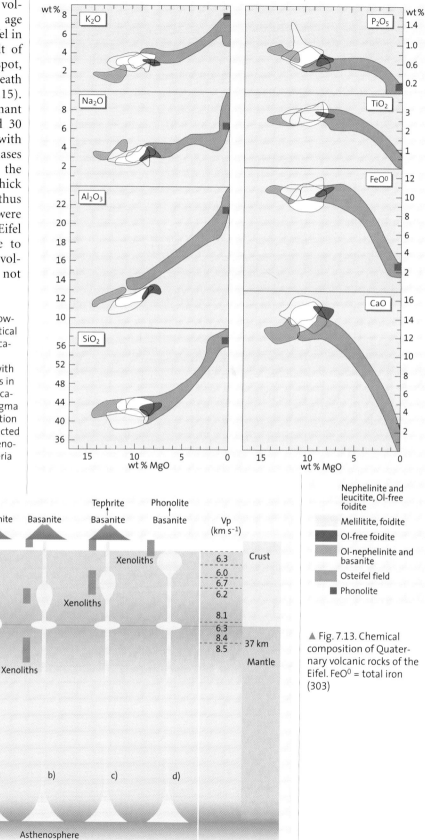

▲ Fig. 7.13. Chemical composition of Quaternary volcanic rocks of the Eifel. FeO⁰ = total iron (303)

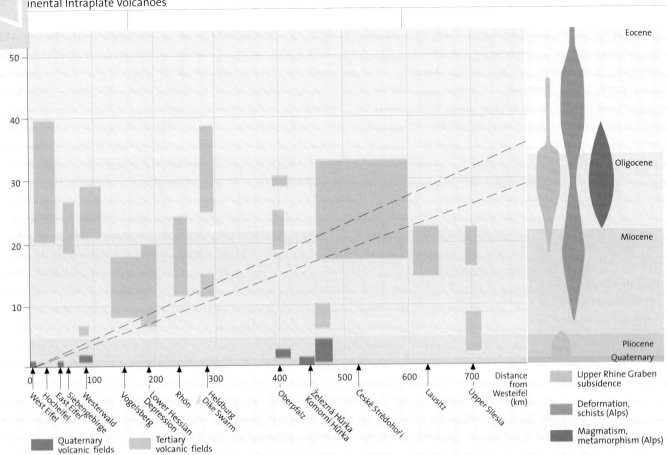

▲ Fig. 7.15. Age and area occupied of young volcanic fields in Central Europe. *Dashed line* trace of age progression of the so-called Eifel hot spot, postulated by (88). The main phases of subsidence of the Rhine Graben and folding and metamorphism in the Alps for comparison (after several sources from 303 and 23)

the volcanic fields represent plumes of small dimensions, as postulated beneath young volcanic fields in central France (127), or a larger plume structure is still a matter of debate (Fig. 7.16).

The Yellowstone Plume

The Yellowstone volcanic system is commonly regarded as a large prototype intraplate plume for continents, comparable to the Hawaii plume system for the ocean basins. Some of the literature, however, neglects the fact that plumes are just models and not physical realities notwithstanding the fact that root zones beneath some areas of high volcanic output down to some 400 km are probably partially melted. For Yellowstone, e.g., powerful arguments are presented for alternative geodynamic scenarios (55). Below, the term plume should thus be understood as meaning melting anomaly.

The birth place of the Yellowstone plume (375) and even the approximate time of birth, manifested in rocks exposed at the surface, are roughly known at present. This contrasts with the early stages of the Hawaii plume that have long disappeared in the depth of a subduction zone in front of northern Kamchatka. Nevada is an Amer-

ican state, well known for its varied underworld activities, also harboring the main testing ground for subterranean nuclear bomb explosions, north of "fun city" Las Vegas. This is also the area where huge caves have been dug in Tertiary ignimbrites (at Yucca Flat), where the long-lived high-level – and low-level – radioactive waste from all American nuclear power plants may one day be deposited (Chap. 13). When the Yellowstone plume started on its long journey, perhaps from the core-mantle boundary, some 3000 km below the surface, is pure speculation. It must have begun hundreds of millions of years ago. A huge, about 17-million-year-old and hundreds of km long dike swarm formed in northern Nevada. In perspective, however, this is small compared to the more than 2000-km-long dikes formed in northern Canada. The Nevada dike swarm is assumed to be the area where the plume was first manifested when it upwarped an area on the Earth's surface about 1000 km in diameter. Today, this area has mostly subsided, or broken up into horst and graben structures. From the Nevada-Oregon border, the American plate migrated above the roughly stationary plume to the southwest, the trace of the hot spot being characterized by voluminous eruptions to the northeast. The roughly

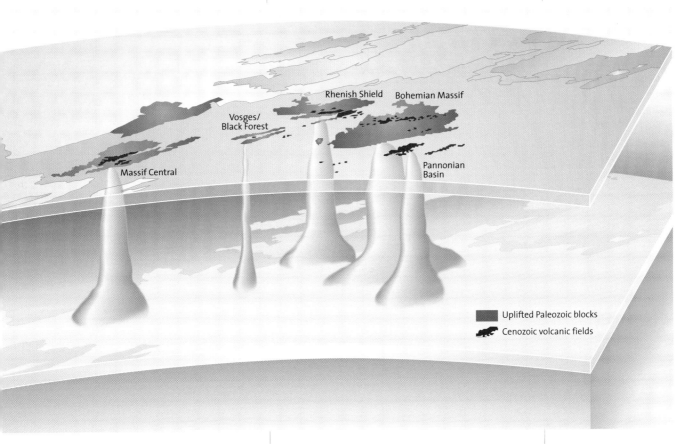

Rhenish Shield Bohemian Massif

Vosges/
Black Forest

Massif Central

Pannonian
Basin

■ Uplifted Paleozoic blocks
🦎 Cenozoic volcanic fields

16–17 million year old Columbia River flood basalts, discussed later on in this chapter, are interpreted as melting products of the Yellowstone plume, as are the more than 400 km long and about 100 km wide basaltic fields of the Snake River Plain. Until about 2 million years ago, the plume produced mainly basaltic magma, only slightly contaminated by digested continental crust. More recently, however, basaltic magmas rarely reached the Earth's surface, but accumulated in the lower crust, where large volumes of crustal material are believed to have become partially melted by the heat of these primitive magmas, thus generating large magma reservoirs (Figs. 7.17, 7.18). Three overlapping calderas are the sources of huge ignimbrites erupted roughly every 700 000 years: the Huckleberry Ridge Tuff (2 500 km^3, 2.0 Ma), Mesa Falls Tuff (280 km^3, 1.3 Ma) and the Lava Creek Tuff (1 000 km^3, 0.6 Ma) (54, 55, 108). These eruptions represent some of the largest Quaternary volcanic eruptions on Earth, dwarfed only by the ignimbrites sourced in the Toba caldera on Sumatra, which erupted ca. 72 000 years ago. Groundwater heated by magmatic fire and mixed with aggressive acids from the magma not only destroyed and altered the rocks (Fig 7.18), generating yellow, hydrothermal-

ly altered rocks, hence the name *Yellowstone*, but also continue to form the basic ingredient for the world-famous geysers. The Yellowstone caldera is thus a hydrothermally active area (Figs. 7.18, 7.19,15.5). The, geologically speaking, extremely short time of human observation and systematic measurements are insufficient, however, to predict if or when further huge eruptions will occur in the future, but evidence is certainly strong that the volcano will reawaken some day. Quite clearly, an eruption of hot pyroclastic flows of Yellowstone-ignimbrite dimensions would result in a disaster with global impact.

Flood Basalts

The tiny scoria cones are peanuts compared to the gigantic piles of lava sheets of the flood basalt provinces and to single vast lava flows, which could make the Guinness Book of Records, at least as far as planet Earth is concerned. Flood basalts represent a very special type of intraplate volcanism. Flood basalt provinces comprise hundreds of thousands of km^3 of magma, covering tens of thousands of square kilometers. This eruptive rate corresponds to that of the entire present mid-ocean ridge system (about 70 000 km) and surpasses that of volcanic hot spots, whose trace

▲ Fig. 7.16. Deep roots (plumes) beneath some young volcanic areas in Central Europe, based on seismic tomography (127)

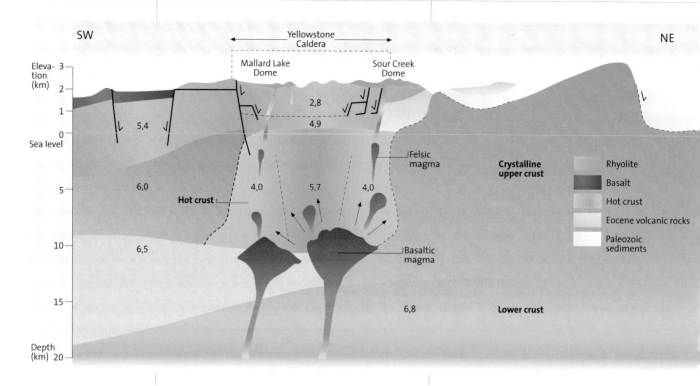

SW

Yellowstone
Caldera

NE

Mallard Lake
Dome

Sour Creek
Dome

Elevation
(km)

Crystalline
upper crust

Rhyolite

Basalt

Hot crust

Eocene volcanic rocks

Paleozoic
sediments

2,8

4,9

5,4

Felsic
magma

Hot crust

6,0

4,0

5,7

4,0

Basaltic
magma

6,5

6,8

Lower crust

Depth
(km)

▲ Fig. 7.17. P-wave velocities (km s⁻¹) and crustal structure beneath Yellowstone Caldera, which erupted three times during the Quaternary. Anomalous lower velocities are interpreted as magma bodies and crystallized but still hot magma chambers (345)

often begins in flood basalt provinces, by a factor of 100 to 1,000. Volcanoes on Mars and Venus, however, belong to a still higher league. Flood basalt fields formed only rarely during Earth history. Their global significance has become appreciated only recently, however (Chap. 14). The step-like erosion due to backweathering of less stable top scoria of lava flows or interbedded sediments (Fig. 7.20), is the reason for the term *trap*, formerly given to these fields, also surviving in the term *Deccan Trap*, a flood basalt field that covers much of central India.

People who drive through some of these immense flood basalt fields may be impressed by their dark and sometimes stark beauty. However, it is overwhelming to visualize that many thousands of km³ of basaltic lava spread within a few weeks to months over thousands of km² and that an entire field, perhaps 1 km thick or more, formed during the geologically incredibly short time of about 1 million years.

Well-known examples of flood basalt fields are the Permian Traps in Siberia, the Karoo basalts of South Africa, the early Cretaceous basalts of Paraná (south America), the Deccan Traps, which cover a large part of India, and the youngest and well-studied Columbia River flood basalt field (Figs. 7.21, 6.32). Single flood basalt lava flows e. g. on the Columbia River basaltic field, have volumes >1000 km³ and cover areas >10000 km². An example for a large single flow is the Roza basalt with an average thickness of 50 m, covering an area

of about 40000 km², representing a volume of about 2000 km³ (368) (Fig. 7.21). This one flow would be sufficient to cover the entire State of New Jersey with a layer of basalt 2 meters thick. Dike systems in the Columbia River flood basalt field are up to 100 km long. Eruptive rates must have reached 1 km³ day⁻¹ km⁻¹ length of a dike, with flow front velocities of several km/h.

The study of continental flood basalt has increased during the last few years because these gigantic, rapidly produced volumes of magma require large thermal anomalies in the Earth's mantle. The volume of new crust generated by continental flood basalts is estimated to amount to $0.1 - 10 \times 10^6$ km³ (61). Oceanic plateaus have even larger volumes, about $10 - 60 \times 10^6$ km³, perhaps because they erupted through a thinner oceanic lithosphere. Age dating has shown that these huge magma volumes formed within 1 – 3 millions of years. Eruptive rates for continental flood basalts are therefore 0.1 – 8 km³/a and for oceanic plateaus 2 – 20 km³/a.

Some of these largest volcanic eruptions on Earth take place when continental lithospheric plates split. An example is the South Atlantic, where the small remainders of flood basalts of the Etendeka plateau in Namibia have the same age as the Parana flood basalts in South America. About 125 million years ago, South America and South Africa were still united before this part of the old Gondwana continent decided to split from south to north in zipper-like fashion. In the northern

◀ Fig. 7.18. Hydro-thermally modified rocks ("yellow stones") along Yellowstone River. Wyoming (USA)

◀ Fig. 7.19. Geyser in Yellowstone National Park

► Fig. 7.20. Columbia River flood basalts along Snake River, Idaho (USA). Individual lava flows are about 20–40 m thick

► Fig. 7.21. Areal extent of Columbia River Basalt Formation and Roza lava flow and its feeder dike system (368)

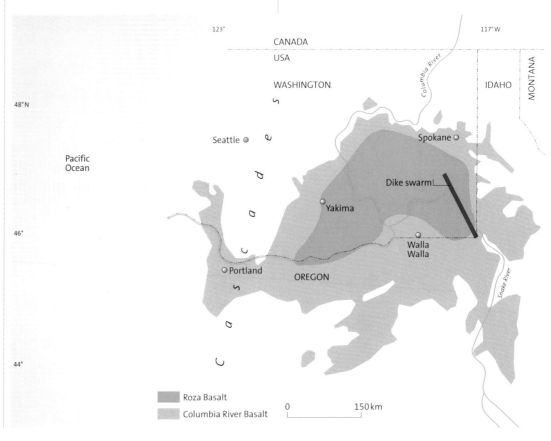

Atlantic, one also finds major remains of a flood basalt province about 60 million years old, which here formed during the much younger separation of continents. This huge former area of flood basalt eruptions is represented today by thick piles of lava along the eastern coast of Greenland, older parts of Iceland, the Faeroe Islands, Scotland, as well as volcanic islands west of Scotland, formerly called the Thule Province. During the last few decades, major packets of basalt lavas with thicknesses in excess of 1–2 km have been detected by marine seismic studies along many passive continental margins and have been called *dipping reflectors*. These thick accumulations of basaltic lava flows apparently formed when the continents drifted apart and some lavas initially erupted subaerially, prior to subsiding below sea level during the increasing separation of the continental plates. There are, as always, exceptions. For example, no rift zones have been detected so far in the northwest of the north American plate (states of Oregon, Idaho and Washington) where the Columbia River basalts erupted mainly between 14 and 16.5 Ma ago.

Flood basalt sequences have recently received much attention because some of them are associated in time with major mass-extinctions. The hottest debate centers around the eruption of the Deccan flood basalts, whose age (65 Ma) exactly coincides with that of the K/T (Cretaceous-Tertiary) boundary (Chap. 14). Could these eruptions be responsible for the mass extinctions, either by excessive warming due to CO_2-emissions or cooling due to SO_2-emissions?

Generation of Flood Basalts

The origin and source of flood basalt magma are highly debated. The discussions center around the question of whether the magmas are generated in rapidly rising mantle plumes or whether they represent asthenospheric or even lithospheric material, partially melted by the addition of heat, supplied by a hot rising mantle plume head.

According to much present thinking, flood basalts are generated when a large plume head impinges at the base of the lithosphere (66). These mantle plumes are probably hotter than the mantle that rises beneath mid-ocean ridges and thus partially melt more strongly during drifting of the lithosphere. In many flood basalt fields, in which magmas have been erupted during very short times (e.g. Columbia River basalts, Deccan Traps), they probably represent melts from a rapidly rising plume.

Some authors think that lithosphere rifts above abnormally hot asthenosphere, which accumulates when the head of a plume intersects a slowly drifting plate. In this model, the hydrated sub-continental lithosphere is thought to become partially melted beneath a hot rising plume, the heat being supplied by the plume, while the magma has its source in the much older lithosphere. Such may be the case where the lithosphere is thicker than about 100 km and flood basalts erupt over a longer period (about 10 million years), as for the Paraná basalts in South America. Here, partial melting may have occurred in a plume after separation of South America and Africa, the trace of the plume being preserved in the seamount chains of the Walvis and Rio Grande Rises in the eastern and western South Atlantic (391).

Summary

All oceanic and continental volcanoes that do not occur along diverging or converging plate boundaries are categorized as intraplate volcanoes. As to be expected, not all volcanoes fit neatly into these categories, examples being Iceland, alkali-basaltic volcanoes behind andesite-dacite volcanic chains, Etna volcano in Sicily or the Afar area on the northern end of the East African Rift.

Volcanoes in the interior of continents are dominantly of basaltic composition and most occur in three types of associations:
1. volcanic fields commonly above uplifted or still rising continental blocks, such as the volcanic fields of the Eifel or the young volcanic fields on the Colorado Plateau;
2. in continental rift zones (and their adjacent shoulders) that can extend over thousands of kilometers some being associated with uplifted blocks, such as the East African Rift system, Rio Grande Rift or Baikal Rift;
3. a different group comprising the mostly tholeiitic flood basalt fields, of which some were formed during rifting episodes. The large flood basalt volumes and the very short duration of their eruption are not well understood. Subsided flood basalt fields are also suspected beneath many passive continental margins, being identified as seismically mapped dipping reflector sequences.

Scoria cone groups dominate in most intraplate fields with lava flows being mostly short. When magma rises through aquifers, eruptions occur in water-saturated grounds, maars or tuff rings or mixtures of these with scoria cones form. Larger central volcanic complexes with greater volumes of more differentiated magmas and longer lava flows are rare.

Intraplate volcanoes play a pivotal role in many current geodynamic hypotheses. They are clearly a surface signal of much wider mantle areas, which reach several hundred kilometers into the mantle and probably represent rising partially melted mantle material. Are the melting anomalies below the lithosphere beneath intraplate volcanoes stationary? Hence, can the direction and velocity (vectors) of the plate be calculated from the traces of a volcanic chain burnt into the moving plates by the melting anomalies? Are oceanic and continental intraplate volcano fields direct evidence for hot mantle plumes (see Chap. 6), rising from the upper or even the base of the lower mantle? Finally, are continental rift zones initial stages for a breakup and drifting-apart of continental lithosphere?

Subduction zones bordering the oceanic plates, topic of Chap. 8, can be geographically defined and geophysically documented much more precisely. These are the zones of the most abundant explosive volcanic eruptions and most volcanoes in this tectonic setting differ drastically from intraplate volcanoes in size, structure, evolution and composition.

Subduction Zone Volcanoes

Lay people asked to name notable or particularly dangerous volcanoes are likely to come forth with names such as Krakatau, Mt. St. Helens, Pinatubo, Mt. Pelée, Vesuvius or Santorini. These are all volcanoes that have grown above subduction zones, and all have produced catastrophic or at least highly explosive eruptions. Why are these volcanoes so explosive? Likewise, a scientist asked which volcanic eruptions had a particularly strong climatic impact will repeat most of these names and might add Tambora, El Chichón, Cerro Hudson and Gunung Agung – all grown above subduction zones. What is so particular for these magmas or the manner of their eruptions that influences atmospheric processes? On the other hand, subduction zone volcanoes will easily win any volcano beauty contest, Mt. Fuji or Mt. Mayon being celebrated examples.

The significance of subduction zones goes much beyond volcanism, however. The negatively buoyant subducting slabs are *the* driving mechanism for plate tectonics. Moreover, the most grandiose Earth factories are located beneath the strings of subduction zone volcanoes. Here, oceanic and oceanic or continental lithosphere interact in a most complex manner, generating continental crust by accretion in some areas, while destroying crust and dragging it into the Earth's interior in other locations.

Strings of large volcanic edifices are the most impressive surface manifestations of deep Earth processes that take place in subduction zones. However, very large and sometimes disastrous earthquakes are even more powerful signals of instantaneous energy bursts within the interior of the globe. Indeed, it was the analysis of earthquakes that led to the discovery of subduction zones.

> It was an eruption, yet no, it wasn't the volcano, the world itself was bursting, bursting into black spouts of villages catapulted into space with himself falling through it all, through the blazing of ten million bodies, falling …
>
> *Malcolm Lowry, Under the Volcano, New York, 1965*

Subduction Zones

During the 1930s, Japanese seismologist Wadati and, independently, American geophysicist Benioff found that the hypocenters of large earthquakes (i.e., the place where they originate) are not distributed randomly at depth. Instead they occur along inclined planes that dip under a continental margin or an island arc (Figs. 8.1–8.7), now denoted the *Wadati-Benioff zone. Deep sea trenches* mark the hinge where the oceanic lithospheric plates disappear at depth. Subduction

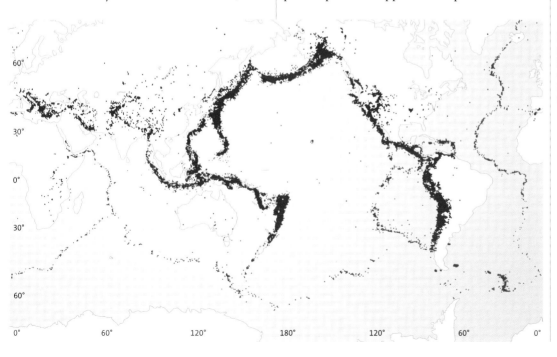

◀ Fig. 8.1. Distribution of large earthquakes (magnitude more than 4.5) between 1963 and 1977 (US Geological Survey Earthquake Information Center)

zones are thus not only characterized by chains of volcanoes, but also by the highest concentration of earthquake hypocenters on Earth. These occur mainly in the uppermost 100 km, but also as deep as >600 km. With the recognition of plate tectonics and seafloor spreading (Chap. 2) the Wadati-Benioff zones define regions along which oceanic lithosphere generated at mid-ocean ridges is subducted into the Earth's mantle below a continental or island arc lithosphere. This is the result of a space problem: the Earth does not expand, so new lithosphere created at mid-ocean ridges must dis-

appear somewhere. As oceanic lithosphere is denser than continental lithosphere – and becomes denser and thicker with age – the downgoing plate is always oceanic.

Volcanic Arcs Above Subduction Zones

About 65% of all subaerial volcanoes that have erupted during the past 10 000 years are located along a belt of active volcanoes circling the Pacific, aptly termed the *Ring of Fire*. These impressive intermittent curved or linear chains of volcanoes called *volcanic arcs* are located landward of deep sea trenches. The circum-Pacific Ring of Fire can be traced from Chile, Peru, Ecuador, Colombia in South America through Central America (Fig. 8.5) and Mexico, the Cascades in Western North America, Alaska, Aleutians, Kamchatka, Kuriles, Japan (Fig. 8.7), Izu-Bonin-Marianas, Tonga-Kermadec to New Zealand, Philippines, Indonesia (Fig. 8.3) and Papua-New Guinea (Fig. 2.1).

About 85% of all historic volcanic eruptions have occurred in volcanoes above subduction zones (338; Fig. 2.6). These comprise almost all well-known and all large explosive volcanic eruptions during the past 200 years: Tambora (Sumbawa), 1815; Krakatau (Sunda Strait), 1883; Santa Maria (Guatemala), 1902; Katmai (Alaska), 1912; Mt. Lamington (New Guinea), 1952; Bezymianny (Kamchatka), 1956; Mount St. Helens (Washington), 1980; El Chichón (Mexico), 1982; small but terrible Nevado del Ruiz (Colombia), 1985; Redoubt (Alaska), 1989/90; Pinatubo (Philippines), 1991 and Cerro Hudson (Chile), 1991. Volcanic arcs above subduction zones also include the Antilles in the Caribbean (Montagne Pelée, 1902 (Martinique); Soufrière Hills (1995–2003) (Montserrat), the young Mediterranean volcanoes of the Aeolian islands including Vulcano, Stromboli

a) Mariana type

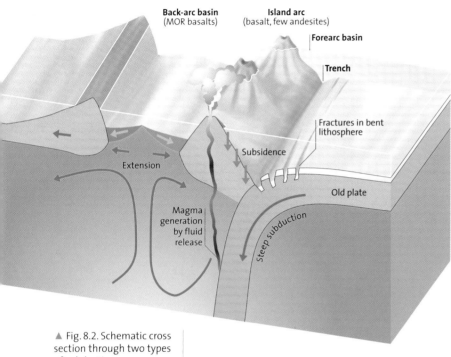

▲ Fig. 8.2. Schematic cross section through two types of subduction zone:
a) Mariana type: steeply dipping oceanic lithosphere subducted beneath younger oceanic lithosphere;
▶ b) oceanic lithosphere being subducted beneath continental lithosphere of the Chilean type (394). *Grey dots* show the position of earthquakes. Main magma types are indicated

b) Chilean type

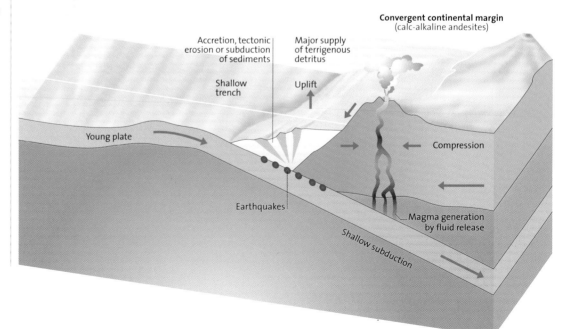

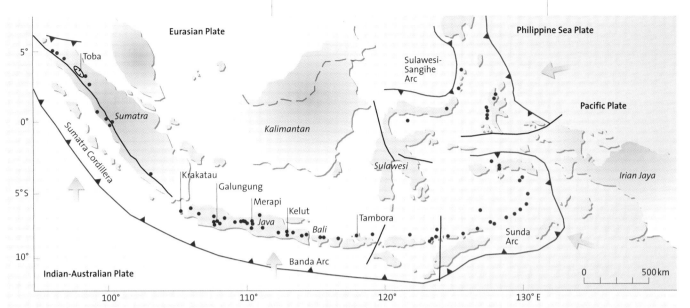

and Vesuvius and the Hellenic island arc (Santorini, 3 400 B. P.), but also young volcanic continental areas, such as Turkey, Armenia or Iran.

Sources of magma and processes of magma formation along converging plates differ fundamentally from those along diverging plates or those within plates (16, 121, 380). From this follows that eruptive processes, mass eruption rates and therefore the structure and size of volcanoes above subduction zones show their own specific characteristics. Volcanic edifices and single deposits, such as gigantic ignimbrite sheets, characteristic of volcanic arcs along convergent continental margins, can have large dimensions and their eruptions are commonly highly explosive. Both characteristics are largely due to the importance of water in the generation of magmas along subduction zones and during the evolution of magma in the magma reservoirs beneath volcanoes. Where does this water originate?

The Japanese geophysicists Uyeda and Kanamori distinguished two types of subduction zones, Chilean and Mariana, based on their contrasting seismicity, dip of Wadati-Benioff zones and other characteristics. The distinction between high stress (Chile-type, large magnitude earthquakes) and low stress subduction zones has been greatly modified subsequently and the problem of why earthquakes along Mariana-type subduction zones are commonly of a lower magnitude is a topic of much current research. Both types of subduction zones are, of course, only end members in a broad spectrum. For example, an island arc-type subduction zone can evolve into a continental-type convergent margin when island arcs become welded together as in Japan and oceanic lithosphere becomes obducted as reflected in ophiolite belts (Chap. 5).

I will briefly discuss these two types of subduction zones, which are also distinguished from each other by their magma composition and therefore the type of volcano; their important geometric and dynamic differences are shown in Fig. 8.2.

Island Arcs

Oceanic island arcs are not very wide, but can be several thousand kilometers in length. They are generated where oceanic lithosphere dips below oceanic or continental lithosphere. One of the best known and largest is the 45-million-year-old Izu-Bonin-Mariana island arc (IBM), which extends for more than 2 500 km from the island of Guam in the south, to Japan in the north (361). In the area of the Izu peninsula and Mt. Fuji it has been colliding for the past 15 million years with the lithosphere of Honshu (Fig. 2.7). The incoming plate in the IBM dips very steeply, from about 50° in the north to near 90° at the Mariana Islands. The reason for this steepening dip is probably the increase in age and therefore density of the subducted plate, whose age ranges from 125 Ma in the north to 170 Ma in the south. The incoming plate is old and cold and the colliding plates are thus not highly coupled with each other. Hence, despite their rapid convergence, low compressional stress occurs across the subduction zone. The lower subducted oceanic plate can be seismically traced more than 1 000 km deep into the Earth's mantle, and probably reaches the Earth's core (126, 395).

Seaward of the Marianas volcanic arc is one of the deepest trenches on Earth, reaching about

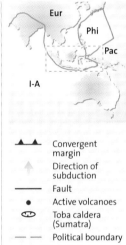

	Convergent margin
	Direction of subduction
	Fault
•	Active volcanoes
	Toba caldera (Sumatra)
– – –	Political boundary

▲ Fig. 8.3. Distribution of major volcanoes and subduction zones in Indonesia (316, simplified from Smithsonian)

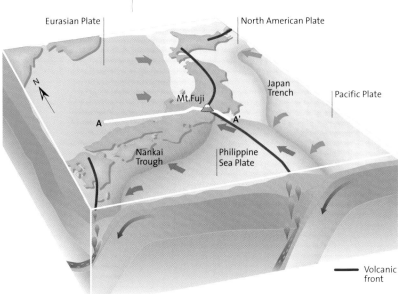

11 000 m b.s.l. The region between the arc and the trench is called the *fore arc*, which in this instance contains only minor amounts of sediment. The sediments are subducted despite their low density (compared to the basaltic crust), perhaps because they are caught in sediment traps in tectonic grabens, which form where the plate starts to buckle down and break. There may also be small frontal sediment prisms that elevate pore pressure and reduce friction. There are no major sediment packets or *accretionary prisms*, welded to the island arc. The fracturing of the plate as it is bent may also allow deep penetration of seawater causing serpentinization of the mantle peridotite. Spectacular examples of the mobility of hydrated (serpentinized) mantle peridotite are the large serpentinite seamounts found in these environments. The easily deformable serpentine layer silicates have become so fluidized that they can

▲ Fig. 8.4. Tectonic setting of Mt. Fuji at the triple junction between the Pacific, Philippine Sea and Eurasian plates. The fore-arc volcanoes southwest of Mt. Fuji, which developed at the triple junction, include the volcanoes Miyakejima and Oshima whose eruptions and deposits are discussed in several other chapters (chapters 4, 9, 13). The line A – A' marks the cross section in Fig. 8.9. Courtesy Japan Geotechnical Consultant Association

▶ Fig. 8.5. Seafloor topography and volcanic arc off Nicaragua and northern Costa Rica. Fractures in bend of slab as it descends beneath Nicaragua may provide pathways for sea water entry causing serpentinization of mantle peridotite. See also Fig. 8.18. Image courtesy Cesar Ranero

◀ Fig. 8.6. Complex Quaternary dacitic volcano Mount Shasta (California). Younger cone Mount Shastina to the right of the main cone. The hilly area between the foreground and the steeper flanks of the cones consists of large blocks of a major debris avalanche, generated during a huge flank collapse of Mount Shasta

extrude to form veritable serpentinite volcanoes on the seafloor.

In island arcs of the Mariana-type, the incoming plate retreats with time, a process called *slab rollback*, while the upper plate advances. This actually causes the upper plate to extend and eventually rupture, most likely along the hot and weak volcanic arc, generating fairly shallow *back arc basins* that are commonly less than 3 000 m deep. This extension causes mantle material to passively well up and partially melt, magmas erupting along spreading centers behind and parallel to the island arc. Magmas of oceanic island arcs are dominantly basaltic (tholeiites and high-alumina basalts), although highly evolved (rhyolitic) and alkaline magmas also occur. The back arc basins vary widely in composition with oceanic tholeiites (MORB) being especially common. As the crust thickens with age, accretion of sediments etc, average crustal densities decrease facilitating the establishment of magma reservoirs to develop where magma can differentiate more extensively.

Convergent Continental Margins

Subduction zones beneath continents, such as those off South America, generally dip much more shallowly (<45°) than those beneath island arcs. These zones are known as *convergent continental margins*. Sediments may be scraped off the incoming lower plate in some areas, forming thick accretionary sediment wedges. The process of accretion was formerly assumed to be the main mechanism by which continents grow in width. Accretion of large prisms of sediments to a continental lithosphere is, however, much less common than formerly thought. Unequivocal examples of active accretion include the Barbados (Carib-

bean), Nankai (Japan), Cascadia (Oregon) and Makran (Pakistan or Iran) margins, totaling only 20 % or less of all subduction zones (407). Much of the trench sediments and debris resulting from mass wasting of the continental slope is subducted. Where accretion occurs, the boundary between the upper continental and lower oceanic plate moves seaward with time. Where sediments are subducted, the boundary moves into the continent by a process called *tectonic erosion*. Large morphological features riding with the oceanic plates, such as ocean island arcs or the much larger oceanic plateaus (Chap. 6), are not subducted but instead are sheared from the oceanic plate and welded to a continent, forming *accreted terranes*. Once a large seamount collides with the upper plate, it may generate major slumps, earthquakes and tsunamis and become an excellent source for trace element-enriched fluids released from the mass of hyaloclastites which form a major portion of many seamounts.

Compression characterizes some subduction zones of the Chile-type. This has several consequences. About 80 % of all large earthquakes (those with a moment magnitude >8) occur in the shallow-dipping subduction zones. Magma cannot rise as easily through a compressed crust as in an extended plate and becomes more easily pooled to form magma chambers. These can be stationary over longer periods of time, leading to higher degrees of differentiation in the magmas. Rising basaltic magmas can also accumulate more easily at the base of the crust; the lower density of the continental crust slows down the rise of the denser basaltic magmas much more effectively than the dense oceanic crust of island arcs. Underplating of basaltic magmas at the crust / mantle density barri-

▶ Fig. 8.7. Decrease in volume of active volcanoes in Japan from the Pacific (volcanic front) to the Sea of Japan (16)

▶▶ Fig. 8.8. Change in volume (in km³) of mafic and felsic volcanic rocks in Japan from Miocene to Present (16)

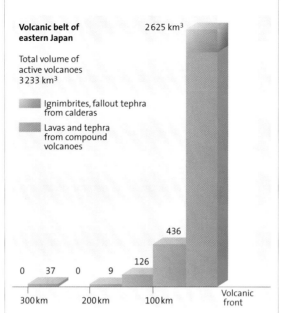

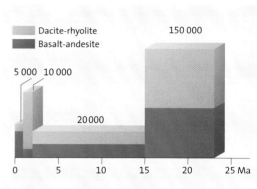

er is the main process held responsible for magma generation by partial melting of crustal rocks.

Volcanic Fronts

Volcanoes in volcanic arcs generally occur only at distances of 150–300 km from the trench and reach their highest spatial density and maximum volume in the *volcanic front* (Figs. 8.7–8.10). As discussed in a later section, rates of fluid release from the downgoing slab, partial melting in the mantle of the upper plate and of magma rise and eruption are at a maximum below such volcanic fronts. A most significant – but poorly understood – geometric aspect is the remarkably constant depth of 100–150 km from the volcanic front to the top of the downgoing slab, approximately the plane of highest seismicity.

The main volcanic zones along the front are 10–50 km wide. The entire volcanic belt in an island arc between the volcanic front and the back arc basin can be up to 300 km wide as in Japan. The composition of magmas inside the volcanic belt changes with increasing depth to the Wadati-Benioff zone (Fig. 8.9).

Several groupings in the distribution of volcanoes can be found. That the distance between large volcanoes is to some degree systematic is often disputed, yet a look at any map shows that volcano spacings in many – but not all – areas are not haphazard. For example, in the northeastern Honshu arc (northern Japan), the mean distance between volcanoes is 23 km but 60–80 km in areas with a lower volcano density. The volume of the larger volcanic edifices in the Aleutians, as well as in Japan, ranges from about 50 to 400 km³. Above gently dipping subduction zones of the Chile-type and well-developed island arc systems such as Japan, the distance between volcanoes is less than above steep subduction zones of the Mariana-type. In the latter case, the volcano density is much more irregular and distances between volcanoes can be as large as 250 km (Fig. 8.10).

The reasons for the regularity in volcano spacings – which also hold for oceanic intraplate volcanoes in some archipelagos (Chap. 7) – are not

▶ Fig. 8.9. Change in chemical composition of basalt magma in a west to east cross section through Japan. Hypothetical depths of origin of magmas are based on the seismically defined Wadati-Benioff Zone dipping east (184). The depths of magma generation are much lower in more recent models

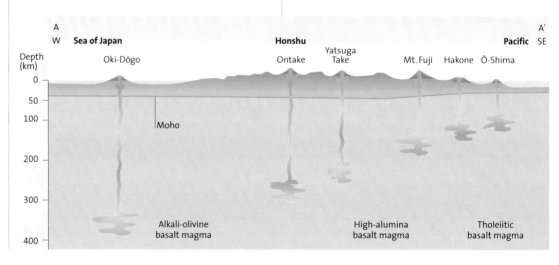

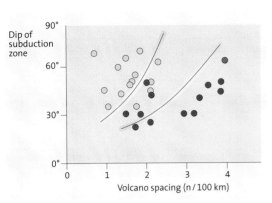

clear. Some authors assume that the regular distance of 60–70 km between volcanoes in the Aleutians is due to the drop-like detachment and rise of mantle diapirs along the zone where magmas are generated – comparable to rising salt diapirs. Other scientists postulate that the compression typical for subduction zones of the Chile-type requires newly produced magma to repeatedly use preheated pathways. In subduction zones of the Mariana type, magma can rise along new pathways more easily in the tensional stress field of the upper plate that stretches as the lower plate rolls back.

Many volcanic fronts are subdivided into 100–300 km long segments. Volcanic fronts may also be offset from each other where a volcanic chain changes direction (Fig. 8.11). Fractures perpendicular to volcanic arcs are reflected in many geological and geophysical parameters; they may form the continuation of oceanic transform faults or older lithospheric fractures.

Small sporadic volcanoes, characteristically scoria cones, occur up to about 400 km landward of the major volcanic belts in the back arc areas of convergent continental margins. The magmas of these volcanoes are more mafic and more alkaline compared to volcanoes along the main front, probably because of the low degree of partial melting at greater depth to the top of the partially dehydrated slab and, in some areas, from asthenospheric mantle little affected by slab fluids.

The elegant form of large volcanic edifices along the volcanic fronts, with their high peaks rising above flat plateaus and their separation from their neighbors by large distances, gives rise to their reputation as majestic. The most famous of these is Mt. Fuji, by its volume the largest among all stratovolcanoes above subduction zones (Fig. 8.12). But one should not forget that even the volume of Mt. Fuji, about 900 km^3, corresponds to that of only a single flood basalt lava sheet (e.g., from the Columbia River flood basalts), or comprises less than 2 vol% of one of the large oceanic volcanic islands. Moreover, many towering volcanoes, e.g., in the highland of Chile, do not rise from sea level but have grown on top of mountains, themselves thousands of meters high. The rocks of many of these giant volcanoes are dominantly moderately chemically evolved andesite and the volcanic edifices rise above flat basaltic shields that represent their initial growth stages (Figs. 8.12–8.16). Some, such as Mt. Fuji, are, however, basaltic to basaltic-andesitic in composition. This temporal succession provides a key for understanding the generation of andesitic and more evolved magmas, especially along convergent continental margins.

◀ Fig. 8.10. Increase in number of large volcanoes in 27 circum-Pacific island arcs for 100-km island arc length with increasing dip of the Wadati-Benioff Zone. *Yellow dots:* island arcs with active back-arc basin (Mariana type); *red dots:* inactive back-arc basins or continental convergent margins (Chilean type) (326)

▼ Fig. 8.11. Subdivision of continental Central American andesite chains in segments along major faults. Small volcanic cones (scoria cones) in the hinterland are associated with tensional fractures (46)

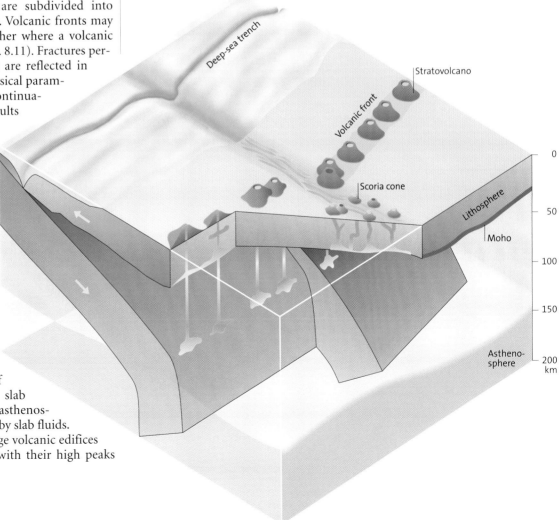

▲ Fig. 8.12. Mt. Fuji, rising to 3 776 m a.s.l., symbol and pride of Japan, last erupted in 1707 and is looked upon as a sacred mountain (see also Fig. 1.1). Its basal diameter is about 40 km. The town of Gotenba in the foreground

▶ Fig. 8.13. Very active Mount Mayon (2,462 m a.s.l.) (Philippines)

◄ Fig. 8.14. Eroded dormant Merbabu volcano (3 150 m a.s.l.) west of very active Merapi volcano near Yogyakarta (Java)

◄ ▼ Fig. 8.15. Arenal volcano (1 633 m a.s.l.) (Costa Rica) since its re-awakening in 1968 one of the most active volcanoes in Central America

▼ Fig. 8.16. Very active San Cristobal (1 745 m a.s.l.), part of the active volcanic arc in Nicaragua. In the foreground the memorial site where ca. 2 000 people were killed in 2000 when a flank of Casitas volcano just east of the San Cristobal volcano complex collapsed following several days of very heavy rain (hurricane Mitch). The floods triggered by the landslide wiped out an entire village

Subduction Zone Magmas

No group of magmas is more complex than that erupting in subduction zone volcanoes, especially those of andesitic and more evolved compositions. Discussions and speculations about their origin have filled thousands of pages in scientific journals and books (16, 121, 380). Hypotheses about the generation of these magmas have contradicted each other repeatedly in the last few decades. Discussions have focussed on types of parent magmas, the mechanisms of generation of magmas, types of source rocks from which magmas are generated and stages and processes of magma evolution.

The prevalence of andesitic magmas in volcanic edifices towering above what are now called subduction zones, had been recognized for a long time and related to a particular type of compressive tectonics. These calc-alkaline magmatic provinces (named because calcium in these rocks is high relative to the alkalis) are also found in the interior of continents, such in the western United States or Turkey, providing evidence for the former existence of active subduction zones now welded into continents.

Key elements in any attempt to understand the origin of subduction-zone magmas include
- the nature of fluids released from the downgoing slab into the overlying mantle wedge,
- the sources of the fluids, i. e. the different rock types making up the slab,
- the age of the overlying asthenospheric mantle and therefore degree of compositional complexity due to past infiltration by ascending magmas,
- thickness and density structure of the hanging plate that control the dynamics of ascending magmas and ponding stages in which magmas can differentiate, transfer heat to crustal rocks to initiate partial melting and promote unlimited scenarios for magma mixing.

Source Materials

Basalts are now known to dominate even in many subduction zone environments. Their abundance had been overlooked in the past as they commonly form the less conspicuous flat shields at the base of stratocones. In other words, the mantle wedge above a down-going slab is a much more important source of subduction zone magmas than formerly believed and

partial melting of slab rocks appears to occur bu rarely. Assuming that the magical depth c 100–130 km below volcanic fronts reflects pres sure-control, slab rocks cannot become partiall melted at this depth because the temperature i too low, an exception being discussed below. Pri mary magmas formed in the mantle wedge abov the slab beneath a volcanic front are thought to b mostly tholeiitic (silica-rich) basalts.

In many volcanic arcs, components of sub ducted sediments are believed to be recognizabl in characteristic chemical signatures (e. g. 91 Diagnostic tracers include not only specific trac elements and radiogenic isotopes (^{87}Sr/^{86}Sr ^{143}Nd/^{144}Nd; ^{207}Pb/^{204}Pb; ^{206}Pb/^{204}Pb), but als other isotopes (δ^{18}O, ^{10}Be/^{9}Be). The isotope ^{10} for example, is produced in the atmosphere b cosmic rays, and is then washed out of the atmos phere and added to marine sediments. This iso tope is characterized by a very brief decay tim (half-life = 1.5×10^6 years) that is useful to estimat the contribution of subducted marine sediment to young lavas in subduction zone volcanoe Other characteristic tracers for slab signatures ar low La/Yb ratios, thought to represent the enrich ment of the mantle in incompatible elements an high Ba/La that may indicate the slab-derive fluid contribution (e. g., 47).

Magmas intruding or extruding along con vergent continental margins are chemically (majo and trace elements) and isotopically much mor variable than magmas of young oceanic islan arcs. Several explanations for this difference hav been suggested. First, the continental lithospher is older than the oceanic one and has thus becom strongly modified compositionally time and agai through infiltration by ascending magmas. Thi leads to an ever-increasing degree of composi tional complexity. Second, continental crust i fundamentally different from oceanic crust i composition and thus physical properties. Thir a thick continental crust can, because of its lowe

▶ Fig. 8.17. Rupture of the Pacific Plate subducted at the Aleutian arc to the east and Kamchatka to the west. The intervening gap characterized by the 1 000 km Bering strike-slip fault may constitute a slab gap for the rise of fertile mantle, potential source areas for the magmas erupted in the giant Kliuchevskoi/Tolbachik volcanic complex (439)

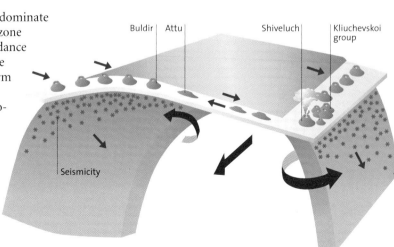

ensity, decelerate or even prevent the further rise f dense basaltic magmas. In many convergent ontinental margins with thick continental crusts, uch as northern Chile, southern Peru or the estern USA, SiO$_2$-rich magmas are especially ommon, both as intrusives (granites) and extruves (e.g. rhyolitic ignimbrite sheets) (Chap. 11). hese SiO$_2$-rich magmas are interpreted to be artial melts of the lower crust. Presumably, the eat necessary to melt the lower crust is supplied y underplated basaltic magmas (150, 154). These asaltic magmas with their chemical mantle sigature can erupt in the same area and mix to arying degrees with magmas derived by melting f the continental crust.

Lateral flow of asthenospheric mantle is inreasingly invoked as a source for magmas or heat n subduction zone settings, especially for those rupting behind the volcanic front. The slab-edge ectonic settings in which some of the largest asaltic volcanoes of the world were produced has een briefly mentioned at the end of Chap. 2. For etter visualization, three examples are shown: 1t. Etna (Figs. 2.7 – 2.9), Mt. Fuji (Fig. 8.4) and he northern edge of the Pacific Plate subducted eneath the Honshu-Hokkaido-Kurile-Kamchata arc (Fig. 8.17).

Another scenario for "fresh" fertile mantle naterial to become partially melted during rise nd decompression is the *break-off* of slabs. In everal subduction zones, seismic tomography of he upper 500 km or so has revealed deep very lense rock bodies interpreted as broken-off slabs vith less dense mantle material filling the gap etween the "near-surface" slab below the volcanic ront and the deeper apparently detached slab Figs. 8.18, 8.19; 282). This may develop when the uoyancy of the incoming plate increases signifiantly, leading to break-off of the dense leading art (282). Such may have been the case in northrn Central America, where major uplift and local eneration of basaltic magma is thought to be due o the buoyant upper-plate response to the influx f mantle asthenosphere, following the break-off nd sinking of the slab. The intrusion, decomression and partial melting of asthenospheric naterial at very complex zones, where several subluction zones meet, or at the edge of a downgoing lab, may explain the highly productive basically asaltic volcano Etna in Sicily (Chap. 2).

Direct partial melts of the basaltic crust of he slab, termed *adakites* for a small island in the vestern Aleutians (Fig. 8.17), are rare and are elieved to be produced when very young and ot oceanic lithosphere is subducted so that the nelting point is reached at relatively shallow lepth.

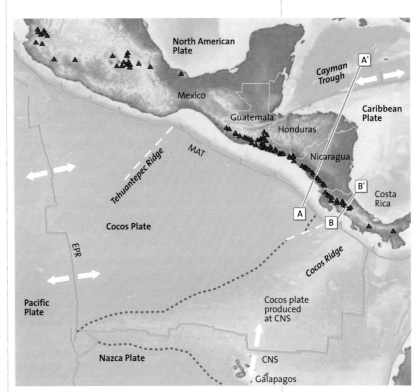

▲ Fig. 8.18. Tectonic setting of northern Central America showing Cocos crust produced at the East Pacific Rise (EPR), Cocos-Nazca spreading center (CNS), triple-junction trace (*heavy dots*), volcanoes (*red triangles*) and Middle America Trench (MAT). Profiles are cross sections through Nicaragua to show slab gap (A – A', Fig. 8.19) and dewatering across Costa Rica (B – B') (Fig. 8.20) (282)

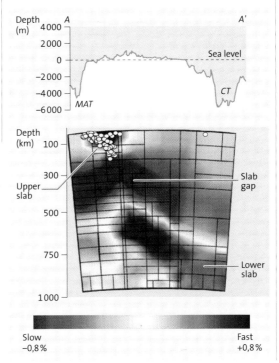

◀ Fig. 8.19. Topography and bathymetry across Central America (A – A', Fig. 8.18) and slab gap between 200 and 500 km. *Blue colors* indicate colder, subducted slab material of Cocos Plate, *red colors* represent hot mantle filling the gap between the broken-off slabs (282)

The Role of Water

That water must play a major role in magma generation and evolution in subduction zones is obvious for several simple reasons. For one, subduction zone volcanoes are much more explosive than most volcanoes in other plate tectonic settings. Indeed, the important role of water, reflected especially in the high H_2O-concentration even of basaltic magmas erupted along subduction zones, is one of the fundamental differences between arc magmas and relatively dry basaltic magmas erupted along mid-ocean ridges and CO_2-rich intraplate magmas. Water contents in melt inclusions in olivine and other phenocryst phases in subduction zones lavas commonly show high H_2O-concentrations. Thirdly, the upper part of the subducted slab is rich in water and water is known to lower the melting temperature of potential source rocks thus facilitating production of magmas (Chap. 3).

There are three different reservoirs for water in a subducting slab. The sedimentary veneer is composed of variable amounts of terrigenous, locally derived volcaniclastic or nonvolcanic debris shed into the trench from the continent and pelagic and hemipelagic biogenic sediments. Water occurs in pores and bound in clay minerals. Hemipelagic clays are thus a much more important source for water than e.g. carbonate oozes. Water in the underlying magmatic oceanic crust is stored in pores and fractures as in the volcanic top layer 2a (Chap. 5) and within secondary minerals, commonly OH-bearing sheet silicates and zeolites formed by reaction of the volcanic glass with sea water. These volcanic rocks also contain various other elements incorporated from seawater, especially K, Rb, Sr and Ba. Large, hyaloclastite-rich seamounts are an especially juicy reservoir of enriched fluids. A third reservoir that has recently come into focus is serpentinized mantle peridotite underlying the oceanic crust.

A topic of much current interest is the time and depth interval of liberation of water from these different reservoirs during the subduction process and the physical processes by which water is removed and by which enriched fluids migrate along the surface of the slab and into the hanging plate. It is generally agreed that hydrous fluids (and in some cases also water-rich magma) are released during subduction and rise into the overlying mantle wedge of the continental or oceanic lithosphere. These not only trigger partial melting in the wedge by decreasing the melting point of peridotite, but also add H_2O-rich solutions or magma and water-soluble elements and isotope ratios to newly formed magmas. Sources and pathways for the fluids are highly debated, however.

The subducted oceanic plate consisting in its upper 5–8 km of igneous oceanic crust and a variably thick layer of sediments is cold. This is reflected in the very low heat flow values along trenches (Fig. 8.20). The slab is in effect a giant cold finger. During subduction, the cold plate is slowly heated by the overlying hot mantle wedge, perhaps also by friction. With increasing pressure and temperature, water is removed in different depth regimes from the three reservoirs (Fig. 8.20) (286). Some 75 % of the chemically bound water in the hemipelagic clay component is lost in the upper 50 km, much of the pore water even earlier. This is quite evident from the abundance of fluids vented along thrust faults when the sediments of the incoming plate become strongly compacted. Basalts of the oceanic crust become metamorphosed, first to greenschist, then amphibolite and finally to the high-pressure rock eclogite. This basaltic crust loses its water mainly between 100 and 140 km depth. Underlying serpentinized mantle loses some 80 % of its water still deeper, between about 130 and 160 km. Thus, dehydration of serpentinized mantle could be a major source for fluids promoting partial melting in the upper plate.

The results of these recent modeling studies are, however, in conflict with conventional scenarios. According to the modeling (Fig. 8.20), much water is removed from the subducting slab at shallow depth and thus at great distance from the volcanic front and the underlying area of melt generation. On the other hand, some element and isotope ratios thought to be diagnostic for slab scenarios are commonly interpreted to have been derived from the sediments. Fluids released from mantle serpentinized during flexural faulting at the outer rise can potentially scavenge elements from the largely dewatered sediment layer and add them to the overlying mantle wedge (286).

The fact that many volcanic arcs first appear along a commonly sharply defined front when the distance to the top of the downgoing slab below is about 100–150 km is not well understood. Most commonly, pressure-control is invoked with the OH-bearing mineral phase amphibole, a major constituent of amphibolite, the metamorphosed equivalent of basalt, becoming unstable at this depth and releasing fluids. Some authors argue that amphibole already breaks down at lower pressure. In detail, however, the generation and vertical and lateral migration of fluids and fluid-rich melts along subduction zones is still enigmatic. Other factors such as increased mechanical permeability along the volcanic front are receiving increased attention.

In many island arcs and some convergent continental margins, the composition of magmas changes systematically inland from the volcanic front. This temporal-chemical correlation was related in the early stages of modern subduction zone research to increasing depth of the Wadati-Benioff zone in Japan (Fig. 8.9). Volcanoes of the volcanic front are characterized by fractionated tholeiites, i.e., SiO_2-rich basaltic magma and their derivative calc-alkaline magmas. With increasing distance from the front, volcanoes become smaller and their spacing increases. At the same time, their composition becomes impoverished in SiO_2, richer in K_2O and more primitive, i.e. less evolved. Based on the present state of knowledge, increasing K and decreasing SiO_2 concentrations with distance from the volcanic front can be explained plausibly by increasing depth to the Wadati-Benioff zone and therefore higher pressure and thus decreasing degrees of partial melting reflected in a higher concentration of incompatible elements. Alkaline basaltic magmas are less dense than tholeiitic ones by about 0.1 g/cm³ because of the higher partial molar volume of alkalis, a difference in density that is accentuated by the apparently much higher H_2O and CO_2 concentration of such magmas (185). These magmas, because of their lower density (ca. 2.62–2.63 g/cm³ at pressures <5 kb), can rise more easily than tholeiites from their source area through the crust without fractionating. Tholeiitic magmas (2.66–2.72 g/cm³), on the other hand, are denser at low pressure than most crustal rocks. For this reason, they are more likely to stagnate at locations of large density contrasts (e.g., the mantle-crust boundary), and fractionate in larger magma reservoirs, so that primary tholeiitic mantle magma rarely rises directly to the Earth's surface. Another reason for the more evolved nature of magmas of the volcanic front of convergent continental margins is the dominance of compression, which increases the likelihood of formation of magma chambers in which primitive magmas can differentiate. In contrast, the hinterland of island arcs is characterized by extension, which facilitates the ascent of smaller volumes of primitive magmas from greater depths.

Summary

Gigantic deep trenches in front of island arcs and convergent continental margins, earthquakes generated above and within down-going (subducted) slabs and chains of highly explosive majestic volcanoes aligned along volcanic arcs are all governed by processes deep in the Earth where oceanic plates sink beneath oceanic or continental lithosphere. Such chains of volcanoes in the northern Andes led Alexander von Humboldt on his famous trip to South America 200 years ago to speculate that they represent fundamental deep Earth seams. This view contrasted with the then prevailing belief, propagated by both volcanists and neptunists, that volcanoes are superficial features fed by near-surface burning coal fires, which melt the surrounding rocks. With the advent of the sea floor spreading and plate tectonic theories in the 1960s, divergent mid-ocean ridges were recognized as the central seams on Earth, along which new lithosphere is constantly generated. Until the 1970s, the plates were thought to be pushed away from the mid-ocean ridges. Today the main cause for the plate motion is considered to be the pull of cold downgoing lithospheric plates that sink beneath continental or oceanic lithosphere on account of their high density.

Most volcanoes above subduction zones are concentrated in volcanic fronts some 150–300 km landward of trenches and about 100–150 km above the seismically defined top of the slab. Magmas erupted in volcanic arcs comprise basaltic and more strongly evolved andesitic to rhyolitic magmas. Andesites and their plutonic equivalents, granodiorites to diorites, were formerly thought to represent partial melting products of the crust. Today, they are chiefly interpreted as differentiated products of basaltic magma that may have become contaminated and mixed with magmas generated by partial melting of the surrounding crust. But the source of most magmas erupted along volcanic arcs must be located at depth in the asthenospheric mantle wedge above the slab. Because of the notorious high explosivity of volcanoes in volcanic arcs which is clearly due to the high water content of the magmas erupted, fluids – and possibly water-rich melts – must play a central role in the generation of such magmas.

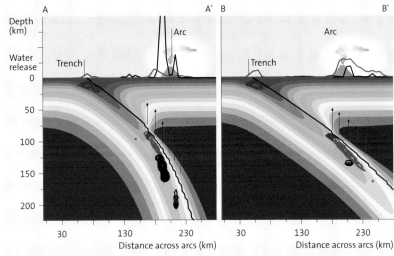

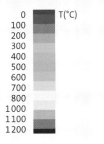

▲ Fig. 8.20. Model of slab dehydration beneath Nicaragua (Fig. 8.18) (A–A' central part) and Costa Rica (B–B'). Color zones of slab and hanging plate indicate modeled temperatures (red = 1 200 °C and *blue* = 0 °C). Curves above surface show water release (red = dewatered sediments from shallow depth (*left*), *pink* from oceanic crust and *black* curves water release from serpentinized mantle). At the steeply dipping subduction zone beneath Nicaragua (A–A') water release from serpentinized mantle is high and occurs at depths between 120 and 200 km, while water release zones beneath the more gently dipping subduction zone beneath Coast Rica overlap, with dewatered ocean crust being highest (286)

Water, enriched in many elements scavenged from seawater, is expelled in cold vents from the deep-sea sediments as they are squeezed landward of trenches where subduction begins. The underlying oceanic crust contains so much water that the entire mass of water in the oceans is recycled into the Earth's interior during the incredibly short time of 0.5 million years, a mere fraction of Earth history. The bulk of this water, however, is also released during subduction. At higher pressure, perhaps at depths just below the volcanic fronts, even serpentinized mantle underlying the oceanic crust may become dehydrated releasing its fluids into the overlying mantle wedge. Such fluids are enriched in water-soluble elements whose character varies with source rock and pressure, highly incompatible elements such as Zr and Nb being retained and liberated only at higher pressure. The fluids are thought to hydrate the overlying peridotite at the base of the upper plate at low temperate and pressure. At higher pressures beneath the volcanic front they are able to lower the melting temperature and trigger the generation of magma in the overlying asthenospheric mantle wedge. Partial melting of slab rocks may occur locally when very young and hot oceanic lithosphere is subducted. Asthenospheric mantle material sucked-in from the side of slabs or in windows in broken slabs and continental crust are other source materials for subduction zone magmas. Enrichment in water is enhanced during differentiation thus contributing to the high explosivity of volcanoes in volcanic arcs. Partial melting and fluid migration beneath volcanic fronts may also play a significant role in the origin of earthquakes in subduction zone settings.

Magmas and thus eruptive rates and processes and, therefore, type of volcanoes differ between young island arcs at one end of the spectrum and thick crust at convergent continental margins at the other. Basalts, andesites and minor rhyolites are characteristic of island arcs where old and dense oceanic plates may dip steeply and recede. Back arc basins opening in the stretched upper plate as a consequence of roll back of the lower plate are dominantly basaltic but commonly with subduction zone chemical signatures. Hot and buoyant young oceanic lithosphere commonly dips shallowly, as along continental convergent margins in South America. Magma ponding at the base of the crust may be responsible for massive partial melting of crustal rocks. Major ignimbrite sheets erupted from large calderas are characteristic for subduction zone settings in convergent continental margins where the crust is especially thick. There are almost unlimited possibilities for differentiation and magma mixing during magma migration through the lithosphere, far more than in diverging plate margins or in the interior of plates.

Volcanoes of volcanic arcs become eroded rapidly – except in very dry climatic zones – because of the dominance of clastic materials, again reflecting their high magmatic water content. These sediments supplied to the deep-sea trenches are mostly subducted. Some are partly recycled to become the roots of new volcanoes above the slab. Accretion of sediments shed from the continental margins or island arcs adjacent to trenches was formerly thought to be ubiquitous and the main process for continental growth. There is now compelling evidence that most sediments are subducted along the ca. 35 000 km global subduction zones. Today, subduction zones are visualized as factories where several types of raw materials are subjected to complex recycling and mixing processes. The major part of the chemically impoverished downgoing slab is dragged down into the Earth's mantle to a depth of at least 700 km. Recent results from mantle tomography show that slabs may sink much deeper, perhaps to the Earth's core. Intraplate ocean island magmas rising to the surface of the Earth far away from subduction zones are nowadays commonly explained by mixing of old possibly pristine mantle material and subducted slab rocks, a geological loop of gigantic temporal and spatial dimensions.

With these speculations on deep Earth processes I will end discussion of large-scale tectonics and volcanism and will now focus on volcanic edifices, their internal structure, their products, and the mechanisms and results of explosive eruptions. Three case histories of subduction zone volcanoes are discussed in later chapters. Pinatubo in the Philippines is a classic example of a subduction zone volcano (251). It is also not unusual for subduction zone volcanoes that a large eruption such as that of 15 June 1991 is preceded by a long period of quiescence – in the case of Pinatubo by more than 500 years. Nevertheless, nobody could predict that Pinatubo would once be the site of one of the largest eruptions of the twentieth century. Chapter 13 gives a summary of the explosive eruption of Pinatubo which resembled that of Mount St. Helens in several ways (Chap. 10) but differed significantly from that of Nevado del Ruiz which resulted in one of the major volcanic disasters in history (Chap. 13).

Volcanic Edifices and Volcanic Deposits

Young volcanoes contrast with most other mountains by forming lovely cones or impressive edifices sometimes rising high above their surroundings. Even small volcanoes, which almost always occur in groups, greatly enhance the attraction of an otherwise morphologically bland landscape. Obviously, areas dotted with young volcanoes are major tourist attractions (Chap. 15). In addition to appreciating the aesthetic appeal of volcanoes, volcanologists, during their professional life, develop the ability to delve beneath a volcano's morphology. They learn how to deduce the internal structure, type of deposit and, therefore, the physical properties of the magma whose eruption makes up the stuff of a volcano (Figs. 9.1, 9.2). Questions that come to mind: what does the form of a volcano tell us about the composition, the gas content and, therefore, the viscosity of the magma? Can we deduce something from the size of a volcano about magma mass eruption rates and thus processes within the magma reservoir, or even the magma production rates in the Earth's mantle? How are different eruptive mechanisms (especially the difference between magmatic and hydromagmatic processes) reflected in the form and architecture of a volcano? How strongly and how often have the volcano flanks collapsed and subsequently rehealed?

In order to answer these types of questions, scientists carry out research along different lines:

- Field and laboratory analysis of the structure of volcanoes and their rocks in natural and artificial outcrops (and in drill holes on land and in the sea);
- Long-term and continuous observation and monitoring of the growth stages of active volcanoes; and
- Theoretical modeling and experimentation in order to better understand fragmentation, eruption and transport mechanisms.

► Fig. 9.1. Major types of volcanic edifices in profile. Pyroclastic cones are mainly of basaltic, lava domes and larger volcanic edifices (*above*) of intermediate to highly evolved composition (340)

Principally, volcanoes consist of three structurally different types of rocks: lava flows, intrusions, and deposits that consist of rock particles, simply called *volcaniclastic deposits*.

Lava Flows

Very few readers will have had the chance to observe "live" flowing lava, with the exception of those who have visited Kilauea Volcano on Hawaii, Arenal Volcano in Costa Rica, Piton de la Fournaise on Réunion, or Etna in Sicily. These four volcanoes differ drastically in size and composition, but active lava flows can be observed on each of them at least every few years. Many more readers have probably walked across cooled lava flows, perhaps balancing precariously over loose blocks. Many others may have seen cross sections of lava flows in road cuts. This is why the natural variety of lava flows has been classified principally by structures on their surface and in cross section. Attempts are increasing, however, to infer the physical properties of flowing lava and the flow processes themselves by structurally analyzing lava flows. These types of analysis are based on observations and field analyses of actively flowing lava, as well as ingenious laboratory experiments, theoretical data, on numerical modeling, and analysis of remote sensing data, such as infrared sensor and others.

Shield volcanoes, typical of ocean island volcanoes, such as Kilauea on Hawaii, were discussed in Chapter 6. Smaller, flat shield volcanoes that interdigitate and form complex basalt plateaus are characteristic

The greater the diversity in construction of a volcano, which is the shell surrounding a channel for molten masses making their way from the inner earth to its surface, the more important it is to precisely determine the characteristics of these constructions.

A. v. Humboldt, Über den Bau und die Wirkungsart der Vulkane, Berlin, 1823
Translated from the German

0   10 km

Composite volcano

Stratocone

Sector collapse and dome

Caldera

Lava dome   Crater row   Scoria cone   Tuff ring   Maar

▶ Fig. 9.2. Alkali basaltic scoria cone on the east slope of Mauna Kea on Hawaii (see also Fig. 6.1)

of rift zones on land, such as Iceland (the type locality for shield volcanoes) or the Snake River Plain in Idaho (Fig. 9.3). The tops of such shields are characterized by small collapse craters or cappings of more viscous late-stage aa lava flows. Shield volcanoes form above eruptive fissures, generally a few hundreds of meters to km long, although exceptionally up to several hundred kilometers. They may also occur above central vents that commonly form during later stages of fissure eruptions. Scoria cones and maars are common in such volcanic fields.

Lava flows of basaltic composition are much more common on Earth than those of other compositions. Moreover, the surface of many other planets is covered exclusively by basalt lava flows and their fragments. Slightly more evolved compositions have been found on Mars, and the famous domes on Venus may also be of more evolved composition.

Pahoehoe Lava

The structural lava type most common on shield volcanoes constructed of low viscosity basaltic magmas is *pahoehoe* lava. These 1–10 m thick flows can travel many km when they are transported through lava tubes. Pahoehoe lavas are very photogenic (Fig. 4.12). It is generally easy to walk on cooled pahoehoe lava flows, with the exception of what has been called *shelly pahoehoe* on Hawaii. The latter consists of large bubbles with thin walls, hostile to the one's shins when breaking through, but which do not survive very long because of their fragile structure.

Pahoehoe lava flows are commonly bulbous with a smooth skin. The hot skin, when still plastic, can form wrinkles known as *ropy lava* (Fig. 9.4). Many types of surface forms have been described, depending on factors such as viscosity, flow velocity, and eruptive rate. This great variety of surface forms can only develop in easily deformable liquids, so it is intuitively easy to understand that pahoehoe lavas characterize melts of low viscosity. When fresh, the surface is very smooth and glassy, but this thin outer glassy film breaks off quickly and is removed rapidly by wind and rain. Pahoehoe lava can change into aa lava when the viscosity and shear stress increase during flow (Chap. 4). Pahoehoe lavas, like other lavas, generally cannot flow very far when they move openly over the Earth's surface. They cool so quickly that increasing viscosity causes them to come to a halt. Some lavas can flow over tens and sometimes hundreds of km due to their transport in lava tubes, below a roof that grows downward from the fast cooling crust. In this way, cooling is greatly retarded. In Hawaii, one can walk over such *lava tubes*, and here and there peek through

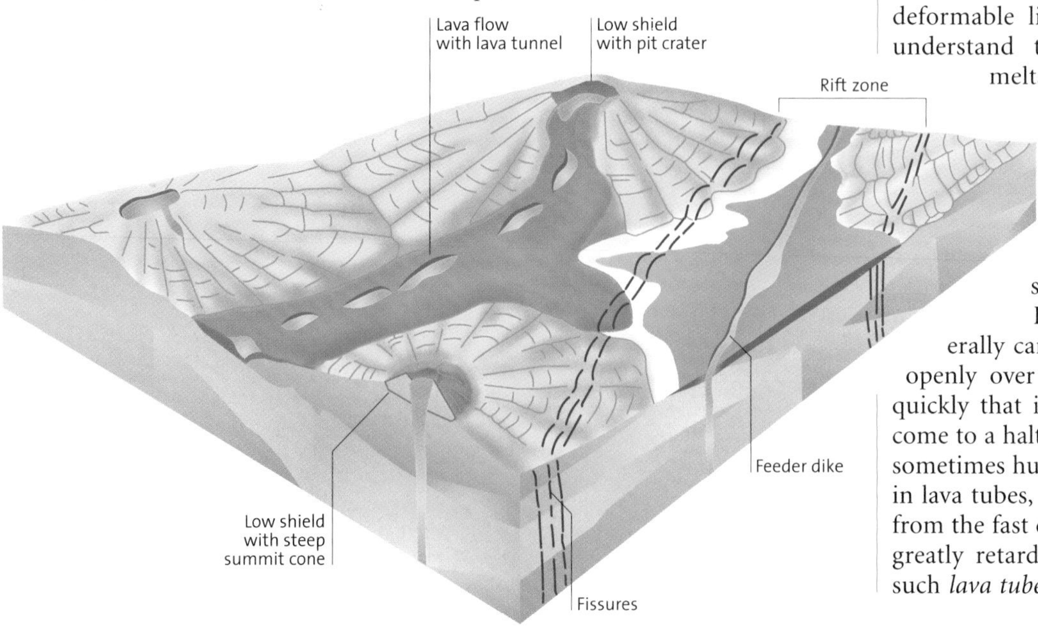

▼ Fig. 9.3. Cross section and morphology of a shield volcano field of the Snake River type (Idaho) (128)

Lava flow with lava tunnel

Low shield with pit crater

Rift zone

Low shield with steep summit cone

Feeder dike

Fissures

windows (called skylights) in the broken roof to watch the lava flowing in the underworld (Fig. 9.5). In cooled lava fields, such as on Hawaii or Lanzarote (Canary Islands), these commonly kilometer-long, drained lava tubes are spectacular tourist attractions. Lava flowing in such tunnels can melt down its bed by a process called *thermal erosion*, some tunnels reaching more than 10 m in height. The walls of such tunnels often show the cross sections of older lava flows and are commonly covered by wallpaper-like melt films, forming smooth skins with drops or stalactite-like lava drops. In Europe, especially illustrative examples of pahoehoe lava include the very fluid lavas of Vesuvius, tholeiitic lavas in Iceland, or the historic and prehistoric lavas on Lanzarote or El Hierro on the Canary Islands.

Pahoehoe lavas commonly move as thin *flow units* because of their low viscosity; several such stacked and interfingering flow units cool together to form complex lava flows. The concept of flow units has, however, assumed a much greater importance in subdividing ignimbrites and understanding their accumulation mechanism than in lava flows (Chap. 11). Gas bubbles in pahoehoe lavas are commonly spherical and are observed in thin, quickly cooled flow units. Pahoehoe lavas also show a great variety of vesicular structures, such as *pipe vesicles*, pear-shaped *vesicle cylinders*, etc. Quarry workers in the large Miocene volcano Vogelsberg in Germany have extracted such porous basalt for centuries to be used in the construction of monasteries or cathedrals, such as that in Cologne. They called this basalt *Lungenstein* (lung rock) because of its peculiar structures and resemblance to the vesicular structure of lungs. When thin lava flows move in a valley or in a lava tube, commonly very photogenic lava rosettes form due to the growth of cooling fractures at right angles to the cooling surface (Fig. 9.6).

Basalt Columns

When thick lava flows or lava lakes cool, fractures form that propagate inward into the cooling lava body, because solid, crystallized and therefore denser basalt has a smaller volume than the melt. The formation of these fractures is energetically most favorable when the horizontal plane through the shrunken basalt body is used optimally. This is the case for hexagonal bodies and is why basalt columns commonly have six sides. However, because only Allah is perfect (the reason why good carpet weavers deliberately introduce a mistake into each article), basaltic columns in nature can have five, six, or seven sides (Fig. 1.4). Heat is not transported away from a stationary lava flow uniformly. The surface of a lava flow cools by

◀ Fig. 9.4. Ropy pahoehoe lava on the south flank of Kilauea volcano (Hawaii)

▼ Fig. 9.5. Skylight in the collapsed roof of a lava tunnel. Red hot lava flowing about 3 m below the surface. East flank of Kilauea volcano (Pu'u 'O'o)

▼ Fig. 9.6. Radially jointed ▼ basalt flow filling a former lava tunnel. The columns have formed perpendicularly to the former cooling surface. The lava rosette has a diameter of about 6 m. Barranco de Agaete (Gran Canaria)

◀ Fig. 9.7. Valley-filling Tertiary basaltic lava flow near Saint Flour (France). The sharp boundary between the larger basal and the smaller upper columns marks the encounter of the solidification fronts during cooling. The fronts rose slowly upward from below and relatively rapidly downward into the interior of the lava flow

radiation and convective rise of heated air, which is constantly replaced by cold air. This is much faster than at the base of the lava flow. Here, heat migrates slowly by conduction into the underlying rock. The temperature maximum thus migrates downward into the interior of a lava flow during the period of cooling. The planes of equal temperature, called *isotherms*, propagate rapidly from the top downward and slowly upward, to finally meet about one-third above the base of a flow. Hence, many lava flows commonly show a razor-sharp boundary between an upper zone of thin irregular columns and a thinner lower zone of thick, more regular columns (Fig. 9.7). A closer look at the surface of the columns reveals small ridges at right angles to their axes. These structures, ranging from a few mm to several cm in width (Figs. 9.8, 1.5), reflect the episodic propagation of fractures during cooling. At the transition zone between a portion of a flow that has cooled enough to be brittle, and the hotter plastic part, stress accumulates until it is episodically released by sudden propagation of the fractures.

Thick, fluid lava flows, such as huge flood basalts, in cross section commonly show a threefold subdivision. A lower, thick *colonnade*, a central zone of thinner columns, called *entablature* (Fig. 9.9), overlain by thicker, more irregular

◀ Fig. 9.8. Quaternary melilite-nephelinite lava tongue about 15 m high. The inclined columns formed at right angles to the cooling surface (a former valley). The columns are marked by subhorizontal ledges that result from episodic advance of fractures between the cooled brittle part of a lava flow and the hot, still plastic interior. Gran Canaria (Canary Islands). Scrope had correctly explained the origin of such inclined columns in 1825 (lower part of figure) (319)

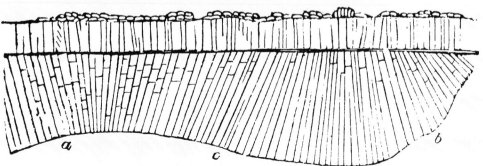

"The columns will be usually more less perpendicular to the surface on which the solidifying process first acts; (as is the case in the desiccation of starch, clay, etc.). Therefore when they are formed upon a convex (as a c) they will diverge towards the exterior of the mass; when upon concave surface, (as c b,) they will converge in the same direction"

▶ Fig. 9.9. Cross section through a basaltic lava flow, flow direction from right to left. The subdivision of the cooling joints is typical for static lava flow sheets. Vesicle structures at the base and in the lower part can be inclined into the flow direction. The tubular (pillow) lavas that develop together with a glass breccia when a lava flow enters water can show fore-set inclination down the transport direction. Palagonite is the typical yellow alteration product of basaltic glass (301) (see also Fig. 9.10)

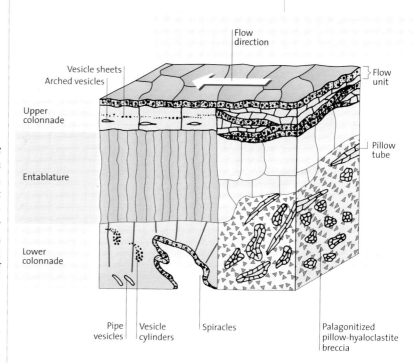

columns near the top (Fig. 9.10). This tripartite subdivision forms because the amount of heat in the interior of a lava flow is initially very large. The temperature difference between the plastic zone and the brittle zone decreases during cooling, because heat is constantly transported away from the entire flow body. The irregular columns of the entablature thus reflect the relatively rapid transition from a period of plasticity into one of brittle deformation.

Aa Lavas

The second major structural type of basaltic lava is *aa lava*, which is more common than pahoehoe on many oceanic islands and continents. It strongly contrasts with pahoehoe because the lava surfaces consist of highly spinose, loose blocks and ridges of lava that have been pushed upward. It can be quite hazardous to walk on these lava flows, as the topmost blocks are not only very loose, but also razor-sharp (Figs. 4.13, 4.14). In cross section, aa lava flows consist of three parts: a basal breccia, a more massive central part, and a capping breccia. Observations of flowing aa lavas clearly demonstrate that the basal breccias form by spinose blocks from the top tumbling down in front of the slowly moving lava flow and being overridden by the flow itself (Fig. 4.2). Laterally, aa lava flows are commonly bordered by irregular ridges or levees that look like lateral moraines in form. In detail, the central, massive part can often be seen to consist of irregularly shaped roundish domains that are slightly vesicular and welded together. Apparently, aa flows move at least in their distal parts by flow of clots or domains of lava that range from individual loose blocks on top to very strongly welded but plastic domains in their interior.

Both types of lava flows are common in Hawaii, hence the terms are of Polynesian origin. Both here and in other active volcanoes, scientists have wondered for decades about the origin of the contrasting flow behavior of basaltic lava. Chemical and mineralogical analyses have shown that both are absolutely identical, so magma composition, and hence compositionally-governed viscosity, cannot be the reason. It is intuitively obvious, and has been observed numerous times in Hawaii, that fluid pahoehoe can change into aa, but never

▲ Fig. 9.10. Thick flood basalt lava flow that entered Miocene Columbia River canyon. Pillow lava in the lower part of the flow dipping down the former transport direction to the left consists of thin lava tubes, some of the quickly cooled lava having become brecciated. The yellow-brown substance (palagonite) is the alteration product of the basaltic glass. The water was so shallow that the lava flow produced its own delta with a later subaerial flow unit traveling over dry ground. Near Sentinel Gap (Washington, USA)

▲ Fig. 9.11. Andesitic lava flow with subvertical ramp structures in its upper part. Crater Lake (Oregon, USA)

andesites in their viscosity) can show more irregular columnar jointing with common subparallel joints (Fig. 9.11). Ramp structures formed by forward thrusting in these flows are quite common, as are top and basal breccias.

These highly viscous lavas form mostly relatively short morphologically prominent flows, a few hundred meters to at most a few kilometers long. Their upper surface consists of relatively smooth, angular, vesicle-poor to vesicle-free blocks, the reason why they are called *block lava* (Fig. 9.12). Rhyolitic and sometimes also dacitic or phonolitic lava flows may consist largely of obsidian. The reason is that the temperature difference between liquidus and solidus in strongly polymerized melts is so small that cooling rates at the surface suffice to cause a fast transition from a molten to the glassy state. The more fluid interior may break through the surface to form more strongly degassed pumiceous obsidian (Fig. 9.13). Differential laminar motion and degassing along planes of motion in these flows are reflected in characteristic layered structures and folds. Typical obsidian lava flows are the ca. 1 000-year-old Rocce Rosso on Lipari (Eolian Islands), several lava flows on Tenerife (Fig. 9.14), Little Glass Mountain in California, and Newberry Crater in Oregon. Obsidian was a precious material for making tools and trading in many early cultures of the Old and New World (Chap. 15).

Domes

Short, thick, sometimes pancake-like lava flows are also called domes. They can sit like huge swallow nests on the lip of craters, point finger-like obelisks into the sky, or mimic thick dollops of toothpaste in the center of a crater (Figs. 6.27, 9.15 – 9.18). Some grow episodically, slowly refilling a crater previously formed by huge explosive eruptions and semicircular rows of domes are common along caldera ring faults. Flank collapses may be caused by the upward push of growing domes (Fig. 9.16). Because dome growth may last several years, changes in form, extrusion rate, stability, and collapse are among the best-studied volcanic phenomena. The evolution of dacitic and andesitic domes has been analyzed in some detail during the past two decades at four sites: the dacitic Mount St. Helens dome (USA, 1980 – 1985), the dacitic Unzen dome in Japan (1991 – 1995), the still growing andesitic dome of Montserrat (Caribbean, 1995 to present) and the ongoing long-term growth of the andesitic dome on Merapi volcano, Java (Indonesia). Other examples include Novarupta (Alaska, 1912), Montagne Pelée (Martinique, 1902), Showa Shinzan (Japan, 1943 – 1945), and Santiaguito (Guatemala).

the other way around. Obviously, viscosity decrease due to cooling appears to be a major factor, but this is only part of the explanation. Peterson and Tilling (261) discovered that the shear stress, causing lava deformation, plays a major role. At high shear stress, e.g., when lava flows down a steep slope, aa can form much more easily out of pahoehoe (at constant temperature and therefore viscosity) than at low strain rate. The overlying principle is that aa lava flows consist of moving lava clumps. Growth of microlites during flow is another major factor in changing pahoehoe into aa lava (50).

Pillow lavas and sheet lavas formed under water were discussed in Chapter 5. Pillow lava can, of course, also form on land, such as when lava flows enter lakes or rivers. Such sequences of lava tubes, locally with detached pillows, surrounded by broken, glassy lava fragments (many altered to yellow palagonite), are common on the Columbia River Plateau (Figs. 9.9, 9.10).

Block Lava

Magmas of felsic, SiO_2-rich (e.g., andesitic, dacitic, or rhyolitic) composition are more strongly polymerized and therefore more viscous. Such lavas can never form pahoehoe structures. The slightly more mafic, SiO_2-poor, andesitic, and some phonolitic lava flows (which resemble

▲ Fig. 9.12. Oblique aerial photograph of Holocene rhyolitic Big Obsidian lava flow in Newberry caldera (Oregon, USA). Typical is the irregular surface and steep front of the lava flow, consisting of obsidian blocks (Fig. 9.13). Photo courtesy Motomaro Shirao

◄ Fig. 9.13. Detail of the surface of the Newberry caldera obsidian lava flow. The quickly cooled surface shows mainly banded obsidian, the white layers representing more porous pumice-like layers

▶ Fig. 9.14. Phonolitic lava flow, broken into huge blocks inside Las Cañadas caldera on Tenerife (Canary Islands). The steep wall in the background represents the scarp of the caldera-like depression, which formed by sector collapse of the north flank of the island about 200 000 years ago. Person for scale (ringed area)

▶ Fig. 9.15. Andesitic dome of Merapi Volcano near Yogyakarta, the most active volcano of Indonesia with the active part visible on the skyline. Tent in foreground for scale

▶ Fig. 9.16. Andesitic Novarupta dome in the center of a crater, the 1912 eruptive center of Plinian fallout and the largest pyroclastic flow erupted in the twentieth century (Valley of Ten Thousand Smokes (VTTS)) (Alaska) (see also Fig. 11.5)

▲ Fig. 9.17. Dome of Unzen volcano, which erupted between 1990 and 1995. Block-and-ash flow and debris flow deposits (*light colored*) fill valleys leading to the east of the volcano. Note protective dikes bordering main channels in the foreground. Behind the town of Shimabara (to the right) excellent view of sector collapse scar of Mayuyama volcano which collapsed in 1792 (see Fig. 13.6), the resulting debris avalanche deposit forming the irregular coastline. (western Kyushu, Japan). Photo courtesy Asia Air Survey

A classical volcano is Showa Shinzan – part of the Usu volcanic complex – on the northern Japanese Island of Hokkaido (Figs. 9.19, 9.20). Crater sediments in this part of Usu volcano started to become uplifted in 1943. Because Japan was preoccupied after Pearl Harbor with other priorities, the postmaster, Masao Mimatsu, in a small village nearby recognized the chance of his life. From June 1944 to September 1945 he documented the constantly changing shape of the growing dome on the paper window of his office. This is a classic example of the initiative of a lay person with excellent observational skills (Fig. 9.20). Precise leveling carried out by Minakami showed that the magma located at a depth of about 100 m caused an uplift of the sedimentary cover by about 650 m (224). The extrusion of the high-silica dacite resulted in a lava dome with an onion-like structure. The Santiaguito dome (Guatemala) has been growing since 1922 on the flanks of the volcano Santa Maria, Guatemala (which had exploded in a large eruption in 1902). This dome had reached a diameter of 1 200 m and a height of 500 m in the first 2 years. In 1967, it had doubled its height. It was extruded in cycles lasting up to 20 years, a cycle beginning with high extrusion rates ($0.5-2.1 \, m^3 \, s^{-1}$) and ending with low rates ($<0.2 \, m^3 \, s^{-1}$). As of this day it is still growing.

Three and a half weeks after the initial large, explosive eruption of Mount St. Helens on 18 May 1980, scientists observed a slowly rising, hot, viscous rock mass in the center of the 700-m-deep crater (which had formed by the decapitation of the volcano). This material, chemically a relatively highly evolved dacitic magma, grew mushroom-like over the vent and changed its form constantly from 13 to 20 June 1980 by rising or laterally swelling by up to 6 m day^{-1}. Eventually, a dome, 300 m wide and 65 m high formed, which then exploded on 22 July 1980. These processes were repeated several times (dome explosions on 8 and 9 September 1980, 18 and 19 October 1980). After the end of the eruptive phase (1985), the final dome was about 300 m high and about 900 m wide (367). The new material makes up only about 3 % of the volume that collapsed on 18 May 1980. Once in a while, the steep flanks of the cooling dome collapse, or a column (several m in diameter and several tens of m high) is pushed out and promptly collapses as it shrinks during cooling. Collapsing domes can generate catastrophic glowing avalanches and hot pressure waves, which in a tragic example led to the death of 43 journalists and volcanologists on 3 June 1991 at the foot of erupting Unzen volcano in Kyushu (Japan). So-called *endogenous domes* (or *cryptodomes*) expand by swelling of highly viscous ($>10^8 \, Pa \, s$) magma emplaced close to the Earth's surface. *Exogenic domes*, however, grow by the piling up of short viscous lava flows and dome flank collapse debris. The distinction between extrusive and intrusive emplacement is commonly difficult to make in older, more strongly eroded domes.

The average volume of a dome is about $10^7 \, m^3$, the growth of a dome lasting several years to decades, episodically accompanied by explosive eruptions. Some domes are extruded as relatively viscous lava spines, such as the famous needle of Montagne Pelée in 1902. This lava spine grew at a velocity of about 10 m day^{-1} and reached a height of more than 230 m. It collapsed into a big rubble pile only forty days after its formation.

Most lava domes occur in subduction zone-settings and range in composition from andesite to dacite (ca. $58-69 \, wt \% \, SiO_2$). In the much rarer intraplate alkaline settings lava domes are commonly of trachytic to phonolitic composition ($55-68 \, wt \% \, SiO_2$). Well-known examples of domes in central Europe are the much-visited Drachenfels near Bonn (trachyte, Miocene), the Middle-

Porfil der Veſtung Hochen Twiel.

Hochen Hewen.

Hochen Stofflen.

◄ Fig. 9.18. Tertiary phonolitic dome Hohentwiel near Singen (Germany). Engraving by M. Merian (1643)

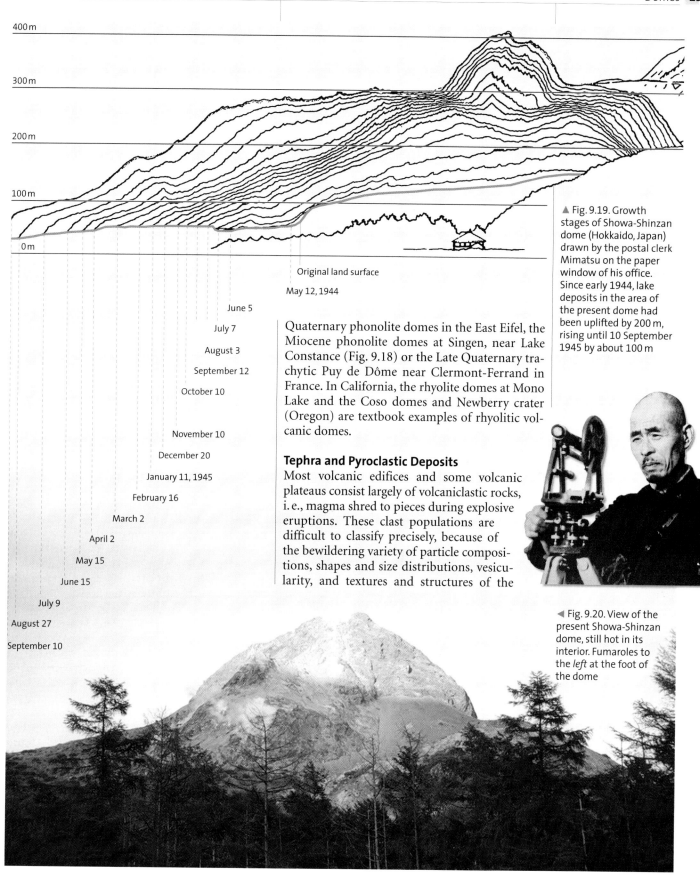

400 m

300 m

200 m

100 m

0 m

Original land surface

May 12, 1944

June 5

July 7

August 3

September 12

October 10

November 10

December 20

January 11, 1945

February 16

March 2

April 2

May 15

June 15

July 9

August 27

September 10

▲ Fig. 9.19. Growth stages of Showa-Shinzan dome (Hokkaido, Japan) drawn by the postal clerk Mimatsu on the paper window of his office. Since early 1944, lake deposits in the area of the present dome had been uplifted by 200 m, rising until 10 September 1945 by about 100 m

Quaternary phonolite domes in the East Eifel, the Miocene phonolite domes at Singen, near Lake Constance (Fig. 9.18) or the Late Quaternary trachytic Puy de Dôme near Clermont-Ferrand in France. In California, the rhyolite domes at Mono Lake and the Coso domes and Newberry crater (Oregon) are textbook examples of rhyolitic volcanic domes.

Tephra and Pyroclastic Deposits

Most volcanic edifices and some volcanic plateaus consist largely of volcaniclastic rocks, i.e., magma shred to pieces during explosive eruptions. These clast populations are difficult to classify precisely, because of the bewildering variety of particle compositions, shapes and size distributions, vesicularity, and textures and structures of the

◀ Fig. 9.20. View of the present Showa-Shinzan dome, still hot in its interior. Fumaroles to the *left* at the foot of the dome

▲ Fig. 9.21. Middle and Upper Laacher See Tephra. The deposits of the Upper Laacher See Tephra begin above the second of 3 brownish ignimbrites and include 1-m-thick fine-grained deposits at the top. These primary deposits are cut unconformably (arrow) by 8-m-thick reworked Laacher See Tephra, forming the upper part of the cliff. Nickenich (Laacher See area, Germany)

deposits. There is a highly specialized terminology for different types of pyroclastic deposits, but only a few names suffice for the purpose of this book.

The term *tephra* (volcanic ash), originally coined by Aristotle and revived by Thorarinsson, is the most convenient term for loose deposits made of volcanic particles because it is independent of composition and grain size (96). Tephra is mainly subclassified according to grain size (Table 9.1). Lithified ash is called *tuff*. A general term for volcanic clastic rocks is *pyroclastic* rocks. These are distinguished from *hydroclastic* rocks, in which fragmentation of the rising magma has been strongly influenced by contact and mixing with external water (Chap. 12). If one is not certain if a tuff is primary or reworked, the term *volcaniclastic* deposit is used. Fine-grained tephra can become lithified quickly to a tuff because volcanic glass, of which most tephra particles consist, is thermo-

dynamically unstable and reacts quickly with ground- or seawater. During this process, some elements, especially alkalis and calcium, are leached from the glass, forming highly concentrated solutions in the pore spaces. New minerals crystallize when such solutions become oversaturated, the framework silicates zeolites, clays and calcite comprising the most common secondary minerals by which loose populations of tephra grains become cemented.

Tephra s.s. is transported away from a vent by many different processes. The simplest cases are ash clouds, from which particles rain down, forming *fallout* deposits. These deposits are characteristically well-sorted (Fig. 9.21) and drape the surface of the land onto which they fall (Fig. 9.22). During collapse of an eruption column or during lateral expansion from the crater lip, particle flows develop that move along the ground, such as the *pyroclastic flows* (Chap. 11) and *surges* (Chaps. 10 and 11). Their deposits are often restricted to valleys (or at least are thicker in valleys) and are poorly sorted. Volcanic debris or mud flows (*lahars*) form, for example, when ash flows enter a river or are deposited on snow or ice-covered slopes, by eruptions through a crater lake or during the remobilization of loose ash following strong rainfall (8, 207) (Chap. 13).

Scoria Cones

Scoria cones are by far the most common type of volcano on land (Chap. 7). They are cone-shaped, relatively regular volcanoes, which may include a crater in their top (Figs. 7.6, 9.2, 9.23–9.27). Their form can be defined by the parameters D_{Ce} (diameter at the base), D_{Cr} (diameter of the crater), H_{Ce} (height) and α (slope gradient) (Fig. 9.27) (268, 434). Basal diameters average about 0.8 km and range from 0.25 to 2.5 km; the average volume is about 4×10^7 m^3, produced at an average eruptive rate of 30 m^3/s. The height varies between 50 and 200 m, with crater diameters between 50 and 600 m, and slope gradients narrowly between 32 to 33°. The dimensions of scoria cones seem to relate to each other in the following way: $H_{Ce} = 0.18\ D_{Ce}$ and $D_{Cr} = 0.40\ D_{Ce}$. The scoria cone Paricutín, which started to grow in a cornfield in Mexico in 1941, reached the unusual height of

► Table 9.1. Classification and nomenclature of volcanic fragmental material

Grain size	Term
< 2 mm	Ash
2 – 64 mm	Lapilli
> 64 mm	Bombs (plastically deformed blebs of magma) Blocks (solid rock fragments)

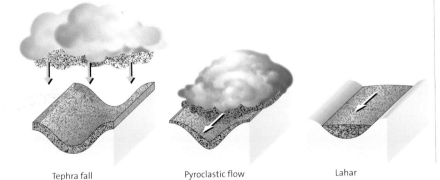

Tephra fall   Pyroclastic flow   Lahar

◄ Fig. 9.22. Schematic of three basic types of volcaniclastic deposits: ash-(and lapilli) fallout tephra forms well-sorted deposits of relatively constant thickness, mirroring irregular morphology. Laterally transported ash flow deposits are poorly sorted, restricted to valleys or form plateaus and thin laterally. Lahars (volcanic debris/mud flow deposits) are poorly sorted, more viscous, restricted to valleys and end more bluntly unless grading into hyperconcentrated stream deposits (303)

▶ Fig. 9.23. Eruptive fissure, lapilli and welded scoria deposits covering the crater row of Duraznero volcano erupted in 1949 (La Palma, Canary Islands)

▼ Fig. 9.24. Cross section through multi-phase scoria cone showing excellent crater unconformity. The bedded, in part phreatomagmatic deposits of the oldest exposed eruptive phase (1) are separated from the main phase of the scoria cone (4) by about 2-m-thick light-colored deposits. The lower part (2) contains a soil horizon, abundant root marks and leaf impressions. The upper part (3) is a phreatomagmatic deposit. The scoria deposits have subsided along several faults towards the left into the interior of the crater. Eppelsberg scoria cone. For details see Fig. 9.25 (Laacher See area, Germany)

◀ Fig. 9.25. Close-up view of old crater rim showing increasing deformation of bedding structures in basal phreatostrombolian deposits (1) towards the center of the crater and change from near-vertical to curved listric faults to the left. Growth faults in main scoria cone phase (4) marked by *arrows*. Most of the crater subsidence occurred during the beginning of cone phase 4, the growth faults (*dashed*) terminating at the end of the scoria cone phase. The crater unconformity and upper part of the sequence are detailed in Figs. 9.26 and 12.5.

▶ Fig. 9.26. Upper eastern wall of Eppelsberg scoria cone illustrating complex and long-term evolution of scoria cones. The massive, bomb-rich scoria deposits of the wall facies of the main scoria phase (4) are overlain by black fallout lapilli layer underlain by thin light-colored phreatic deposits (5), bedded phreatostrombolian deposits (6) capped by a brown massive paleosol (7), once covered by a dense forest. The vertical pipes (encircled) are the molds of former trees (see also Fig. 12.5). This forest was rapidly covered by thinly bedded gray to dark phreatomagmatic tuffs (8) that are more coarsely bedded in the upper part. Holes (arrow) are evidence of broken off and horizontally transported tree trunks. Another black basaltic subplinian lapilli fallout deposit (9) near the top of the quarry wall beneath reworked volcaniclastic and glacial loess deposits (10) is capped beneath the vegetation by light-colored Plinian fallout of Laacher See volcano (11) erupted 12 900 years ago

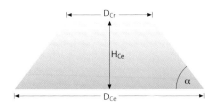

▲ Fig. 9.27. Morphological parameters of a scoria cone. D_{Ce} Basal diameter; D_{Cr} crater diameter; H_{Ce} height of cone; α slope gradient

610 m (201). Eruptive rates in scoria cones are usually highest on the first day, e.g., from about 10^4 to 10^6 m³/day and decrease thereafter. Thus, a height of 100 m can be reached in one day of eruption. About half of the scoria cones cease to grow after 1 month, 95 % in less than 1 year of activity, but in some cases they can remain active as long as 15 years or even longer. Cerro Negro in Nicaragua (Fig. 7.7), which began to grow in 1850, is still active and erupted several times during the past decade. Following the initial intense explosive phase (due mainly to a concentration of volatiles in the upper part of a magma column) (Chap. 7), lava flows issue from many scoria cones during late stages of activity.

The higher the eruptive rate, commonly the larger the total volume of a scoria cone. The volume of most scoria cones varies from 10^4 to 10^8 m³, that of associated lava flows being commonly about one order of magnitude higher. In other words: scoria cones generally represent only a very small fraction of the total mass of erupted magma.

Growth stages of scoria cones have been studied at Paricutín, Heimaey, Stromboli and Etna, classic studies including those by McGetchin et al. (215). These authors postulated that a scoria cone starts to become unstable during its later stages of growth, with loose material sliding down the cone flanks to form an apron which superficially resembles talus deposits that are reworked after volcanic activity has ceased (Fig. 9.28). Many Quaternary scoria cones in the Eifel volcanic field, well-exposed due to extensive quarrying, show the stages postulated by McGetchin et al. (Figs. 9.29, 9.30). Nearly all of these scoria cones consist of several eruptive centers, nested within each other or overlapping laterally. Eruptive centers can be active practically synchronously and one vent may erupt material in a hot state, while a nearby vent produces hydroclastic deposits.

Nearly all scoria cones in the Eifel start with phreatomagmatic deposits in their initial stage (Chap. 12). The basal deposits are commonly fine-grained and some are rich in accretionary lapilli, typical of phreatomagmatic deposits. These are overlain by coarse-grained, very poorly sorted and extremely xenolith-rich breccias (Fig. 12.6) alternating with fine-grained, massive tuff layers, in which the volcanic particles are angular and glassy and contain very few vesicles. These initial deposits tend to pinch out, however, a few 100 m away from the central vent. Their grain size and bed thickness decrease rapidly. The particles in the deposits of these initial maars or tephra rings have commonly been transported by base surges. The deeper levels of scoria cones, well exposed in older crater areas of the Tertiary volcanic areas in central Europe, show evidence for the basal phreatomagmatic and some of the later pyroclastic eposits to have subsided along normal or listric ~~f~~aults into the basement (Figs. 9.24, 9.25). The ini~~t~~al craters have thus formed by caldera-like col~~la~~pse into the pre-volcanic surface because a sig~~n~~ificant mass deficit had formed below the sur~~fa~~ce.

In well-exposed scoria cones (Fig. 7.6) one ~~c~~an distinguish two major facies: the crater facies ~~a~~nd the wall facies. These are separated from each ~~o~~ther by several semi-circular faults or funnel-~~s~~haped surfaces, along which material has slid ~~d~~own into the crater (Fig. 9.24).

The pyroclastic cone deposits, dominantly ~~fo~~rmed ballistically and by lava fountains, change ~~t~~heir character quickly away from the central vent ~~a~~rea. The wall facies above the basal breccias and ~~t~~uff layers commonly consists of bombs and frag~~m~~ents thereof (Fig. 9.30). These are generally ~~w~~elded together close to the central vent area, but ~~b~~ecome increasingly non-welded with distance. ~~S~~ingle, less vesicular, spindle-like bombs can be ~~f~~ound in many cones. The very outer facies is ~~o~~ften dominated by fragments of bombs or lava ~~b~~lebs, which have formed by bombs bouncing off ~~t~~he flanks of the volcano. This process is well-~~i~~llustrated in films or videos of recently formed ~~s~~coria cones, such as Paricutín or those of the ~~g~~reat Tolbachik eruptions in Kamchatka. These ~~p~~rimary deposits are then covered by re-deposited ~~t~~ephra, making up the outer apron.

Crater walls are commonly not well-devel~~o~~ped in the central part of scoria cones deposits, ~~b~~ut can be recognized by unconformities between ~~t~~he welded deposits dipping outward and those ~~d~~ipping into the central crater. While the crater of ~~t~~he initial maar phase is often significantly larger ~~i~~n diameter than that of the scoria cone grown ~~i~~nside the maar, the crater area decreases in diam~~e~~ter as the cone grows.

The lower crater floor facies commonly con~~s~~ists of very solid lava rocks, made up of rapidly ~~e~~jected lava fountains lava blebs that coalesce to ~~f~~orm *spatter deposits* or *agglutinates*, many of ~~w~~hich resemble compact lavas (Figs. 9.31, 9.32). ~~T~~he very tough agglutinate blocks remain in the ~~c~~enter of many quarries following the termina~~t~~ion of quarrying operations, because they ~~r~~equire larger machines for crushing or have to ~~b~~e blasted and are thus of less commercial value ~~t~~han the loose lapilli and bombs. The central ~~c~~ompact lavas in some volcanoes represent solidi~~f~~ied small lava lakes.

Agglutinates are also common in more evolved ~~v~~olcanoes, where they form massive and bedded ~~d~~eposits that may extend from the summit down ~~t~~he flanks as on Komagatake volcano (Fig. 9.30). In ~~b~~asaltic or phonolitic lava fountains, i.e. low vis-

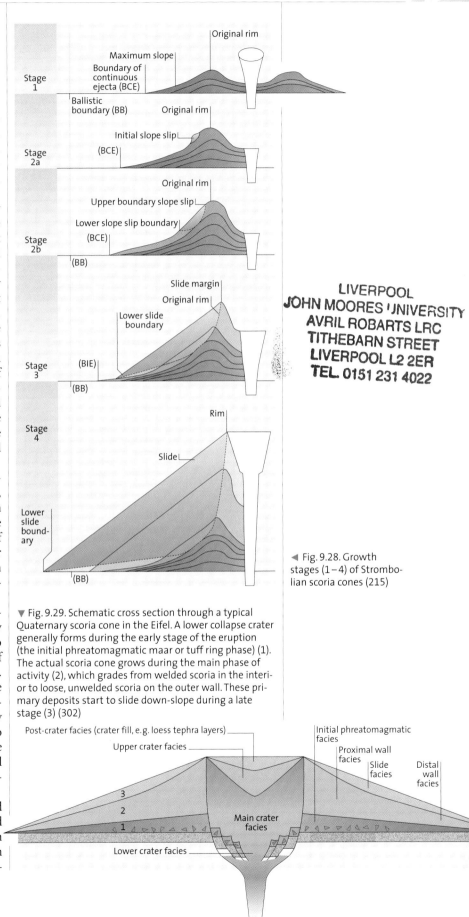

◄ Fig. 9.28. Growth stages (1–4) of Strombolian scoria cones (215)

▼ Fig. 9.29. Schematic cross section through a typical Quaternary scoria cone in the Eifel. A lower collapse crater generally forms during the early stage of the eruption (the initial phreatomagmatic maar or tuff ring phase) (1). The actual scoria cone grows during the main phase of activity (2), which grades from welded scoria in the interior to loose, unwelded scoria on the outer wall. These primary deposits start to slide down-slope during a late stage (3) (302)

▲ Fig. 9.31. Strongly-welded dacitic lava fragments (agglutinates) erupted in 1929. Crater rim of Komagatake volcano (Hokkaido, Japan). Block 1 m across

▲ Fig. 9.30. Typical wall deposits of a scoria cone consisting of bombs (large bomb, 1 m in diameter), bomb fragments and vesicular lapilli. Rothenberg volcano (Laacher See area)

▶ Fig. 9.32. Transition from strongly welded (lower part of photograph) to extremely welded lava-like agglutinate. Late Quaternary Wartgesberg volcano (Eifel, Germany)

cosity melts, such spatter deposits may grade imperceptibly into lava flows that are actually fed from fire fountains. Such lava flows represent an extreme variant of pyroclastic rocks (Fig. 9.32).

The youngest deposits in many crater fills are composed of both large, round bombs that may exceed 0.5 m in diameter, and smaller lapilli. These deposits are the only ones to which the widely misused term *agglomerate* may strictly apply. The final deposits in many scoria cones are commonly black, well-sorted fallout lapilli, which can extend laterally for several km and are basaltic sub-Plinian deposits, i.e. cover an area up to ca. 500 km². The crater area is commonly red or purplish red due to oxidation by hot gases streaming preferentially through the central part of the edifice. The outer walls of scoria cones are commonly black and non-oxidized. Many dikes may still be recognizable and solid at deep levels of well-exposed scoria cones. They breach out upward and appear as films of lava, forming large, funnel-like zones that are filled-in by fallout and back-slid material. From these upper zones alone they cannot be easily recognized as dikes, unless the lower parts of scoria cones are exposed. Crater fills, such as loess, paleosols or tuffs derived from

◀ Fig. 9.33. Crater-fill deposits of 200 000-year-old scoria cone Kollert (Laacher See area). The crater-fill consists of black basaltic lapilli deposits, erupted in nearby scoria cone, reworked glacial loess deposits and reworked tuff and lapilli that slid down into the crater from the upper part of the cone. Bones of mammals and tools of stone age people are common in these deposits, testifying to early settlement of the area

other volcanoes commonly form magnificent environmental or volcanic records of the last few hundred thousand years (Fig. 9.33). Maars, tuff rings and tuff cones are discussed in more detail in Chapter 12.

Stratovolcanoes

Stratovolcanoes have volumes up to about 800 km^3 (Fig. 9.34) and a lifetime commonly between 10^5 and 10^7 years. This means that a much larger variety of intrusive and extrusive rocks and chemical compositions can be found in stratocones than in scoria volcanoes. Hence, they are difficult to define precisely. Because of their complex and long evolution, they are also called *polygenetic or composite volcanoes*. Typical stratocones are the andesitic to dacitic volcanoes above subduction zones (Chap. 8; Figs. 8.3, 9.1). They are also common in the alkali-rich (tephritic, phonolitic, trachytic) volcanoes on many oceanic islands in the Atlantic, such as the Canary Islands (some examples being the young Pico de Teide on Tenerife or the Roque Nublo volcano on Gran Canaria, Fig. 9.35). Mount St. Helens, Vesuvius, and Mount Fuji – the latter two of mafic composition – are classic examples of stratocones.

The central core of these volcanoes is commonly dominated by intrusive rocks, whose slow ascent in the form of needles or domes has been documented repeatedly in historic eruptions as discussed above. The dominantly intrusive core zones, from 500 m up to several kilometers wide, are surrounded by many different generations of lava flows, many brecciated, and a great variety of pyroclastic and reworked epiclastic rocks (Chap. 11), whose percentage increases with increasing distance from the central vent area. Coarse-grained pyroclastic rocks, which form by repeated collapse of growing domes (block-and-ash-flows, Fig. 9.36) dominate in the central area of dacitic and andesitic stratocones. Pumice-rich pyroclastic flow deposits, which form during ash fountaining or collapse of eruption columns, become more common with distance because of their high mobility. Large central volcanoes are often associ-

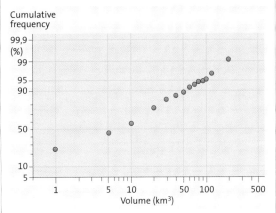

Cumulative frequency

◀ Fig. 9.34. Cumulative frequency of volumes of late Quaternary stratovolcanoes in Japan (16)

ated with calderas and are surrounded by ignimbrite plateaus (see below).

The form of stratocones, especially their particularly high height/width ratio, depends largely on the viscosity of the erupted magma, which in turn is a function of its SiO_2 concentration or the crystallinity. The high water content coupled with the high viscosity of the andesitic to dacitic magmas are the reason for their common explosivity and the fragmentation of lava flows on the steep volcano flanks. Also, the slopes become oversteepened due to dome intrusion. Because volcaniclastic materials may make up more than 90 percent by

▲ Fig. 9.35. About 500-m-thick flank deposits of eroded Pliocene Roque Nublo stratocone. The section shows alternating lava flows, debris flow deposits and a few fallout layers. Near Tejeda (Gran Canaria, Canary Islands)

◄ Fig. 9.36. Oblique aerial photograph of block-and-ash flow deposit with steep front emplaced in February 1976 on Augustine Volcano (Alaska). For close-up of one of the large blocks see Fig. 11.25, and details in Figs. 11.23 and 11.24. In the left background, hilly deposits of sector collapse debris avalanche deposit that continues into the sea

eruptions. The melting of a glacier during the eruption of Nevado del Ruiz (5 500 m a.s.l.) in Colombia on 15 November 1985 (Chap. 13) is an example of a rather small eruption of hot tephra onto a glacier and snow, whose melting initiated a chain of events that ended in disaster. Such hydroclastic eruptions are possibly the trigger for many more purely magmatic explosions, a topic discussed in more detail in Chapters 11 and 12.

Flank Collapses, Debris Avalanches and Debris Flows

Volcanic *debris avalanches* remained almost unrecognized until the collapse of the northern flank of the volcano during the initial eruption of Mt St. Helens on 18 May 1980 (Chap. 10). All large edifice-building volcanoes are now known – or suspected – to episodically experience sometimes huge *sector collapses*. Such flank collapses are also common in many oceanic volcanoes (182, 239, 328, 329, 330) (Figs. 9.37–9.41). The collapses of volcano flanks result in debris avalanches that

volume, these volcanoes are like sponges and can hold huge volumes of water within in their highly porous deposits. Tall volcanoes, such as those dotting mountain ranges in South America or the Cascades in western North America, are commonly glacier-clad. Interaction of rising magmas with external water is – apart from the high water contents in subduction zone magmas (Chap. 8) – another major factor for the high explosivity of stratocone

may grade into debris flows which can reach giant dimensions. Debris avalanches are extremely mobile and can flow with very high velocity and for distances of more than 10 km on distal slopes of less than 1°. Their deposits, many tens of m thick, can cover more than 1 000 km². Sector collapses often generate horseshoe-like scars on the flanks, some of which may reach caldera dimensions. More than 20 large flank collapses have formed

▲ Fig. 9.37. Thick debris avalanche deposits formed by late
▲ stage sector collapse of the Pliocene Roque Nublo stratovolcano on Gran Canaria. Height of Roque Nublo monolith 80 m. The roughly 200-m-thick deposits forming the plateau consist of giant displaced masses

▲ Fig. 9.38. Middle Miocene (10.2 Ma) debris avalanche deposits consisting of competent dark domains of fragmented phonolitic lava flows and light-colored plastically deformed soft pumiceous deposits. Upper Fataga Formation. Lower Arguineguin canyon (Gran Canaria, Canary Islands)

▶ Fig. 9.39. Marine channel between the islands of Gran Canaria (*left*) and Tenerife (*right*) (Canary Islands). The scarp of the 600 000-year-old flank collapse near Güimar is shown on the south flank of Tenerife, while the fan of debris avalanche deposits resulting from this collapse is shown on the seafloor. Young, 600-m-high volcano Hijo de Tenerife in the center of the channel (182)

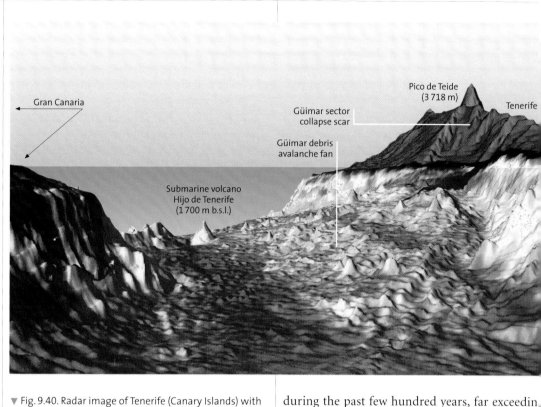

Gran Canaria

Pico de Teide (3 718 m)

Tenerife

Güimar sector collapse scar

Güimar debris avalanche fan

Submarine volcano Hijo de Tenerife (1 700 m b.s.l.)

▼ Fig. 9.40. Radar image of Tenerife (Canary Islands) with several scarps resulting from flank collapses: in the south Güimar, in the northeast Orotava (ca. 600 000 years old) and in the north the scarps of the Las Cañadas flank collapse. The Cañadas Volcano has grown over the erosional remnant of three older basaltic shield volcanoes: the Roque del Conde Massif (ca. 10 – 11 Ma), the Teno mountains in the northwest (ca. 6 Ma) and the Anaga peninsula (ca. 3 – 5 Ma) in the east. Sector collapse scar outlines based on P Navarro (pers. comm.) and (412)

Las Cañadas landslide (0.2 Ma)

Güimar landslide (0.6 Ma)

0   10   20 km

during the past few hundred years, far exceeding the number of true caldera collapses. In some volcanoes, debris avalanche deposits dominate the volcaniclastic aprons surrounding the central edifice. Flank collapses probably represent the most important process modifying the growth and evolution of volcanoes in all tectonic environments.

Sector collapses are the result of structural instability of a volcano. Oversteepening of the flanks caused by intrusions, weakening of the core of a volcano by hydrothermal alteration (aggressive acid gases), dike intrusion into water-rich deposits increasing the pore pressure, or repeated dike intrusions into rift zones are commonly invoked trigger mechanisms for flank collapse (Fig. 13.7). Sometimes, as in 1980 on Mount St. Helens, such flank collapses can trigger, or at least precede, the explosive eruption of a rising magma column, suddenly depressurized by deapitation of the volcano. Andesitic and dacitic stratocones with heights of more than 1 000 m, steep flanks and slope dips that may exceed 30° are particularly prone to collapse. In the central Andes, the percentage of volcanoes that have collapsed correlates positively with altitude. Less than 10 % of volcanoes less than 500 m high have collapsed, compared to over 75 % of volcanoes whose height exceeds 2 500 m (99). Structural factors, such as steep flanks with alternating competent lava flows and unconsolidated clastic deposits are important predisposing factors for flank collapses. Debris avalanches on oceanic islands are discussed more fully in Chapter 6. On Tenerife, for example, the central part of the island collapsed about 200 000 years ago. The famous Cañadas caldera walls were probably formed mainly by such huge slides (413) rather than by classic collapse into a magma chamber as sometimes postulated (210). The significance of debris avalanches as volcanic hazards is discussed in Chapter 13. The caldera type locality – subject of the next section – the beautiful deep hole Taburiente on the Canary island of La Palma (Fig. 9.41) – shares the fate with many type localities, being now convincingly re-interpreted as due to lateral flank collapse.

Calderas and Caldera Volcanoes

Calderas represent some of the most impressive volcanic landforms, sometimes of huge dimensions, and many of the most beautiful lakes hosted by volcanoes are water-filled caldera basins (Figs. 9.42–9.47, 9.51, 15.22, 15.23). A question often posed to geologists: what is the difference between a crater and a caldera? The answer is easy from a purely descriptive point of view, when considering the diameter. It is more difficult when trying to visualize the substructure and reconstruct the processes by which morphological depressions are generated in the top of a volcano. There are several quite different processes by which such holes in the ground are formed. Moreover, there is no clear-cut distinction between what is called crater and caldera. *Craters* have diameters ranging from a few tens to at most a few 100 meters, while the diameters of most calderas range up to tens of km (in rare cases more than 50 km).

Funnel-shaped craters are generated during the growth of tephra-dominated volcanoes as in scoria cones by ejection of particles and a driving gas phase through a hole, the conduit, and fallout around the vent. The funnel shape of the depressions is the result of erosion of the tephra walls when the eruptive mixture is decompressed when leaving the conduit and expands, especially during

water-magma interaction, as well as by episodic downward sliding of the loose material of the unstable inner walls. *Pit craters*, common on Kilauea volcano on Hawaii (the type locality), are sub-circular, steep-walled holes, up to 1.5 km wide and 300 m deep (Chap. 3; Fig. 3.7, 3.8), generated by collapse. Most lava lakes of Kilauea volcano have formed by the filling of older pit-craters with new lava, most of them having been completely infilled during the past three decades. *Maars*, kettle-like

► Fig. 9.41.
Morphology of La Palma (Canary Islands) as presented by Leopold von Buch (37). The large hole in the northern, about 1 to 3-million-year-old shield volcano (Caldera de Taburiente) cut by Barranco de las Angustias is the type caldera. Today, it is explained by sector collapse, generating submarine debris avalanche deposits and by subsequent erosion (Fig. 6.29). The late Quaternary and historic (last eruptions in 1949 and 1971) basaltic eruptions are restricted to the southern, about 30-km-long ridge, which represents a rift zone

▶ Fig. 9.42. (a) Index maps of Miyakejima volcano (Japan) at northern end of Izu-Bonin chain

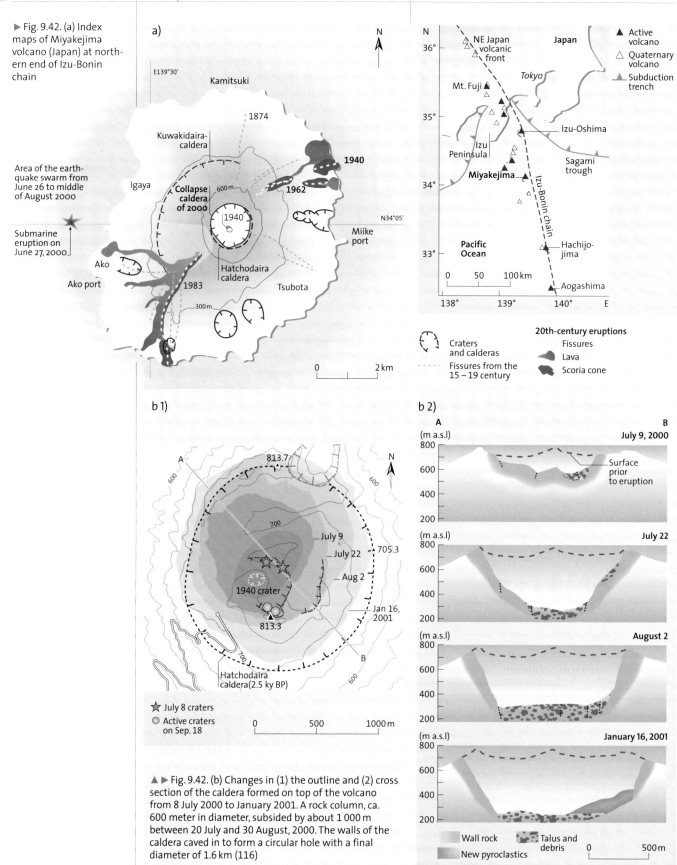

a)

E139°30'

Kamitsuki

1874

Kuwakidaira-caldera

Area of the earthquake swarm from June 26 to middle of August 2000

Igaya

Collapse caldera of 2000

600 m

1940

1962

1940

Miike port

N34°05'

Submarine eruption on June 27, 2000

Ako

Ako port

1983

Hatchodaira caldera

Tsubota

300 m

0   2 km

N

N Japan
36°   NE Japan volcanic front

Mt. Fuji   Tokyo

35°

Izu Peninsula   Izu-Oshima

34°   **Miyakejima**   Sagami trough

Izu-Bonin chain

33°   Pacific Ocean   Hachijo-jima

0   50   100 km

138°   139°   140°   E

▲ Active volcano
△ Quaternary volcano
◄ Subduction trench

Craters and calderas
Fissures from the 15–19 century

20th-century eruptions
Fissures
Lava
Scoria cone

b 1)

A
813.7

600   600

700

July 9
July 22   705.3
Aug 2

1940 crater

Jan 16, 2001
813.3

700   600

B

Hatchodaira caldera (2.5 ky BP)

☆ July 8 craters
○ Active craters on Sep. 18

0   500   1000 m

N

b 2)

A   B
(m a.s.l)   July 9, 2000
800
600   Surface prior to eruption
400
200

(m a.s.l)   July 22
800
600
400
200

(m a.s.l)   August 2
800
600
400
200

(m a.s.l)   January 16, 2001
800
600
400
200

Wall rock   Talus and debris
New pyroclastics   0   500 m

▲ ▶ Fig. 9.42. (b) Changes in (1) the outline and (2) cross section of the caldera formed on top of the volcano from 8 July 2000 to January 2001. A rock column, ca. 600 meter in diameter, subsided by about 1 000 m between 20 July and 30 August, 2000. The walls of the caldera caved in to form a circular hole with a final diameter of 1.6 km (116)

▶ Fig. 9.43. Aerial photograph of collapsing caldera wall on top of Miyakejima volcano erupted in July-August 2000. Photo courtesy Asia Air Survey

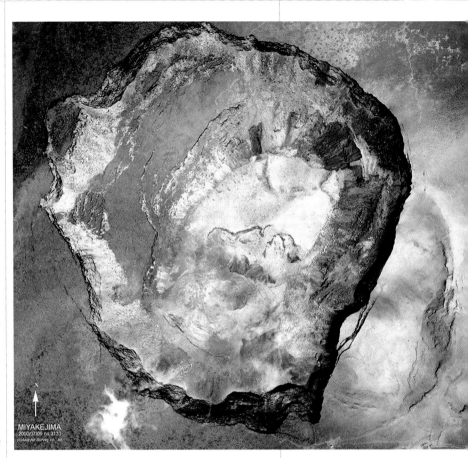

holes (Chap. 12), whose diameter may exceed 1 km, are often called explosion craters. The main process of their formation is, however, collapse.

Calderas on Basaltic Volcanoes and the 2000 Eruption of Miyakejima Volcano

Large calderas on top of basaltic shield volcanoes, such as Kilauea or Mauna Loa on Hawaii, have diameters up to several km and are characterized by vertical walls. Kilauea caldera, e.g., is about 165 m deep with an outer diameter of about 6 km. It is widely believed that these calderas form by incremental collapse of the roof of a volcano. The cause is thought to be a decrease in magmatic pressure below, e.g., when the magma has erupted on the flanks of the volcano and not at the top, creating a void below the structurally unstable roof above the central shallow magma reservoir. Calderas have also been recognized on the tops of a number of submarine basaltic volcanoes by high-resolution side-scan sonar mapping (Chap. 6).

The eruption of Miyakejima volcano in 2000 and the interpretation of a preceding earthquake swarm enabled scientists to reconstruct the processes leading to formation of a caldera on top of a basaltic volcano in remarkable detail (116) (Figs. 9.42–9.44). Miyakejima volcano (814 m a.s.l. prior to the 2000 eruption and extending some 1000 m to the seafloor), an active basaltic andesite stratovolcano island about 200 km south of Tokyo, is part of the Izu-Bonin volcanic chain. From 26 June to 8 July 2000, magma migrated northwestward below the sea floor, starting from a reservoir below the summit as indicated by seismicity. The dike reached a length of 10 km and a width of a few m, representing a magma volume of at least $1-2 km^3$. In other words, a very large volume of magma had apparently escaped laterally from the main reservoir beneath the center of the island. A very minor submarine eruption occurred on 27 June about 1 km off the western coast of the volcano. Collapse of the summit of the volcano began on 8 July accompanied by a minor phreatic eruption and lasted until the middle of August. Phreatic and phreatomagmatic eruptions resumed on 10 August and lasted until late September. By far the largest eruption occurred on 18 August after subsidence had stopped, producing

▶ Fig. 9.44. Schematic cross section of top of Miyakejima volcano showing upper part of conduit system, collapse blocks, rising of volatile-rich and sinking of dense degassed magma (290)

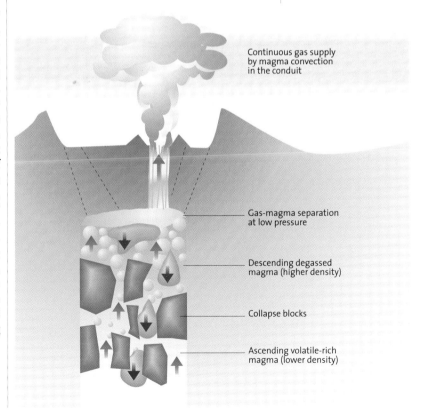

Continuous gas supply by magma convection in the conduit

Gas-magma separation at low pressure

Descending degassed magma (higher density)

Collapse blocks

Ascending volatile-rich magma (lower density)

▶ Fig. 9.45. Holocene 7 700-year-old Crater Lake caldera (Oregon, USA) (ca. 10 km diameter). The andesitic volcanic island Wizard Island in the western part of the lake (*center* of photograph) grew after caldera collapse

▼ Fig. 9.46. Santorini caldera (Aegean Sea) formed about 3 500 years ago. The light-colored pumice deposits that resulted from this eruption overlie the caldera margin on the opposite side of the bay

8×10^9 kg of tephra with 40 % juvenile material. The total volume of tephra erupted from the volcano in 2000, dominantly lithics and dense juvenile bombs, amounts to 1.1×10^7 m^3 (DRE). The growth rate of the caldera was 1.4×10^7 m^3/d and the final volume of the caldera was about 6×10^8 m^3. In other words, the volume of erupted material was exceedingly small, 1.1 % (!) of the void created by caldera collapse. Clearly, eruption of magma and lithics could not have triggered caldera collapse.

The collapse process itself appears to have consisted basically of two parts. A central stoping rock column 600–700 m in diameter is thought to have subsided piston-like, bounded by outward dipping faults. Synchronous with central piston subsidence was significant lateral enlargement of the surface basin by caving-in of the vertical walls along listric faults, as shown clearly by aerial photographs (Fig. 9.43). The final caldera had a diameter of 1.6 km in September, the floor being 450 m deeper than the former land surface. Because the floor of the caldera did not change much in depth after the first few days (while lateral caving-in continued for over a month), subsidence of the central rock column was calculated as ca. 40–53 m/d assuming 40 days of subsidence from July to August. The total subsidence then amounted to 1.6 to 2.1 km (!) to account for the volume of the caldera depression.

◀ Fig. 9.47. Wall of Miocene caldera (14 Ma) about 20 km in diameter in the center of a basaltic shield volcano. The dark horizontal shield lavas (*left*) were more strongly eroded than the thick silicified and therefore erosion-resistant ign-imbrite cooling units accumulated inside the caldera basin. The thick ignimbrite in the center of the photograph with large caves in its upper part is about 80 m thick. The green rocks dipping to the right in the lower part of the photograph are hydrothermally altered pumice lapilli deposits. Black-and-white line marks caldera wall. Los Azulejos below Montaña Horno (Gran Canaria, Canary Islands)

The origin of the void that caused summit collapse is easily accounted for by lateral drainage of magma at depth as indicated by robust seismic evidence. Support of the roof of the magma reservoir was then lost and a stoping rock column subsided into the partially emptied reservoir. This created a shallow near-surface void that may have been steam-filled to account for the initial phreatic activity. Intrusion of magma into the stoping column caused the major phreatomagmatic explosion on 18 August. The volume of fresh magma at depth must still be large and possibly convecting to account for the very high SO_2 emission that continues until now. This has prevented the return of some 5 000 people evacuated from the island following caldera collapse and subsequent strong sulfur degassing (Chap. 4).

The calderas at the top of basaltic shield volcanoes are constantly modified by many eruptive phases and grow upwards during long periods of activity, several hundred thousand years as in Hawaii. The mass of magma erupted in these volcanoes during eruptions is small. Ash-flow calderas, on the other hand, are very large, their form and eruptive products having been generated during a few major events.

Ash-Flow Calderas

Large calderas, also called *ash-flow calderas* (193), are subcircular collapse craters, up to about 20 km in diameter (rarely up to 60 km). Many have morphological basins, partly infilled during subsequent eruptions, in which central volcanoes may grow (Figs. 9.45–9.49). Many of the so-called *somma volcanoes* are, however, actually formed in the head scarp of flank collapses. Calderas were once classified according to type locality, as was done for different types of eruptions, and terms such as calderas of the Krakatau type, Valles type, etc. were distinguished from each other. Today, a more rational classification is attempted, based on form, size and inferred mechanism of formation (Fig. 9.48) (194). There are, of course, several transitions and intermediate types between specific end members and for some, such as piecemeal calderas and downsag calderas, there are only very few well-documented examples.

The subsidence of a caldera basin is generally visualized as a relatively coherent to piece-meal sinking of a more or less circular rock plate above a void formed at a depth of very few kilometers (Fig. 9.48). The void is generated when several to more than 1 000 km³ of magma are suddenly and rapidly erupted, mostly along ring fractures, dominantly as pyroclastic flows. In some cases, a void or caldera forms when magma has become siphoned away from a sub-terranean magma reservoir to an eruptive site that may be several km away. A well-known example is the collapse of Katmai volcano in 1912, where the eruptive site

▶ Fig. 9.48. Different types of subsidence mechanisms of calderas, depending on the depth and diameter of magma chambers. The dashed lines show postsubsidence depths above the magma chambers (194). The cause for the trapdoor caldera could have been an asymmetric pluton and for the downsag caldera and the funnel caldera deep-seated, small volume plutons

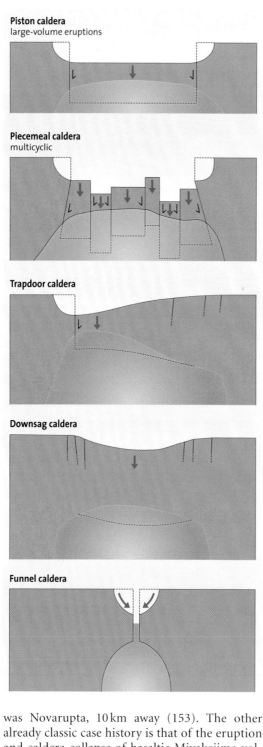

Piston caldera
large-volume eruptions

Piecemeal caldera
multicyclic

Trapdoor caldera

Downsag caldera

Funnel caldera

was Novarupta, 10 km away (153). The other already classic case history is that of the eruption and caldera collapse of basaltic Miyakejima volcano in 2000 described in detail above.

Because the walls of subsidence calderas may be very steep directly after the roof has sunk, they repeatedly collapse, forming breccias of huge landslide blocks mixed with the erupted ignimbrite or the sediments formed in the basin. In other words, the impressive walls of calderas with their smaller or larger embayments form after the initial subsidence and are thus several 100 m to kilometers outside the actual faults. Ignimbrites, which can extend several km to more than 100 km away from a caldera, may also accumulate inside the basin and reach thicknesses of more than 1 000 m (while the *extracaldera* sheets are rarely more than 100 m thick). Some large calderas are actually the roof regions of larger granitic to granodioritic plutons that are exposed where the upper few kilometers have been stripped away by erosion (Figs. 9.49, 9.50). Some calderas are associated with hydrothermal ore deposits (Chap. 15).

A typical cycle of caldera subsidence and ash flow eruptions is shown in Fig. 9.46. Prior to collapse, several smaller stratovolcanoes of intermediate composition may develop above smaller, near-surface magma reservoirs. These herald the later, huge magma chambers sometimes of batholitic dimensions. The slow emplacement of large magma bodies causes ring fractures to form. Following the eruption of ash flows and syngenetic caldera collapses, the basin fills with ash flows, which cover the roof of the volcano. Prior to many ash flow eruptions, a compositionally zoned magma chamber has developed by differentiation of the magma, magma mixing and exchange of material with the wall rocks. Several calderas have distinct eruptive cycles with compositional zonation developing in the magma reservoirs during non-eruptive intervals. Eruptive periods are sometimes separated by periods of quiescence, lasting from 10 000 to several 100 000 years. During later stages of caldera collapse (Figs. 9.48, 9.49), a central area is uplifted by emplacement of new magma or rebound (called *resurgence*) and fragmented into smaller domains. Dome intrusions and accumulation of sediments in the more or less circular morphological moat between the central uplift and the collapsing margins terminate a caldera cycle. Some of the best-studied calderas are Valles caldera, Long Valley caldera, Crater Lake caldera, and the Tertiary calderas in Colorado in the western US, the caldera of Taupo volcano in New Zealand, several calderas such as Hakone caldera in Japan, Santorini caldera in the eastern Mediterranean (194) (Fig. 9.46) and the Miocene Tejeda caldera on Gran Canaria (Fig. 9.47). One of the largest active caldera systems on Earth is the Yellowstone caldera in Wyoming (USA). The seismic roots of this giant volcano magma system can be followed to depths of over 200 km (Chap. 7). The accumulation of magma beneath this caldera is the heat engine that drives the activity of the largest active geyser field on Earth (Figs. 7.17, 15.5). The majestic beauty of

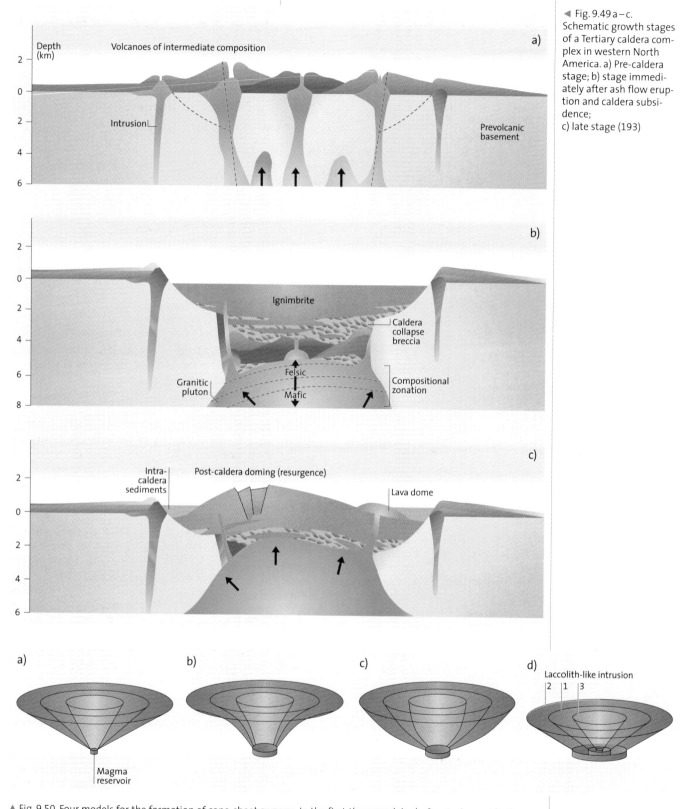

◀ Fig. 9.49 a–c.
Schematic growth stages of a Tertiary caldera complex in western North America. a) Pre-caldera stage; b) stage immediately after ash flow eruption and caldera subsidence;
c) late stage (193)

Fig. 9.50. Four models for the formation of cone-sheet swarms. In the first three models a) after Anderson b) after Philipps and c) after Gudmundsson, a central focal point is assumed. d) The well-exposed cone-sheet swarm in the center of Gran Canaria (see Fig. 6.26) can be explained plausibly by a laterally extensive magma reservoir (laccolith) as base for the dike swarm with irregular temporal succession (2–1–3) of the about 500 cone-sheet intrusions (299)

▶ Fig. 9.51. Mural showing Tianchi crater lake (Lake of Heavenly Peace) that fills the caldera of Baitoushan (also named Paektusan) volcano, which formed during a large eruption about 965 A.D. View from the North Korean to the Chinese side with exit of the lake

caldera lakes is a major aesthetic and thus touristic highlight in many volcanic areas and not without appeal to politicians (Fig. 9.51; Chap. 15).

Summary

Volcanoes consist of lava flows, clastic deposits and intrusions in different forms and proportions representing the entire spectrum of magma compositions. Lavas and intrusions dominate in submarine volcanoes that form the mosaic of the oceanic crust, and in fields of flood basalt volcanism. The number of intrusions increases relative to extrusive and clastic rocks toward the core of a volcano. Scoria cones, maars, domes and caldera volcanoes are the morphologically most clear-cut volcanoes, which have been discussed in more detail in this chapter. Domes are typical for the later stages of large dacitic-rhyolitic but also phonolitic volcanic eruptions and can grow to several 100 m. They seem to try to heal the wound generated during the beginning of an eruption. Processes by which pyroclastic deposits are generated, which form the bulk of most subaerial volcanoes (except shield volcanoes and flood basalts), are discussed in more detail in Chapters 10–12.

Present-day research interest in topics discussed in the next few chapters also reflects the attitude of scientists trying to decipher the mechanisms of individual developmental stages in a volcano before they dare to say something about the final complex edifice. A more holistic view in understanding of volcanoes will develop once we know more about the mechanisms of eruptions, intrusions and constructive and destructive processes. This is similar to the developments in biology, where scientists refocus their attention again at a plant as an entity, following–and in parallel with–decades of physiological and biochemical research on specific aspects.

Strombolian, Hawaiian and Plinian Eruptions and the Mount St. Helens Eruption 1980

Europeans who choose not to spend their money on New Years Eve on fireworks, can instead book a flight to Naples and take a 12-h boat trip to the Aeolian Islands. They can climb the steep slopes of 960-m-high Stromboli volcano and enjoy natural fireworks every 10–20 min, all year-round. Except during the last week of December 2002 when part of the island collapsed into the sea and magma escaped from the flank (Sciara di Fuoco) instead of on the top. The ancient Romans, pragmatic as they were, probably did not think much about the source of the gases, the gas pressure, or the velocity with which particles of magma are thrown out of the several craters in Stromboli, but simply called Stromboli the *Lighthouse of the Mediterranean*.

Quite naturally, eruptions of this permanently active volcano have been called *Strombolian*. The father of volcanic eruption narratives was Pliny the Younger, to whom we owe the graphic description of the demise of Pompeii. He is now remembered by the term *Plinian*, used for eruptions like those of Vesuvius in A.D. 79. The island of Vulcano, type locality for all firespitting mountains, exploded strongly several times in the years 1888–1890 inspiring the name *Vulcanian* eruptions. Other common terms coined include *Hawaiian*, for the spectacular lava fountains of Kilauea volcano, and *Merapian*, for hot pyroclastic block flows, which form during the collapse of rising viscous lava domes.

Today, scientists try to quantify physical and chemical processes within the interior of a volcano, which, in combination with environmental factors, govern eruptive processes. The older descriptive terms for different types of eruptions have thus lost some of their significance, especially since there are also Strombolian types of activity in Kilauea volcano, etc. New classifications are based on different types of fluid regimes, which govern the transport of fluids which contain bubbles (322).

Pyroclastic Fragmentation

The two most important boundary conditions that control the *pyroclastic* fragmentation of a magma into particles are (1) the formation of a magmatic foam and resulting tearing apart of the unstable thin walls of melt in the foam, and (2) the brittle fracture of a magma when the tensile stress of a melt is exceeded (see also Chap. 4). An explosively erupting volcanic system can be subdivided in a simplified way into several levels.

In the upper part of a magma column that contains variable amounts of dissolved volatiles, a zone develops where bubbles start to nucleate, perhaps on crystal faces. In other words, volatile elements previously dissolved in the melt form a free gas phase (Fig. 10.1). Gas bubbles begin to grow when the magma becomes oversaturated in one or more of the volatile components. Much of the decompression takes place when the magma rises, or when a gas species, such as H_2O, is enriched during magmatic differentiation and formation of H_2O-free crystals above its saturation point. Degassing begins at variable depths controlled not only by different magma compositions, but also by the type of volatile species (H_2O, CO_2, SO_2, etc.). The rise speed of an ascending magma depends strongly on its composition and hence its viscosity. Broadly speaking, there is no significant volatile oversaturation at low magma rise speeds (< ca. 0.1 m/s). At rise speeds >10 m/s degassing becomes pronounced when the magma has reached the upper conduit (85). At rise speeds between these two extremes, bubbles form at varying depths.

The dynamic processes during the growth of bubbles are complex (271) and depend on
- viscosity
- diffusivity
- temperature
- lithostatic pressure
- concentration of volatiles (e.g., H_2O)
- separation of bubbles from the melt

It is not only the formation of bubbles that leads to an increased rate of magma ascent. It may simply be the increasing buoyancy of a magma that precedes the phase of bubble formation. Negatively buoyant magma rise may represent the initial stage of many large explosive eruptions. Increasing buoyancy can be caused by a reduction

> ### Ke pah'ù nei ka honua – the Earth explodes
> *Last words of the tribal chief of Vulcanesia before the island was shattered into pieces and became engulfed by the waves of the Pacific ocean in August 1882*

in density that develops during differentiation. Alternatively, and possibly most commonly, it can also be driven by the increasing pressure in a magma reservoir, when newly arrived magma batches push the resident magma out. This has been documented repeatedly during eruptions of Kilauea volcano on Hawaii and also in many other volcanoes, including Mount St. Helens (see below). Basaltic magma that intrudes an older, more strongly differentiated and cooler magma, will heat the vintage magma, causing release of volatiles. This process commonly acts as the actual trigger of an explosive eruption as at Pinatubo 1991 (Chap. 13). The actual explosive acceleration can then be triggered by two overlapping processes:

- oversaturation of magmatic gases (Chap. 5),
- interaction of the rising magma with external water, resulting in a phreatomagmatic eruption (Chap. 12)

Rising viscous magmas can fragment when the bubble volume reaches about 65% along a *fragmentation zone* (109). Pumice with a higher vol-

▼ Fig. 10.1. Schematic of magma reservoir – eruption column system. Shearing and vesiculation increase during decompression as the magma rises. The two-phase gas-particle mixture generated at the fragmentation front may leave the vent with supersonic velocity and may erode the vent. The eruption column rising above the vent consists of a short gas thrust part driven by volatile expansion, the main convective stem incorporating air, and the umbrella region expanding laterally at the level of neutral buoyancy (in part after 352)

ume fraction of bubbles has expanded after fragmentation. The actual fragmentation is thus not only the result of strong expansion by formation of bubbles but of *film thinning*, which develops by shearing of the rising melt. The irregular distribution of shearing stresses in a conduit causes different degrees of fragmentation, commonly found in pumice deposits and within single pumice clasts. When the bubbles burst – possibly in a relatively sharply defined *fragmentation zone* – the gas pressure of the bubbles will drop rapidly and the rise velocity will suddenly increase. Degassed rhyolitic melts become rigid very quickly at viscosities in excess of 10^7 Pa s. Tubular bubbles, common in Plinian pumice deposits, possibly reflect high shear stresses in the rising magma.

Highly viscous silicate melts, e.g., dacites and rhyolites, can fragment by *brittle fracture* during sudden decompression under different conditions (5). Firstly, when tensions accumulate during propagation of a *rarefaction* wave (which migrates into the erupting magma during an explosive eruption; 172, 173). In this case, stresses in excess of the tensile stress of the magma are created and the wave propagates faster than the magma can react by viscous deformation. Hence, the magma fragments by brittle fracture. In the second case, a fragmentation wave is generated as rapid magma decompression affects the pressurized bubbles in the melt. The pressure difference between the bubbles and their surroundings is often high

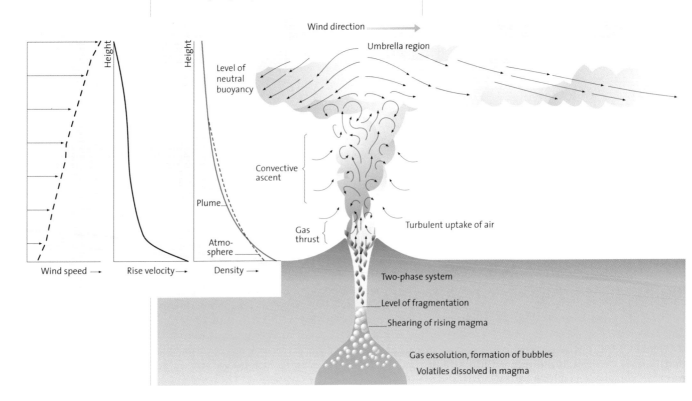

Wind direction

Umbrella region

Level of neutral buoyancy

Convective ascent

Plume

Gas thrust

Turbulent uptake of air

Atmosphere

Two-phase system

Level of fragmentation

Shearing of rising magma

Gas exsolution, formation of bubbles

Volatiles dissolved in magma

Height

Height

Wind speed →   Rise velocity →   Density →

enough behind the rarefaction wave that magma fragments by bursting of bubbles. In the third case, the now vesiculated and fragmented magma is no longer impermeable. Hence, the friction caused by fast flow of decompressed gases through irregular pore spaces can contribute to fragmentation.

If the bubbles burst in a relatively sharply defined fragmentation zone, the gas expands rapidly, depending on the load pressure, leading to strong acceleration. Above the fragmentation zone, the bubbly melt transforms into a suspension of melt particles in gas. The rise velocity will increase drastically because the viscosity of the suspension has decreased by several orders of magnitude, and the expanding gas-particle system can reach supersonic velocity (421). The mixture of gas and particles ejected from a crater is called an *eruption column*. Many types of eruptions can be characterized by their particular eruption column structure and dynamics.

Eruption Columns

Eruption columns consist of blebs and fragments of magma (molten or quenched solid particles), crystals, and country rock (lithic) fragments, dispersed in a continuous gas phase. This mixture is shot out of a crater vertically or subvertically with velocities up to several 100 m/s and a total density that is higher than that of the surrounding atmosphere. The jet will incorporate relatively cool atmospheric air on its sides. This air is mixed into the eruption column, becomes heated by the pyroclasts, and expands, decreasing the density of the rising mixture. At the same time, friction between the outer boundaries of the eruption column and air causes drag and initiates gravitational fallout of particles.

Sparks and Wilson (351) subdivided eruption columns into two parts. The lower part, called the *gas thrust region*, is where a mixture of pyroclasts and gas is jetted several 100 m to a few km into the atmosphere by initial acceleration. *Nozzle velocities* (i. e., the maximum velocity to which the mixture of pyroclasts and gas can become accelerated by the expansion of the magmatic gases) range from slightly over 100 m/s in Strombolian to more than 600 m/s in major Plinian and Vulcanian eruptions (421). The nozzle velocity depends mainly on the gas content of a magma and therefore on the explosive pressure in the fragmentation zone. The dense gas thrust is soon decelerated by losing momentum.

The gas thrust region can incorporate up to four times its mass of cold air at its margins. If enough air can be incorporated to decrease the bulk density of the jet to below that of the surrounding atmosphere, the mixture then rises as a *convective eruption column or plume* (356). Column rise will be accelerated by removal of large clasts, which decreases the overall column density aiding transition to convective rise. This transition is reflected in a drastic change in morphology from the strongly focused columnar gas thrust, to a cauliflower-shaped expanding convective eruption column. This convective plume has a density lower than the surrounding atmosphere, despite its load of pyroclasts, because the hot gas thrust has heated the entrained air. The positive buoyancy can transport convectively rising eruption columns to heights exceeding 30 km, until they reach air of the same density which is the *level of neutral buoyancy* (LNB). The eruption column will then expand laterally into the *umbrella region*, also called *mushroom cloud*, although the central accelerated part will overshoot for some distance until its lateral parts descend into the umbrella region (Figs. 10.1 – 10.4). The degree of lateral wind drift of the rising eruption column depends on the wind velocity, which varies significantly with height, latitude, season, weather pattern etc, as well as the *mass eruption rate* or the rise velocity of the eruption column (Figs. 10.5 – 10.6). The wind drift is most effective at the tropopause, which is characterized by strong, relatively constant wind currents (*jet stream*).

When an emerging eruption column does not incorporate enough air or not mix sufficiently with the air to reduce its total density to below that of the surrounding atmosphere, the rising jet will decelerate until a height where the velocity reaches zero. Because the plume density at this height is still greater than that of the surrounding atmosphere, the particle-gas suspension will fall back to the Earth's surface. In this way, the eruption column, or marginal parts of it, collapse, and the jet transforms into a *tephra fountain*, where descending pyroclastic mass flows form (Chap. 11).

The height of a convective eruption column is mostly determined by thermal energy which is a function of the mass eruption rate, i. e. the mass and magma temperature of vesicular lava blebs, crystals and gases, transported per unit time through a conduit. The total height is then determined by the buoyancy of the plume. It also depends on vent dimensions, and other parameters. Obviously, the processes are much more complex in nature because water vapor condenses (149), larger particles sink, conduits become unstable, and a whole range of other environmental factors come into play.

Pyroclastic eruptions and eruption columns range across a wide spectrum. Below, I will discuss two basic types, one representing basaltic and the other highly evolved magmas.

◄ Fig. 10.2. Eruption column rising above Mount St. Helens during 18 May 1980. Photo US Geological Survey J Postman

▼ Fig. 10.4. Frequency of different heights of eruption columns between 1975 to 1985 and heights of tropopause for equatorial and high latitudes

▲ Fig. 10.3. An approximately 14-km-high eruption column rising above the 3 108 m high Redoubt volcano (Alaska) on 21 April 1990. This spectacular eruption column was generated about 4 km away from the vent by gases and fine ash particles rising above the surface of hot pyroclastic flows that had spread along the foot region of the volcano. The upper umbrella has formed at 12 km height at the boundary between troposphere and stratosphere while the lower umbrella is at >5 km in the troposphere. Photo US Geological Survey courtesy C Nye

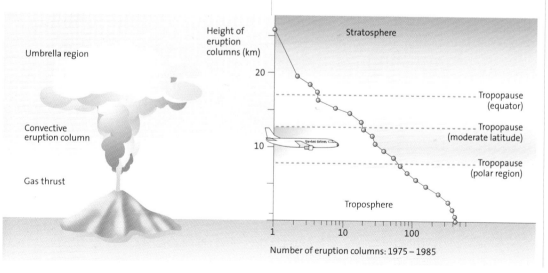

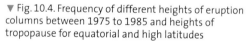

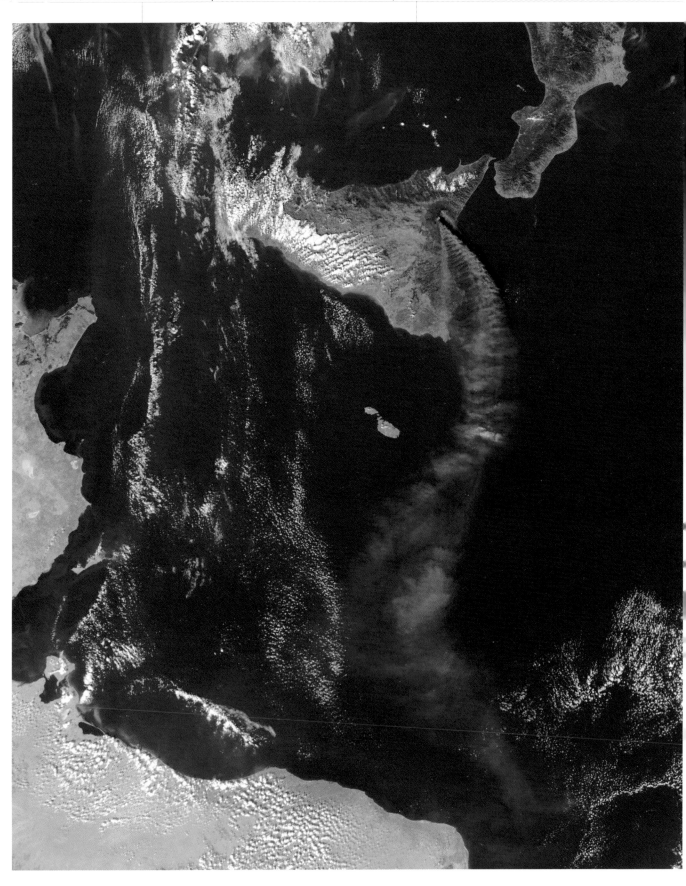

©esa 2001 - processed by ESA/ESRIN

◄ Fig. 10.6. Close-up satellite view of Etna volcano showing ash plume blown eastward and lava flows (*red*) issuing from several vents. Photo Landsat 5 TM processed by ESA/ESRIN

Strombolian and Hawaiian Eruptions

Fire fountains of low viscosity basaltic lava represent one end of the spectrum of eruption columns (Fig. 10.7). Basaltic magmas can erupt along fissures, but the magma will become concentrated into approximately cylindrical conduits during the course of an eruption, because this is the energetically most favorable ratio between volume and surface area. Basaltic lava fountains are comparatively low and their fragments are relatively large (centimeters to meters). The fragments lose little of their heat and accumulate rapidly around a conduit. Magma blebs may also coalesce quickly when falling to the ground, to eventually move away as lava flows. At higher gas content and larger eruption velocities, these fountains can grow to several hundred meters in height. The higher a fountain is, the greater is the cooling of particles during travel through the air. Particles on the outer envelope of such fountains cool especially fast and are deposited as scoria. Continuous and high lava fountains are called *Hawaiian*. Large lava volumes can be contained within Hawaiian lava fountains, which can rise up to 500 m above the ground.

Stromboli volcano (Aeolian Islands), ca. 220 km southwest of Naples, was called Lighthouse of the Mediterranean in ancient times, not only because it was permanently active, but also because it erupted at regular intervals. This type of activity is still typical although more energetic eruptions producing lava flows or flank collapses occur occasionally, such as in December 2002 and January 2003, for the moment interrupting or terminating crater activity. At intervals of 10–20 min, lava blebs are thrown out of several small orifices in the main crater of Stromboli. Rising gases tear apart the cooler, more viscous skin of the slowly rising lava column, which is pressed upward by large expanding gas bubbles. During these explosive fragmentations, lava blebs are thrown out vertically and laterally from the crater. Only a few small particles (ash and small lapilli) are produced. Because the large gas bubbles burst only a few tens of meters below the Earth's surface, their energy is low and lava blebs are thrown at most a few hundreds of meters. Lava flows are occasionally produced at Stromboli, but the normal eruptive mechanism is eruption of

◄ Fig. 10.5. Satellite view of the ash cloud erupted from Etna volcano (Sicily) on 21 July 2001. The direction of the plume changes with that of the prevailing winds from east to south reaching northern Africa. The distance from Etna to Africa is 500 km. Image courtesy Jeff Schmaltz. MODIS Rapid Response Team at NASA GSFC (MODIS: Moderate Resolution Imaging Spectrometer on Aqua Satellite)

gases with relatively little melt. This style of eruption is called *Strombolian*.

Many different physical parameters govern the eruption of basaltic magma (422, 423). An example for the changes in velocity of basaltic magma as it rises through fissures 0.2–0.6 m wide is presented in the following calculations. A basaltic magma can reach an overpressure up to 200 bar due to its low density compared to the surrounding rock (lithostatic pressure), which causes the magma to rise. This magma (density 2.8 g/cm³, viscosity 10³ Pa s, H_2O content 1 wt %) rises at a mass transport rate of $2-3\times10^8$ g/s through a 6-m-diameter circular conduit in crustal rocks (with density of 3 g/cm³). Gas bubbles will begin to form about 800 m below the Earth's surface. The rise speed is now about 2.4 m/s and will increase to 18 m/s as the conduit diameter narrows to 3 m, until a depth of about 100 m. At this depth, the gas bubble fraction will exceed 75 % of the total volume. The magma is then torn apart and forms a gas-particle mixture, which results in an extreme reduction in the viscosity of the rising system. The strong acceleration leads to a transition from subsonic to supersonic velocity. Assuming that the diameter of particles is less than 2 cm, the gas-particle mixture will reach the Earth's surface with a velocity of 160 m/s. Assuming that the melt may already be torn apart at some 65 vol % bubbles (see above), the depth at which a two-phase system forms will increase accordingly. In newer models (see above) film thinning i.e., liquid stretching and disruption control melt fragmentation more than bubble bursting.

The most important difference between the two classic forms of basaltic eruptions, Strombolian and Hawaiian, appears to be the rise speed of the magma relative to the rise velocity of the bubbles forming within it. When the velocity of the rising magma is less than 0.1 m/s, the ascent rate of the rapidly growing bubbles determines the system behavior. Growth rates of bubbles by diffusion and decompression is a topic of much current research interest. A basaltic magma with 0.7 wt % CO_2 will form bubbles with diameters of about 1 mm in about 40 h at the base of the Earth's crust (30 km). Bubbles 1 mm in diameter about 1.5 km below the Earth's surface, will reach a diameter of 1 m in 3 to 15 h in the rising magma column by diffusion and decompression. Large bubbles rise faster than small ones, overtake them, and can coalesce. This volume increase, in turn, accelerates the rise speed to such a degree that a chain reaction can develop, leading to an explosion. In Strombolian eruptions, the magma column rises at relatively low velocities of order 0.5 m/s; velocities of lava particles thrown out of the crater are around 100–400 m/s and overpressures in the bubbles range from 1 to 4 bar.

▶ Fig. 10.7. Basaltic lava fountain rising to 200 m during 31 December 1969 eruption of Mauna Ulu (Kilauea volcano, Hawaii)

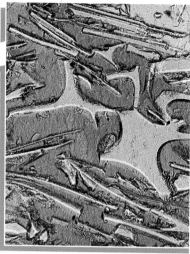

▲ Enlargement of central part of accretionary lapillus, showing rhyolitic glass shards (light-colored), about 50 μm in diameter, embedded in calcite cement

content and viscosity of the melt, both being much higher in silicic than in basaltic magmas. Other factors include the conduit morphology, conduit radius, eruptive rate and volume of a magma, all of which are larger than in most basaltic eruptions, the flood basalts excepted (425).

To better understand different stages in the evolution of a Plinian eruption, let us assume the following, simplified average conditions (425): magma temperature 850 °C, density 2.3 g/cm³, gas content (H_2O) 5 wt %, viscosity 10^5 Pa s. We assume particle sizes generated during fragmentation in the upper part of the magma column (as discussed earlier in this chapter) to be small

▼ Fig. 10.13. Proximal pyroclastic deposits from Laacher See volcano (Eifel, Germany), erupted 12 900 years ago. The lowermost light-colored fallout deposits are overlain by a light-colored, about 12-m-thick ignimbrite, which had become eroded (*arrows*) by subsequent ash flows (eruptive and erosional phases having alternated repeatedly). Erosion was concentrated in a paleovalley, shown in cross section in the left side of the large wall of the quarry. Thinner, darker-colored ash-flow deposits within and above three well-sorted pumice fallout layers (*BA*) are pyroclastic flow deposits that thin in the overbank facies to the *left* and thicken towards the central axis of the paleo-valley (*right*). Flow direction towards the observer. Wingertsberg (Laacher See area, Germany)

◀ ▲ Fig. 10.12. Photomicrograph (diameter about 2 cm) of a rhyolitic tuff with y-shaped glass shards. An accretionary lapillus consisting of coarse glass shards in the center is surrounded concentrically by a thin rim of fine-grained ash. The tuff has been cemented by large calcite crystals (colorful in polarized light) formed during diagenesis. Miocene Ellensburg tuff (Columbia River plateau, Washington, USA)

BA

Ignimbrite

▲ Fig. 10.14. Volcanic arc forming the Cascade volcano chain between northern California and southern British Columbia. Mount St. Helens in southern Washington is highlighted

crater lips. The diameter could be 1.6 m at a depth of 320 m and may be enlarged to 15 m on the surface, whereby the exit velocity would increase to 275 m/s and the exit pressure decrease to 8 bars.

The Mount St. Helens eruption of 18 May 1980 was the first Plinian eruption documented and analyzed in detail (42, 195, 387). The analysis of the composition of the magma and its gas content allowed a new level of precision in information about the state of the magma prior to and during the eruption. The eruption of El Chichón volcano in Mexico, during which about 2 000 people perished, occurred two years later. However, this was not studied to the same detail because the volcano erupted in a remote area. The major importance of the El Chichón eruption lies largely in its sulfur-rich magma composition and the climatic impacts caused by stratospheric aerosols (Chap. 14).

The Eruption of Mount St. Helens on 18 May 1980

Mount St. Helens in southern Washington State (USA) was the most beautiful Cascade volcano prior to the destruction of its summit on 18 May 1980. It is part of the impressive chain of young andesite/dacite volcanoes, many about 50 km apart, extending from northern California into southern British Columbia (Canada) one of the most impressive volcanic chains on Earth (Fig. 10.14). Mount St. Helens is the most active volcano in this chain, as demonstrated by the abundance of its young ash layers found in soils, lake sediments, and in bogs in the northwestern United States and southwestern Canada. Based on this evidence, Crandell and Mullineaux (68) predicted that Mount St. Helens would probably erupt during the next 100 years, perhaps within the twentieth century, even though there was not a shred of evidence of an impending eruption when they published their classic paper in 1978:

"During the last 4 500 years, Mount St. Helens has never been inactive for more than 500 years. In fact, it is the most active volcano in the Cascades. The volcano's behavior pattern suggests that the present quiet interval will not last as long as 1 000 years. Instead, an eruption is more likely to occur within the next 100 years, and perhaps even before the end of the century. Because of the variable behavior of the volcano in the past, we cannot be sure whether the next eruption will produce lava flows, pyroclastic flows, tephra or volcanic domes or some combination of these."

This clear and brave forecast became a milestone in the history of volcanology, because the volcano generously proved them right and erupted in a dramatic series of events barely 2 years later (195, 387).

enough to be in thermal equilibrium with the accelerated gases. The velocity is so high that no energy is lost, nor do significant numbers of gas bubbles form during the transport, which lasts only a few seconds. Under this set of conditions, the first bubbles would form at a depth of 5 km. Bubbles would reach a volume of ca. 77 % at 1.6 km below the Earth's surface and then form a gas-particle dispersion. The velocity at the surface would be about 160 m/s, the pressure 56 bar. The original eruptive fissure will soon become funnel-shaped as the high-pressure suspension erodes the

10

◄ Fig. 10.15. Swelling of the northern flank of Mount St. Helens between 27 April and 17 May 1980 (381)

▼ Fig. 10.16. Eruptive phases of Mount St. Helens from 18 May 1980 to July 1980 (234)

The first signs of an impending eruption of Mount St. Helens (which had been quiet since 1857) included a period of abundant small earthquakes. These began on 16 March 1980 and were followed by a larger earthquake of magnitude (M) 4.7 on 20 March at 15:47 h. Between 27 March and 18 May, several small eruptions occurred. These, however, were probably purely steam explosions (phreatic eruptions, Chap. 12), during which fragmented country rocks were erupted but no new magma although small particles of quenched juvenile magma are commonly difficult to identify. Beginning about mid-April 1980, the northern flank of the volcano expanded by up to 2.5 m/day, probably because a rising magmatic intrusion was pushing the northern flank outward (Fig. 10.15). On 18 May at 08:32 h local time, the bulging northern flank of Mount St. Helens became destabilized by an M 5 earthquake. The entire top of the volcano (about 12.7 km³) collapsed and cascaded down the northern flanks of the edifice at a speed of 200–250 km/h (Fig. 10.16).

A new crater, 700 m deep and open to the north, had been carved out of the towering cone. The rock avalanches accumulated to form a 70-m-high plateau at the foot of the volcano and filled the canyon of the Toutle River to a depth of ca. 50 m for 20 km to the northwest (Fig. 10.17). The debris avalanches were accelerated so enormously during collapse of the top of the volcano that they were able to race up more than 400 m on mountain slopes as much as 6 km away from the former top of Mount St. Helens. The level of Spirit Lake to the north of the volcano rose by about 60 m when huge masses of debris cascaded into it (Fig. 10.18).

Fortunately, "only" 60 people lost their lives. If the entire area around the volcano had not been cordoned off, perhaps thousands of people could have perished. Nobody could have predicted the exact time or style of the gigantic eruption.

Several different processes interacted and followed each other, resulting in a complex sequence of events during the eruption on 18 May. Rising magma caused expansion and destabilization of large rock masses in the volcano. Large earth-

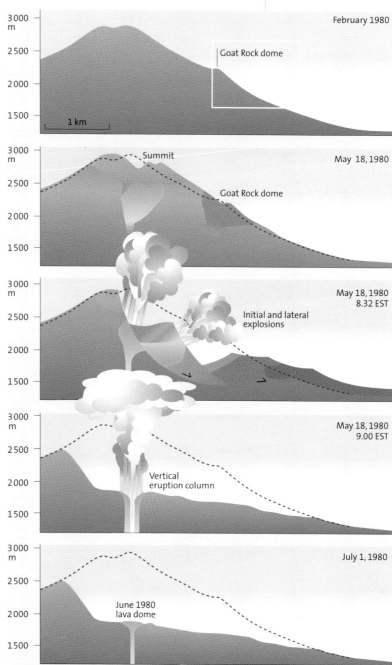

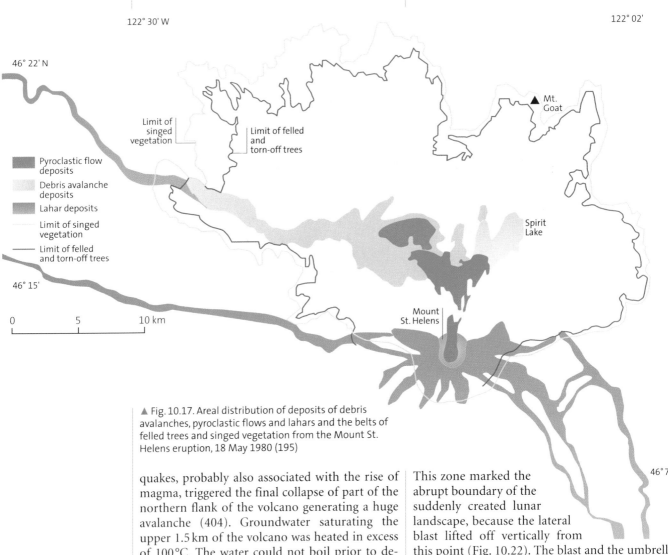

▲ Fig. 10.17. Areal distribution of deposits of debris avalanches, pyroclastic flows and lahars and the belts of felled trees and singed vegetation from the Mount St. Helens eruption, 18 May 1980 (195)

quakes, probably also associated with the rise of magma, triggered the final collapse of part of the northern flank of the volcano generating a huge avalanche (404). Groundwater saturating the upper 1.5 km of the volcano was heated in excess of 100 °C. The water could not boil prior to decapitation of the volcano, however, because of the high lithostatic pressure. When the northern flank suddenly collapsed, the overheated water flashed into steam, thereby developing enormous explosive energy. Steam expansion led to further fragmentation of rocks deeper in the volcano, including the upper part of the rising magma intrusion. A giant lateral blast expanded northward at 100–400 km/h (Chap. 11), destroying everything in an area of about 550 km² (Figs. 10.19–10.22). The enormous power of this blast broke off and splintered all trees 1–2 m in diameter in the densely forested area and uprooted many trees (Fig. 10.19). Trees were felled in a zone 10–15 km away, aligned with the blast direction (Fig. 10.20). The blast energy of the laterally transported ash was sufficient to topple over large tractors and other logging equipment, many of which were also buried by the ash. Pine needles were only singed by hot gases on the margin of the devastated area.

This zone marked the abrupt boundary of the suddenly created lunar landscape, because the lateral blast lifted off vertically from this point (Fig. 10.22). The blast and the umbrella cloud are discussed in more detail in Chapter 11.

Collapse of the volcano's northern flank, which formed the lid to the rising magma column, resulted in a sudden pressure reduction of the underlying magma. This allowed the gases in the rising magma to expand, burst, and generate an eruption column that rose at about 09:00 h local time with an average velocity of 50 km/h, to a height of 25 km (143). The umbrella region expanded at 09:20 h when the column rose to 14 km. Column height oscillated between 14 and 19 km above the Earth's surface during the nine subsequent hours. About 20 pyroclastic flows were spawned by the collapsing eruption column during 18 May (Figs. 10.23–10.24). They were channeled by the huge notch generated by the initial collapse of the northern sector of the volcano. Pyroclastic flows transported about 80 % of the total erupted volume of c. 0.25 km³ (43). The main mass of tephra in the eruption column was

▲ Fig. 10.18. Deposits of Mount St. Helens eruption of 18 May consisting of large debris avalanche blocks and hot pyroclastic flow deposits or units, in which blocks of glacier ice or pockets of lake water generated steam explosions and subcircular craters. The left part of the photograph shows rafts of trees torn off by the initial blast and blown into Spirit Lake. The lake level rose by 60 m after the eruption as a result of the influx of voluminous debris avalanches

◄ Fig. 10.19. Stumps of fir trees, remaining after trees had been torn off by the powerful blast of the Mount St. Helens eruption on 18 May. Blast direction from left to right

► Fig. 10.20. Trees blown down by the initial blast of the Mount St. Helens eruption of 18 May, 1980. Note cars (arrow) for scale

▼ Fig. 10.21. Deposit of the initial blast of Mount St. Helens eruption of 18 May 1980. The deposit consists of about 50% of gray juvenile dacite lapilli mixed with multi-colored fragments of older rocks and pieces of wood from trees felled by the pressure wave

blown by the prevailing winds to the east over the Northamerican continent (Figs. 10.25, 10.26). Ten kilometers east of the volcano, the thickness of the ash layer was 50 cm, but decreased rapidly farther eastward. The base of the new crater was about 1 000 m deeper than the former top of the volcano (Fig. 10.27).

The eruption of Mount St. Helens led to the foundation of a new observatory, the David Johnson Cascades Volcano Observatory, named after a volcanologist who was killed by the initial blast. A monumental publication that was produced in the incredibly short time of 18 months after the eruption (195) presented the results of the first basic studies of the eruption.

▼ Fig. 10.22. Margin of the area devastated by the blast of the Mount St. Helens eruption of 18 May 1980. The blast expanded into the dense forest, foreground and opposite slope of Clearwater Creek. The zone of felled and oriented trees is bordered by a brownish zone, in which the trees are still standing, while the needles were singed by the hot blast

◄ Fig. 10.23. Pyroclastic flows descending Mount St. Helens during an eruption of June 1980. Photo courtesy PW Lipman, US Geological Survey

▶ Fig. 10.24. Canyon eroded in pyroclastic flow deposits erupted from Mount St. Helens in June 1980

Analysis of the Eruption Dynamics

The dynamic changes of the physical and chemical properties of the volcano-magma system prior to, and during, the eruption of Mount St. Helens on 18 May have been studied extensively (41–43, 287, 297 among others). The results are based on geophysical, geochemical, and experimental petrology data, as well as theoretical studies, in a comprehensive quantitative analysis of processes that occur during the eruption of an explosive volcano.

The diameter of the magma chamber prior to eruption was about 1.5 km at a depth of 7 to 13 km, the zone free of earthquakes (Fig. 10.28). The total magma volume at this depth amounted to about 10–20 km³. A small portion of the magma erupted on 18 May in Plinian eruptions had the following properties prior to eruption (Table 10.1):

▼ Table 10.1: Physical parameters of Mount St. Helens eruptions of 18 May 1980 (42) (see also Fig. 10.29)

Composition:
dacite; 40% crystals (plagioclase, hypersthene, amphibole, titanomagnetite, ilmenite); 60% glass (melt)

Water content of magma column at a depth of 7 km prior to eruption:
4.6 wt.% H_2O (analyzed in melt inclusions in phenocrysts)

Temperature:
920 to 940 °C (determined using coexisting Fe/Ti-oxides)

Viscosity:
2.3×10^6 Pa s = magma + crystals

Volume of magma at a depth of 7 to 13 km assuming a vent diameter of 1.5 km:
10 to 20 km³

MSR (magma supply rate) = laminar rise of magma up to the zone of fragmentation by degassing.
MSR = 1.9×10^7 kg s^{-1}

MDR (magma discharge rate) = fragmented material.
Low viscosity, turbulent transport. At MSR 1.9×10^7 + 15% rock fragments = 2.28×10^7 kg s^{-1} MDR

Velocities:
U_1: ascent velocity below the zone of gas saturation: 1 m s^{-1} assuming a vent diameter of 47 m
U_2: ascent velocity between zone of gas saturation and zone of fragmentation: 1.1 to 2.5 m s^{-1}
U_3: exit velocity (all H_2O liberated): 330 m s^{-1} assuming a vent diameter of 105 to 135 m

Erupted volume on 18 May 1980:
1.3 km³ tephra + 15% rock fragments = 0.213 km³ magma (DRE = dense rock equivalent)

Eruption on 18 May 1980:
8:30 until 17:30 = 9 h

Eruption column:
height 25 km at 9 h, thereafter oscillating between 14 and 19 km

Type of erupted material:
8:30 h debris avalanche und blast deposit. Thereafter fallout and 18 pyroclastic flows (flow units)

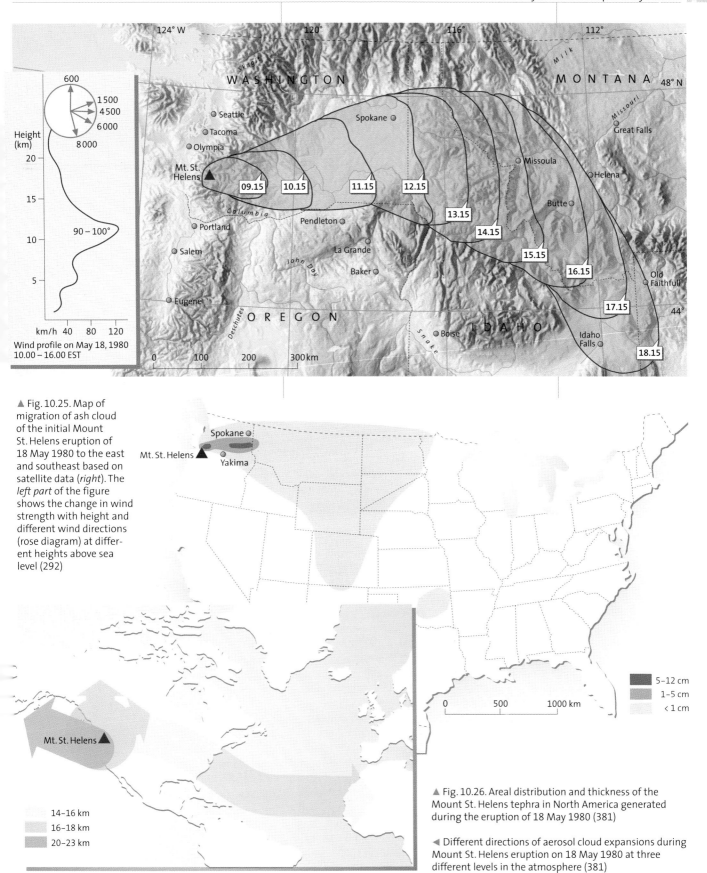

▲ Fig. 10.25. Map of migration of ash cloud of the initial Mount St. Helens eruption of 18 May 1980 to the east and southeast based on satellite data (*right*). The *left part* of the figure shows the change in wind strength with height and different wind directions (rose diagram) at different heights above sea level (292)

▲ Fig. 10.26. Areal distribution and thickness of the Mount St. Helens tephra in North America generated during the eruption of 18 May 1980 (381)

◄ Different directions of aerosol cloud expansions during Mount St. Helens eruption on 18 May 1980 at three different levels in the atmosphere (381)

▲ Fig. 10.27. Crater generated by the large eruption of Mount St. Helens on 18 May 1980. The difference in height between the crater floor and the original top of the volcano amounted to about 700 m

The magma was undersaturated with respect to H_2O, i.e. the partial pressure of H_2O was about 0.5–0.7 of the lithostatic pressure. This is evidenced by the observed phenocrysts at H_2O saturation of the magma $(P_{H_2O} = P_{total})$ being unstable based on experimentally determined phase equilibria. The phenocrysts crystallized in a magma that contained H_2O and CO_2, at a pressure around 220 ± 30 MPa, corresponding to a depth of about 7.2 ± 1 km. This depth estimate is remarkably similar to that deduced from the hypocenters of earthquakes beneath the volcano registered on May 18 (7–9 km). These earthquakes were probably generated at the top of the rising dacitic magma.

The rate of magma supply to the level of fragmentation corresponded to the mass eruption rate, or the rate of the material erupted from the crater during the main, 9-hour-long Plinian phase on 18 May. This was shown by the relatively constant height of the eruption column throughout the day. A magma supply rate of 1.49×10^7 kg/s for 18 May, was calculated based on the duration of the eruption and the amount of juvenile magma erupted (about 0.25 km^3 DRE) (43). This eruption rate lies within the range of the mass supply rate 5.0×10^6 to 2.1×10^7 kg/s, calculated using the height of the eruption column (measured by radar) based on (424). From these data and assuming a viscosity of 2.3×10^6 Pa s calculated at 4.6 wt % H_2O and 40 vol % crystals, a rise velocity of the non-fragmented magma (U_1) from 1 m/s at a conduit diameter of 47 m was derived. This in turn corresponds well to a rise velocity of 0.6–0.7 m/s inferred from seismic data (297).

The exit velocity of tephra particles can be calculated according to (419) with the following parameters: 0 = effective density of the gas, U_0 = eruptive velocity, B = product of radius and density of the largest particle, extrapolated to the vent and C = friction coefficient (close to 1, when U_0 is appreciably below the speed of sound).

Using the known grain-size distribution, and assuming that all H_2O in the melt (4.6 wt %) was in a gas phase at the eruption of the magma with 40 vol % crystals (corresponding to c. 2.3 wt % H_2O in the erupted total tephra gas mixture),

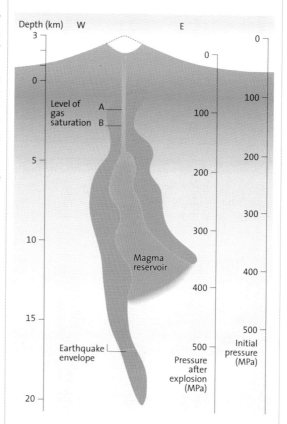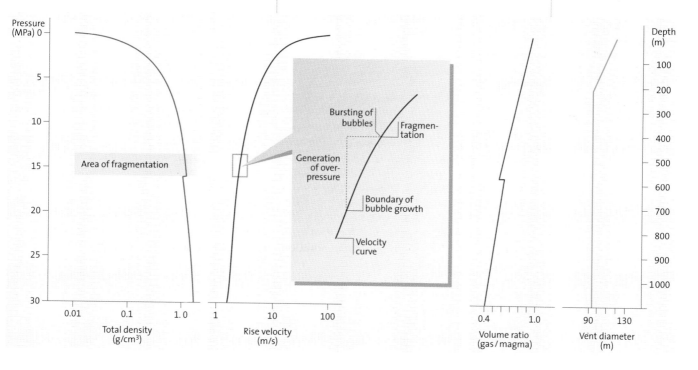

implies nozzle velocities of 200–330 m/s. Carey and Sigurdsson reconstructed the dynamic evolution of the Mount St. Helens magma on 18 May based on the following data.

At a depth of >4 560 m (125 MPa), the magma probably rose because of its low density, since it was undersaturated with respect to H_2O. Whether or not the rise was caused by absorption of water or enrichment of water during crystallization, is unknown. At about 4 560 m, the magma became water-saturated so that initial bubbles could nucleate. However, only after rising to a depth of 580 m below the surface (16 MPa), did the magma contain 75 vol% vesicles (which, according to Sparks (351), is the critical boundary for fragmentation of a melt by bursting of bubbles, a condition now interpreted differently see Chap. 4). In the depth range between 4 560 and 580 m, the velocity of the rising magma column increased from 1.1 to 2.5 m/s, not considering the drastic increase in viscosity of the melt containing bubbles.

At a depth of <580 m, overpressure was generated in the bubbles due to the difference between the internal bubble pressure and the decreasing lithostatic pressure. This overpressure increased until the tensile stress of the bubble walls was exceeded. The interval between the end of bubble growth and the critical overpressure was, according to Carey and Sigurdsson, between about 510 and 560 m (Fig. 10.29). On the other hand, the depth extent of this fragmentation zone and the evolution of the complex processes during tearing-apart of the melt are still highly specula-

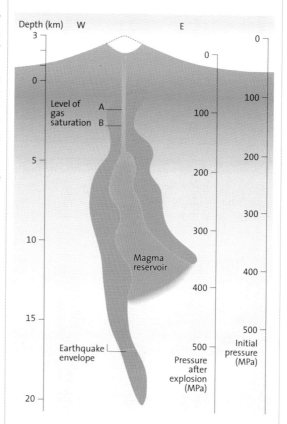

▲ Fig. 10.28. Postulated magma reservoir with envelope of earthquake hypocenters beneath Mount St. Helens after the eruption of 18 May 1980. Interval between the termination of bubble growth and critical overpressure between about 510 and 560 m below the Earth's surface (*A* and *B*) (42, 297)

▼ Fig. 10.29. Change of critical parameters of the shallow dacitic magma during the 18 May 1980 eruption of Mount St. Helens (42)

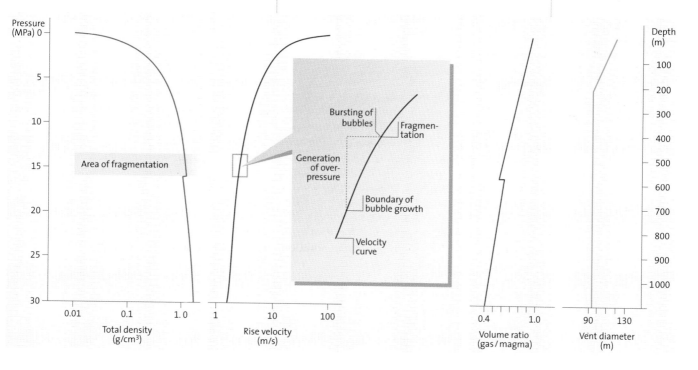

tive. Density and viscosity of this system drastically decreased above the zone of fragmentation. Gas formed a continuous phase between the magma blebs. The flow regime changed from a laminar into a turbulent one, and the velocity increased rapidly. This also accelerated erosion of the crater walls. In total, about 15 wt% country rock fragments were mixed into the dacite tephra.

In order to reach an eruption velocity of 200–300 m/s (which can be calculated from the areal distribution of the tephra layers and the amount of gas liberated) the crater diameter must have been between about 105 and 135 m, at an ambient pressure of 0.1 MPa. The vent, in which a dome started to grow immediately after the Plinian eruption, had a diameter of about 200 m after 21 May (234). Hence the vent must have become enlarged following the Plinian phase, e.g., by slumping of the crater walls.

The most important trigger of degassing and pyroclastic fragmentation was apparently the sudden decrease in pressure in the upper part of the magma column following collapse of the roof of the volcano. The zone of fragmentation therefore migrated suddenly downward by about 1 km, equilibrating with a new lithostatic gradient. A huge magma volume thus suddenly became oversaturated in the gas phase. Perhaps the relatively high mass eruption rate and the great height of the eruption column (24 km) during the first half hour of the eruption on 18 May can be explained by this sudden disequilibrium at the beginning of the explosive discharge.

The scientific harvest of the 18 May eruption also comprises a better understanding of volcanic blasts and airborne umbrella clouds that rise from blasts and pyroclastic flows (see Chap. 11).

Summary

The three most common types of pyroclastic eruptions are those called Strombolian, Hawaiian, and Plinian, each being characterized by a particular style of eruption. Several different types of eruptive styles may occur, however, during the course of a single eruption. Intermittent Strombolian eruptions, resulting from the episodic rise of large gas bubbles in low viscosity basaltic magma, lead to the creation of scoria cones. The impressive basaltic lava fountains, which at Hawaii, Etna (Sicily) or Cerro Negro (Nicaragua), can reach heights that exceed 100 m, may be caused by the sudden emptying of the upper part of a magma chamber in which gases had accumulated during extended residence. They may also start in deep-seated magma reservoirs, while more strongly degassed lava in shallow reservoirs may be erupted less explosively. Fire fountains in Hawaii and Etna commonly result in lava flows. A rising magma can become fragmented pyroclastically when the bubble volume has reached about 65 vol%, whereby the fragmentation is governed largely by shearing in the rising silicate melt.

Plinian eruption columns consist of two parts. The lower gas thrust region, driven by the acceleration of the decompressed gases, rises only a few hundred meters. The bulk of a spectacular eruption column (which can rise to > 40 km) is called the convective region. A column rises because of the positive buoyancy of the mixture of gas and particles, plus incorporated and heated ambient air. One of two main depositional mechanisms of tephra in Plinian eruptions is fallout of pumice, lapilli or, at larger distances, ash to form widespread tephra lobes. Pyroclastic flows that result from the collapse of dense and unstable parts of eruption columns can travel laterally over the ground away from the volcano. Such pyroclastic flows are the topic of the following chapter.

The eruption of Mount St. Helens triggered a major breakthrough in volcanology. In the 1970s and early 1980s, questions concerning the cause and course of explosive eruptions had become a central topic in physical volcanology. Modern methods for analyzing and monitoring active volcanoes (especially seismically, geodesically and gas analyses) had been available since the 1960s. They were mainly developed at very active Kilauea volcano on Hawaii, because of the excellent logistics. The scientists of the US Geological Survey, who had learned their trade on Kilauea volcano, were able to apply their experience to the eruptive processes of Mount St. Helens, and other researchers could build on those results. A second case history, that of Laacher See volcano, which produced the most widely dispersed, geologically young Plinian tephra deposit in central Europe, will be discussed in the following chapter.

Pyroclastic Flows, Block and Ash Flows, Surges and the Laacher See Eruption

World-famous volcanic eruptions that have become embodied into our cultural heritage include the A.D. 79 eruption of Vesuvius, the 1883 eruption of Krakatau, but especially the eruption of Montagne Pelée on 8 May 1902 at 07:52h on the Caribbean Island of Martinique. This disaster, the major natural catastrophe that heralded the twentieth century, is a symbol for the destructive potential of nature, as well as for the hubris of man, who fails to take the power of nature seriously. A famous example of the folklore of volcanic disasters (sometimes only loosely based on facts) is the story of one of the few inhabitants of St. Pierre, Augustus Ciparis, who survived the devastation of the town being incarcerated in the dungeon of the town jail (Fig. 11.1). Witness reports include observations of survivors on ships anchored in the harbor of St. Pierre, some of which were capsized due to the enormous force of the blasts. Scientists, who had rushed to the scene, for the first time observed and described glowing clouds speeding down the flanks of a volcano with velocities exceeding 200 km/h greatly expanding in transit. Lacroix (188), who studied the effects of the eruption in detail, called them *nuées ardentes*, a term used first in the nineteenth century by Fouqué, but now replaced by the terms pyroclastic flow or pyroclastic density current.

The utter destruction of the city of St. Pierre was mainly the result of powerful hot blasts that carried little ash (Fig. 11.2), something that is often overlooked. The glowing avalanches themselves, carrying the bulk of the material from the collapsing dome of the volcano, were restricted to the valley of the Rivière Blanche, 3 km northwest of St. Pierre (Fig. 11.3).

It was not until many years later that scientists realized that nuées ardentes are actually composed of two parts. A basal *ground-hugging avalanche*, in which most particles are transported, is

> Surrounded as it is by high and rugged mountains, the most striking feature of the conformation of the Valley of Ten Thousand Smokes is the flatness of its floor. One could ride a bicycle for miles along its smooth surface, and there are many places between the lines of activity that would be ideal landing fields for airplanes.
>
> *Robert F. Griggs, The Valley of Ten Thousand Smokes, Washington, 1922*

No airplanes have, to my knowledge, landed in the VTTS as it is commonly known

◀ Fig. 11.1. Prison cell in the center of St. Pierre (Martinique) in which the celebrated Cyparis survived the hot blast of May 1902

► Fig. 11.2. Deposits of several hot density currents that swept over St. Pierre between May and August 1902. Currents traveled from left to right destroying the upper structure of a house of which some walls remained (*left*). Accumulation of broken tiles and piles of charcoal in the lee of the house (*arrow*).

► Fig. 11.3. Block-and-ash flow deposit in the Rivière Blanche canyon. The flow has eroded and uplifted a chunk of muddy deposit rich in plant material (to the right of scale *arrow*) laid down by several mudflows that rushed through the valley during the early phreatic stages of the eruption in 1902

hidden by an overriding billowing *cloud* of hot gases and fine ash particles, which can rise tens of km into the atmosphere (Figs. 10.23, 11.4). The glowing avalanches themselves are generally restricted to valleys but the hot pressure waves (blasts) traveling ahead of them can expand sideways and race across the slopes of valleys and the

flanks of the volcano. Such hot blasts form one of the most formidable hazards associated with explosive volcanic eruptions.

Hot *pyroclastic density currents* have become a prime object of volcanological research. They represent one of the major volcanic hazards and their velocities can exceed 300 km/h. Some travel far-

◀ Fig. 11.4. Pyroclastic block flows, generated on 3 June 1991 during collapse of part of the dome of Unzen volcano. The pyroclastic cloud expands over the margins of the town of Shimabara (houses in foreground) (Kyushu, Japan). Photo courtesy Nagasaki Photo service

◀ Fig. 11.5. Flat-topped ignimbrite sheet erupted in 1912 filling the ca. 20-km-long Valley of Ten Thousand Smokes (VTTS) (Alaska)

ther than 100 km and cover vast areas the size of some of the smaller US states (Fig. 11.5). Moreover, they often speed down the slopes of a volcano with little noise, despite their deadly destructive power. Pyroclastic density currents of various types have occurred during more than half of all present-day explosive volcanic eruptions. Deposits such as those produced by the many eruptions of Montagne Pelée throughout 1902 that destroyed St. Pierre, have only small volumes, much less than 1 km³. Huge pyroclastic flow deposits are known from the geological past with volumes exceeding 5 000 km³, ten times the volume of large stratocones. These large-volume pyroclastic flows

◀ Fig. 11.6. A light-colored flow unit of ignimbrite A (Gran Canaria) overlain (*upper part* of photograph) by darker, more mafic flow unit. The compacted, fiamme-like pumice lapilli differ in composition. The dark fiamme are trachytic and the light-colored ones rhyolitic. Barranco Medio Almud (Gran Canaria, Canary Islands)

are associated with calderas, up to tens of km in diameter, which develop above large, commonly compositionally zoned magma reservoirs (Chap. 9; Figs. 3.13, 11.6, 11.7).

In this chapter, I will provide a brief glimpse into the nature and origin of pyroclastic flows *sensu lato* and their deposits, with a detailed discussion of the late Quaternary eruption of Laacher See volcano in Germany, in many respects a classical example of a Plinian explosive eruption.

Some Historical Notes

Soon after Lacroix described what he called nuées ardentes, his model was applied immediately to pumice-rich, massive deposits, obviously transported along the ground (the Brohltaltrass) that had been generated during the Late Pleistocene Laacher See eruption (400) (Fig. 10.13). The origin of these deposits had been debated throughout the nineteenth century, with most workers having assumed that they represented volcanic mud flows (lahars). The name *trass* was applied in the scientific literature of the nineteenth and early twentieth century for similarly massive, pumice-rich flow deposits. Despite the pioneering work of Lacroix and the application of recent eruption models to late Quaternary deposits, and despite the famous observations by Perret (260) (who built himself a hut on the slope of Montagne Pelée to closely observe nuées ardentes for more than 10 years), it was not until the 1960s when the worldwide occurrence and importance of pyroclastic flows and their deposits became recognized.

The early research on the deposits of pyroclastic flows is one of the fascinating stories in the science of volcanology. Von Fritsch and Reiss (406) described strange volcanic rocks with flame-like structures from the area around Arico (Tenerife, Canary Islands). In hand specimen, these had the appearance of pyroclastic rocks consisting of individual particles, although their relationship in the field clearly suggested that they had been transported along the ground, similar to lava flows. In their own words, translated from the German:

"*The eutaxites* (the name given by Fritsch and Reiss to these rocks) *show clear flames, expressed by the alternation of different materials, and for this they have the look of clastic rocks. The field relationships of the eutaxite masses known to us, quite clearly show that these interesting rocks flowed along the ground*" (pp. 414, 417).

Indeed, the Arico ignimbrite shows many aspects of pyroclastic flow deposits that are now known to be especially diagnostic (Fig. 11.8a–l). These authors also noted the occurrence of similar rocks in Indonesia, Azores, New Zealand, and other areas, now known to be classical areas of ignimbrites. Abich (1) called similar rocks covering vast plateaus in Armenia *tufolavas*, because they resembled both tuffs and lava flows. Wolf (431) compared eruptions of pyroclastic flows that had issued from Cotopaxi volcano (Ecuador) with overflowing boiling rice. Ten years after the eruption of Montagne Pelée, about 12 km³ of magma erupted from the volcano Novarupta (Alaska) in the form of pyroclastic flows, the most voluminous eruption of the twentieth century (Fig. 11.5). The famous explorer Griggs, who made it to the remote area with an expedition in 1916, named the flat area underlain by the hot mass, from which rose thousands of fumaroles, *Valley of Ten Thousand Smokes (VTTS)*. Most of the fumaroles had died out by 1923 (94, 151) (Fig. 11.9).

Marshall (209) recognized that widespread sheets of rhyolites in New Zealand, previously interpreted as lava flows, consisted of welded glass shards and explained their origin by fire clouds, i.e., fallout of hot magma particles. The term *ignimbrite*, coined by Marshall, is the term most commonly used today for such deposits. Recognition of the Bishop Tuff as an ignimbrite was another step forward in the study of ash flow tuffs (120). It covers an area of over 1 000 km² and erupted 760 000 years ago from the Long Valley Caldera (California). Seismic unrest that started

▲ Fig. 11.7. Unwelded compositionally zoned Quaternary KP-4 ignimbrite (ca. 120 000 old) erupted from Kutcharo caldera (Hokkaido, Japan). The light-colored basal material is rhyolitic, the dark part containing some mixed rhyolite-andesite lapilli

n 1983 in this volcanic structure may possibly erald another eruption (Fig. 11.10). Milestone apers by Smith (346, 347) and Ross and Smith 285) on the Bandelier Tuff in New Mexico mark he beginning of modern dedicated studies of ash ows and their deposits (Fig. 11.11). Many empirical, theoretical, and experimental studies published during the past 40 years illustrate that pyroclastic flows and their deposits have not lost their ascination to scientists (Fig. 11.12).

Research into pyroclastic flows and their deposits is still vigorous. Central questions include: o chemical and mineralogical gradients and layrs deduced from widespread compositional stratification in the deposits reflect a particular compositional architecture in shallow magma reservoirs? Or are the compositional changes largely the esult of processes during the evacuation of a magma reservoir, in the conduit, during multiple pisodes of roof subsidence of calderas or during mplacement? Which are the most important rocesses in the eruption column that causes it to ollapse or be manifested as low ash fountain? Can different mechanisms for the origin of pyroclastic flows and modes of transport be distinguished from each other? For example, how many ypes of pyroclastic density currents (surges sensu ato) can be distinguished from each other? Are most pyroclastic flows highly dilute or do they move as denser grain flows or do both modes grade into each other with distance from source? Which are the most important intrinsic and extrinsic factors that control the steady or unsteady eruption of huge masses of hot particles and gas and how are the different eruption scenarios reflected in the flow behavior of pyroclastic flows and thus their deposits? What is the origin of distinct eruption-controlled flow units and how are they related to episodes of roof subsidence during caldera formation?

A well-documented general discussion of the many terms and genetic concepts for the origin of pyroclastic flows and their deposits covering the last 100 years would require a long chapter on its own. Major difficulties arise from the difference between the interpretation of older deposits and attempts to decipher processes that govern the transport and sedimentation of actual pyroclastic density currents. Observing and analyzing processes within fast-moving, hot and extremely dangerous pyroclastic flows in which the major bedload is enveloped by billowing clouds of ash and

▶ Fig. 11.8a – l. The structural and compositional characteristics and facies changes with topography, thickness (temperature) and distance of the classical Quaternary phonolitic Arico ignimbrite on Tenerife (Canary Islands) span much of the diagnostic features of ignimbrites worldwide. Proximal and intermediate facies:

▲ a) basal Plinian fallout pumice lapilli deposit with a lower light-colored and an upper darker slightly more mafic layer. The top of the fallout sequence has been slightly eroded by surges, the channel being filled by cross-bedded surge deposits.

▲ b) Lower flow unit with a poorly welded base overlying remnants of eroded fallout and surge deposits is separated sharply from more mafic flow unit (hammer on boundary). Clear regionally mappable eruption-controlled flow units can be traced from near the source to most distal outcrops and reflect brief pauses in the eruption

▲ c) Local basal lag deposit consisting of boulders of light green phonolitic lava picked-up when the density current raced over very irregular terrain across the steep slopes of the island. Note large black obsidian (now hydrated) blebs

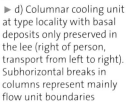

▶ d) Columnar cooling unit at type locality with basal deposits only preserved in the lee (right of person, transport from left to right). Subhorizontal breaks in columns represent mainly flow unit boundaries

▲ e) Moderately welded ignimbrite at type locality showing broad compositional range in juvenile lapilli types (green: more evolved, dark: more mafic) and rock fragments

▲ f) Strongly welded ignimbrite showing two flow units differing in grain size and composition, the lower lapilli being more evolved

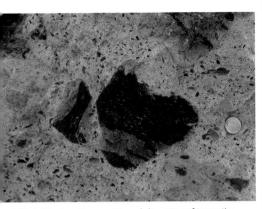

g) Distal facies: Compositional diversity of juvenile lapilli in distal poorly welded but zeolitized deposit showing banded pumice with darker more mafic and light green more evolved phonolite. Green lapillus at top is more homogeneous

▲ h) A lower light colored planar-bedded to massive entirely pumiceous but commonly slightly welded density current bed is regionally overlain by a lithic-enriched layer deposited at the base of the first major flow unit.
Initial fallout pumice has been largely eroded. Plant fragments are inclined downcurrent (to the right of hammer handle)

▲ i) Local lapilli-pipes in the near-shore area mostly restricted to the lithic-rich basal layer may have formed where the hot flow covered wet ground

▶ j) Poorly welded massive ignimbrite with light-colored base and unwelded pumice lapilli in massive matrix

▼ k) Layered structure in distal ignimbrite near-shore representing flow-induced surges during last stages of flow

▲ l) Accumulation of rounded pumice lapilli at the top of the unwelded near-shore distal ignimbrite

► Fig. 11.9. Degassing channel (fossil fumarole) in the upper part of the VTTS ignimbrite (Katmai National Park, Alaska, USA)

▼ Fig. 11.10. Light-colored rhyolitic fallout lapilli, overlain by reddish ignimbrite with basal pumice concentration of the Bishop Tuff erupted 760 000 years ago from Long Valley caldera (California). The caldera has become seismically active since 1983. Near Bishop (northern California, USA)

gas is clearly impossible. Moreover, methods and approaches used by scientists in their study of pyroclastic flow deposits vary widely and there is no consensus in opinions concerning major aspects of pyroclastic flow systems.

Terminology

The terminology of ground-hugging flows of hot, variably inflated, lava particles and variable amounts of magmatic gas, steam, and ingested air has been extraordinarily confusing. The main reason is the fact that scientists coined genetic terms for something they inferred to have happened– understandable in view of the problem of observing and analyzing pyroclastic flows in transit. In some cases, it took decades to realize that concepts of earlier scientists were based on incomplete or incorrect perceptions of processes. For example: when Marshall (209) coined the term *ignimbrite*, he did not realize that the hot lava particles that welded to form a compact rock could not have fallen from the sky. If this had been the case, particles would have cooled and been unable to weld together upon landing. They must have been

▲ Fig. 11.11. Columnar, welded Quaternary ignimbrite (Bandelier Tuff), underlain by light-colored fallout pumice deposits. Near Los Alamos (New Mexico, USA)

◄ Fig. 11.12. Trachytic ignimbrite erupted in A.D. 965 from Baitoushan volcano (Korean side). Pinnacles represent more strongly cemented ignimbrite with fossil fumaroles in the upper part. Ignimbrite fills the valley of Yalu River defining the border between China (*left*) and North Korea (*right*). See also Fig. 9.51

emplaced by ground-hugging flows, a process that is strongly heat-conserving. The term *nuées ardentes or glowing clouds*, applied by Lacroix to the dramatic pyroclastic density currents racing down the flanks of volcano Montagne Pelée in 1902, was very graphic but equally misleading. The bulk of the material transported by these flows, the hidden bedload, traveled largely in the valleys while the spectacular glowing cloud expanded across the hills and grew into the sky, yet carried most material beyond the volcano slopes.

To make matters even more complicated, there has been a tendency even in the expert scientific literature to use identical terms for the eruptive processes, the particle population that flows, and the deposits generated once the flow has stopped. This lumping of terms glances over the fundamental differences between eruptive, transport, and depositional processes and the final laid-down product.

Today the scientific community has largely agreed to use the term *pyroclastic flow* – or *pyroclastic density current* – for inflated mixtures of hot volcanic particles and gas that flow in variable

concentrations and varying velocity along th ground. There are three end member types o pyroclastic flows and their deposits.

Pyroclastic flows proper are composed of low density vesicular volcanic particles – pumice – tha range in size from ash to lapilli, accompanied b variable amounts of crystals and lithic fragment and are buoyed-up by gas. The deposits of suc flows are called *ignimbrites* in most countries, th term being restricted to welded deposits in som Historical examples are the deposits produce during the eruptions of Krakatau (1883), the VTT (1912), and Komagatake (Japan) (1929). Man pyroclastic flows were generated during the erup tions of Mount St. Helens (1980) and Pinatub (1991). When more than 50% of the particles i such flows are smaller than 2 mm, they are calle *ash flows* and the deposit may be called an *ash flo deposit*. When the ash or lapilli particles are ho enough to weld together, the rocks are calle *welded tuffs* or *welded ignimbrites*. It must be re membered, however, that grain size and degree o welding in the deposits may change significantl both vertically and with distance from the vent.

The deposits of glowing avalanches that cam to rest in the valley of Rivière Blanche close to th city of St. Pierre in 1902 are not ignimbrites prop er in the above definition. They are composed o poorly to moderately vesiculated blocks of lav and ash-size particles (Fig. 11.3). This type of flo is generally of small volume and, because of th

▼ Fig. 11.13. Miocene rhyolitic ignimbrites that fill the highly irregular primary surface of a very viscous thick rhyolitic lava flow consisting of several flow units and flow breccias. The lower ignimbrites terminate on the upper slopes of the primary depression, while the uppermost ignimbrite sheets cover the smoothed-out terrain. South coast of Gran Canaria (Canary Islands)

high density of the large clasts, comes to rest within a few kilometers of the vent. Many such deposits form by gravitational collapse of slowly ascending viscous lava domes, breaking through and rising above the ground. The cumbersome name *block-and-ash flow* (BAF) is generally used for such glowing avalanches, or, when laid down, *block-and-ash flow deposits*. The simpler term *pyroclastic block flow* is also in use (303). Because these were the flows that were generated at Montagne Pelée, the flows are sometimes called nuées ardentes or nuées and the deposits *nuées ardente deposits*. Block-and-ash flows are very common and commonly hazardous. Volcanoes with historical examples, apart from Montagne Pelée, include the very active volcano Merapi (Java) (Fig. 9.15), Unzen volcano (Japan) in the 1990s, Santiaguito in Guatemala in eruption since 1922 and, beginning in 1995, Soufrière volcano on the small island of Montserrat in the Caribbean (Fig. 1.17).

Pyroclastic flows *sensu lato* are associated with a third type of pyroclastic particle transport recognized since the 1970s, the *surges*. This is a complex group of low density flows that defies precise definition, even more so because of highly variable properties. Surges commonly leave only thin, ephemeral deposits that often are quickly washed away by erosion. A common denominator is that these are very low concentration flows with a high gas (and air) component that move fast and turbulently, are often highly destructive, and accompany–or precede–both pyroclastic flows *sensu stricto* and block-and-ash flows. *Base surges*, the study of which actually initiated recognition of hot surges, form a major class by themselves. They are not transported and deposited in a hot and dry state and are one of the most distinctive modes of transport of particles during phreatomagmatic eruptions, when rising magma physically and violently interacts with external water (Chap. 12).

Ignimbrites
Structures and Fabrics

Ignimbrites are composed of glass shards, pumice lapilli, crystals, and rock fragments, with ash commonly the dominant particle size (i.e., <2 mm ø). Typical is their massive character, in which obvious bedding and good internal sorting of particle sizes are lacking or subtle. This poor sorting (Figs. 11.6–11.8) is the result of *mass flow* transport of a mixture of small and large particles. If the same particle population had fallen through an air column it would have separated into different grain size ranges at different distances from source and formed well-sorted deposits.

Ignimbrites deposited at low temperatures (<500–600°C) are not welded and form relatively

unconsolidated massive sheets when young. Pyroclastic flows deposited at higher temperature are often sintered or welded, and can even crystallize during cooling to form lava-like rocks (Figs. 11.13 –11.15). No wonder that most high temperature ignimbrites had been interpreted as lava flows for more than 100 years! Welded pumice lapilli commonly form distinct, sometimes black, blebs (*fiamme*) in a fine-grained ash matrix (Fig. 11.8). From a materials perspective, ignimbrites can be worked much more easily than lava flows and are quarried in many countries (Fig. 11.16). Ignimbrites are easily cut, the blocks being of light weight and well-insulating.

Most ignimbrites are composed of several *flow units*, or individual flow deposits that represent distinct depositional events following each other within minutes, days, or longer time intervals. When cooling together, they form a *cooling unit* (346) (Figs. 11.8, 11.17, 11.18). Ideally, ignimbrites show a characteristic succession of layers:

- Initial Plinian fallout in many but not all
- Basal surge layer, laid down by a turbulent surge advancing ahead of the main flow, commonly eroding friable underlying deposits such as initial fallout (Figs. 11.8, 11.19);

▼ Fig. 11.14. Photomicrograph of a strongly welded Miocene compositionally zoned (trachytic-basaltic) ignimbrite (P1, Gran Canaria). The light-colored crystals are feldspars, the blue crystal is amphibole and the brownish irregular particles are glassy fiamme, strongly deformed when deposited hot. The glass shards below the feldspars in the upper part of the photograph have become strongly welded to an almost homogeneous glass, in which the glass shard evidence has disappeared. Diameter of photomicrograph 2 cm

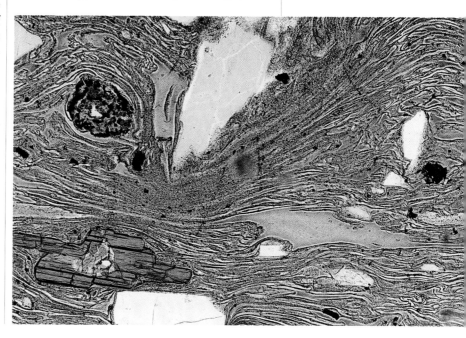

▲ Fig. 11.15. Columnar structure in highly welded, lava-like Miocene rhyolitic ignimbrite VI (Gran Canaria, Canary Islands)

◄ Fig. 11.16. Mining of poorly welded Miocene ignimbrites with a mechanical saw equipped with horizontal and vertical saw blades. Las Banderas ca. 25 northeast of Managua (Nicaragua)

◄▼ Fig. 11.17. Flow units of unwelded Laacher See ignimbrites. Dark rock fragments are enriched at the base and large, light-colored pumice lapilli in the upper part of a flow unit. Wassenach (Laacher See area, Eifel, Germany)

▼ Fig. 11.18. Schematic cross section through an ignimbrite flow unit (104)

Fine-grained ash	
Pumice accumulation	
Central zone	Flow unit
Lapilli pipes	
Enrichment in rock fragments	
Fine-grained base	
Basal surge layer	
Basal fallout layer	

Main body, in which dense rock fragments may locally become concentrated at the base during flow in the distal facies and light-weight pumice accumulates near the top (Figs. 11.8, 11.18)

A fine-grained ash layer on top of the main ignimbrite body, generated when small ash particles lofted by escaping hot gases from the moving or arrested ash flow are sedimented when the turbulence of the ash cloud has dissipated.

Distribution

Volcanic rocks that occur in the form of vast sheets, often forming plateaus that characterize and dominate a landscape, fall into two groups. One is represented by the flood basalts, such as on the Columbia River Plateau (Chap. 7; Fig. 7.20), the other by ignimbrites, a classical example being

dated fallout tephra deposits. Landscapes characterized by ignimbrite plateaus include many Tertiary to Recent volcanic areas around the Pacific from New Zealand through Indonesia, Japan, western North America (Figs. 11.5, 11.10, 11.11) and Central (Fig. 11.16) and South America. They also occur in the young volcanic areas between the African and Eurasian plates from Iran in the east through Turkey, with the famous cave-dotted Göreme deposits, to Italy in the west. Plateau-forming ignimbrites are also common on ocean islands, such as Tenerife and Gran Canaria (Canary Islands) (Figs. 11.6, 11.8, 11.13–11.15). In Europe, widespread ignimbrites were formed during the Permian, their red cliffs dominating the landscape in part of the Oslo Fjord (Norway), Bolzeno (Italy), and in France at the western end of the Alps. Indeed, it was the plateau-forming character of widespread sheets of

▲ Fig. 11.19. Base of Quaternary rhyolitic ignimbrite KP-4 showing basal fallout lapilli, eroded and partly reworked by subsequent turbulent surges (arrow), their deposits being cross-bedded and planar. These are overlain by the massive main body locally beginning with a fine-grained base. 5 km east of Kutcharo caldera (Hokkaido, Japan)

▼ Fig. 11.21. Model of eruption of ash flows from a chemically and mineralogically zoned magma chamber with synchronous caldera subsidence (346)

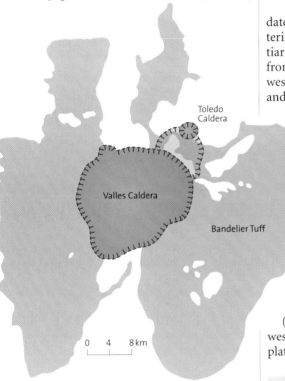

▲ Fig. 11.20. Distribution of plateau-forming Quaternary rhyolitic ignimbrite of Bandelier Tuff (New Mexico) around Valles caldera and the slightly older Toledo caldera (347)

the impressive ignimbrite ridges in the Los Alamos area in New Mexico (Fig. 11.11). Pyroclastic flows, some with volumes of hundreds or thousands of km³, are highly mobile because of their very low viscosity, and therefore can cover areas of hundreds to thousands of square kilometers. Welded or partially welded ignimbrites form compact, hard, columnar-jointed rocks that weather in prominent cliffs when underlain by less consoli-

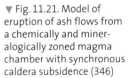

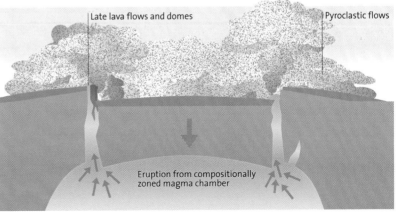

Late lava flows and domes — Pyroclastic flows

Eruption from compositionally zoned magma chamber

▶ Fig. 11.22. Important parameters that help to define the conditions for collapse (*above the diagonal line*) of a convecting eruptive column (*lower right*) (425). The lower tephra deposits of Laacher See volcano fall into the area of convection, that of ash flows of the middle Laacher See tephra in the area of collapse. At the observed eruption of Mount St. Helens on 18 May 1980 the physical parameters correspond to the ash flow eruptions in the collapsed area. Data of LST (31), of Mount St. Helens (43)

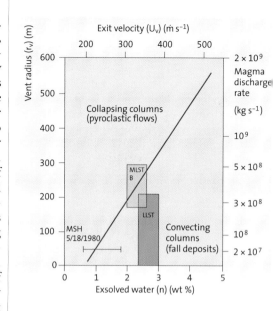

ignimbrites that helped to decipher their origin. Lava flows of the same chemical composition, such as rhyolites and dacites, are, with few exceptions, so viscous that they form only relatively short stubby lava flows (Fig. 9.12). Huge volumes of rhyolitic ignimbrites occur in some large igneous provinces, such as southern Africa or southern America (LIPs, Chap. 7). They are so strongly welded that the discussion as to whether or not they are true ignimbrites is still continuing.

Most large ignimbrite volcanoes consist of several widespread sheets erupted over several million years, such as those making up the Yellowstone Plateau. Large-volume pyroclastic flows erupt along more or less concentric fissures, along which the crust collapses into the partially evacuated magma reservoirs to form large calderas. These collapsed roofs may subside hundreds of meters, over an area with diameters that may exceed 20 km. Fortunately, huge plateau-forming pyroclastic flows have not erupted in historical time. If erupted today, they would be catastrophic and completely devastate and bury areas of hundreds to thousands of square kilometers.

Small-volume valley-hugging pyroclastic flows are erupted each year from many volcanoes. For example, during the late stage of the famous eruption of Vesuvius (A. D. 79), small pyroclastic flows buried the town of Herculaneum. The deposits of these ash flows are still interpreted as having been laid down by mudflows even in some more recent non-technical literature. The famous trass in the Laacher See area in Germany (Fig. 11.17) was also once thought as having been deposited from mudflows.

Origin
The main problems concerning the origin of pyroclastic flows – some under discussion for more than 100 years – are (1) their generation at the source, (2) their mode of transport and (3) deposition and – since the late 1960s – (4) the generation and evacuation of compositionally zoned magma reservoirs.

The British traveler Anderson and the geologist Flett, who arrived in Martinique on 14 May 1902 to study the catastrophic destruction of the city of St. Pierre, witnessed the third major nuée ardente eruption of Montagne Pelée on 6 June. They correctly reasoned that the city St. Pierre was destroyed by ash clouds moving along the ground and not, as some people believed, by ash rains. They further assumed that a mixture of gas and tephra, formed deeper in the conduit, had risen to the lip of the crater from where it descended the flank of the volcano because of gravity. The American geologist Jaggar, founder of the famous

volcano observatory on Kilauea volcano, instead had postulated that the erupted mixture was first transported to great height and then collapsed to become laterally diverted when it hit rising newly erupted material. Lacroix modified this idea by assuming a laterally directed explosion, because the glowing clouds were not restricted to valleys. The origin of this lateral explosion was strongly debated and the model of lateral blasts was not discussed seriously for many decades, apart from the huge eruption of volcano Bezymianny in Kamchatka (24, 123). Today, several modes of origin for pyroclastic flows *sensu lato* are distinguished from each other: (1) lateral eruptions such as those of Mt. St. Helens (Chap. 10), (2) the collapse of eruption columns, (3) ash fountains to be discussed in more detail below, and (4) pyroclastic block flows that form by collapse of viscous domes, also treated below.

Wilson et al. (425) defined several boundary conditions under which a rising convecting eruption column (Chap. 9) transforms into a collapsing one. Firstly, a convecting eruption column may not develop when the mixture of tephra and gas of the gas thrust of an eruption column is initially so dense, or does not mix sufficiently with air, that the total density is higher than that of the atmosphere. Alternatively, a convecting eruption column may be transformed into a collapsing one during an eruption. The most important factors that determine whether or not an eruption column develops a convecting phase are shown in Fig. 11.22. Variables include: the gas content of a magma (n), which influences the nozzle velocity (U_v), the radius of the vent (r_v), and, mainly dependent on these factors, the mass eruption rate ($\dot{m}$). The fields subdivided by a diagonal line in

Fig. 11.22, show areas of convecting and collapsing eruption columns. An eruption column, for example, can collapse when the vapor phase or free magmatic gas content decreases during the course of an eruption. This situation can occur when volatiles have become strongly enriched in the upper part (*cupola*) of a magma column. The nozzle velocity does not necessarily decrease in harmony with the decreasing gas content, because gases in the lower part of a magma column are at a higher pressure. However, decreasing nozzle velocity and/or an enlargement of the diameter of the vent, and therefore an increase in eruptive rate, may also cause the column to collapse at relatively constant volatile contents. In nature, the boundary conditions between convecting and collapsing columns change during an eruption in complex ways. A reconstruction of the real sequence and interplay of processes is thus impossible. Nevertheless, the diagram proposed by Wilson et al. is a good approximation and has been refined more recently (e.g., 248; 435). While the models of Wilson et al. assume that the bulk mixture can be treated as a single phase *pseudogas*, newer models allow fallout of individual clasts. A sharp distinction between convecting and collapsing columns is no longer maintained.

Transport and Cooling of Pyroclastic Flows and Their Deposits

The interpretations of how closely the particles in a pyroclastic flow are spaced have varied widely over the last few decades. Highly inflated flows characterized by very low particle concentrations were envisioned initially. Subsequently, many scientists invoked relative high concentration flows advancing in a more laminar mode much of the evidence being based on the poor grain size sorting. Plug flow was envisioned as a possible end stage of transport. Deposition of particles from the base upwards is discussed as an alternative. At present, prevailing opinion holds that flows move in a turbulent manner, most being strongly inflated over much of their transit along the ground. Clearly, flows strongly change their character with distance from vent and as the supply at source waxes or wanes. In any case, discussion on this very complex set of problems will continue for some time with quantitative modeling increasingly replacing field studies.

The low degree of friction, one of the reasons for the incredible velocity of pyroclastic flows and their ability to spread laterally, can best be explained by particles being buoyed up by rising gases. The nature and origin of the gas phase is a matter of debate. The principal sources of gas are (1) degassing of magmatic particles in transit,

possibly not a major source, (2) gases incorporated during collapse of the eruption column, (3) air ingested and heated at the front of the moving pyroclastic flows or (4) vaporized moisture from the ground, possibly only locally important.

Another debate is concerned with the details of the mode of deposition. Several models are under discussion: En masse freezing, layer-by-layer deposition from the base upwards and emplacement by flow units. In any case, as the gas phase and fine ash continue to escape from the flow by elutriation, internal friction increases, velocity diminishes and material is deposited in the higher concentration bedload or in a more massive plug, as is likely in many block-and-ash-flows. Fundamental is the observation that most ignimbrites when studied closely consist of several flow units that can be mapped over large distances and, in well-exposed areas, be traced to the source caldera (Fig. 11.8). These *eruption-controlled flow units* must record discontinuities in the evacuation of the magma reservoir such as successive phases of roof subsidence. The younger flow units in a cooling unit are generally more mafic and also compositionally mixed, such mixing possibly being induced by incremental roof collapse into the reservoir.

If the deposit is very hot and thick, cooling to ambient temperature can last for decades or hundreds of years. For example, when Griggs and his companions entered the VTTS in 1916 and noted the thousands of rising fumaroles, they thought that a magma chamber existed beneath the steaming pyroclastic flow deposits (called *sand flow* by them; 129). Later it was recognized that the gases were derived from the flow itself, which was overlying glacial gravels and not a hot plutonic body. Channels, along which the gases escaped and through which fine ashes were blown into the atmosphere, are preserved as hydrothermally altered pipes (Fig. 11.9). In older and more strongly eroded deposits, fossil fumarolic pathways reflect characteristic *degassing channels* (Figs. 3.13, 11.12). More recent ideas about the transport mechanisms of pyroclastic flows are discussed at the end of this chapter.

Compositional Zonations

One of the main areas of scientific inquiry into the petrology of igneous rocks is the origin of chemically and mineralogically zoned magma reservoirs. Many of these studies have been based on the analysis of ignimbrites because these rock bodies result from the fast evacuation of very large volumes of magma from the top down (Chap. 3). It is basically the decrease of gas content during the course of an eruption that causes

and crystallized. This plug may then become blown out in an explosive Vulcanian explosion (Chap. 12) to be followed by the newly arrived more volatile-rich magma. The study of ash flow deposits thus allows to document the derivation of highly evolved magmas from more primitive ones in detail.

Pyroclastic Block Flows and Their Deposits

Volcanologists now become well-informed by rapid information distribution through the internet or via the monthly bulletins of current volcanic eruptions, such as the Smithsonian Global Volcanism Network. However, it is still not commonly recognized that most pyroclastic flows generated each year around the globe consist of relatively dense blocks of lava and coarse-grained nonvesicular ash particles rather than pumice and fine ash (Figs. 11.23–11.30). The deposits of pyroclastic block flows have received much less attention than the pumice-rich ignimbrites, mainly because their small volume does not harbor as much petrological information, representing only a tiny fraction of a magma column. Moreover, they are very coarse-grained and not amenable to detailed grain size studies. Also, many are not associated with large Plinian explosive eruptions.

▲ Fig. 11.23. Hot steaming surface of pyroclastic block flows erupted in 1976 on Augustine volcano (Alaska)

► Fig. 11.24. Steep, about 2-m-high front of pyroclastic block flows generated in 1976 by dome collapse. Augustine volcano (Alaska) (see also Fig. 9.36). Person as scale (*circle*). Depositional fan begins at mouth of steep chute (Fig. 11.23) leading to growing dome

Plinian eruption columns to collapse after having transported material to great heights (Chap. 10). The mass of volatile elements dissolved in a magma column increases with time in the roof region as a result of prolonged differentiation (150). The reverse case also occurs when a non-erupted magma plug in the conduit is degassed

◄ Fig. 11.25. Huge stranded dacite block at the foot of Augustine volcano (Alaska), left by draining of hot pyroclastic block flow. Fumarole degassing above previously deposited hot block flow deposit in the foreground

Glowing avalanches that result from collapsing domes are sometimes called *nuées ardentes of the Merapi type*. They have been most frequently observed and studied at this very active volcano in Java, which has been erupting almost continuously for centuries. Larger eruptions or phases of increased activity occur two to three times per decade and people have perished repeatedly during Merapi's eruptions, including about 70 fatalities in November 1994. A major recent eruption occurred in July 1998, and no end of Merapi's persistent eruptive activity is in sight.

The deposits of pyroclastic block flows are commonly poorly sorted. Grain sizes are distributed bimodally to polymodally and large blocks may exceed 5 m in diameter (Fig. 11.25). The composition of the particles is commonly almost monolithologic, as they are usually derived from a compositionally relatively homogeneous small-volume dome. Particles are angular to slightly rounded, their density is high with fine, irregularly shaped vesicles and a glassy matrix containing variable amounts of phenocrysts and microlites. Most compositions are andesite or dacite. The deposits fill valleys, in which the surface of the deposits is relatively flat, but they form pronounced hourglass-shaped fan deposits on gentle slopes (Fig. 11.24).

Thin, fine-grained ash deposits associated with the pyroclastic block flows are deposited by

▶ Fig. 11.26. Pyroclastic block flows – right side of volcano flank – on the slope of an andesitic dome growing since 1995 on the Caribbean island of Montserrat. See also Fig. 1.17

▶ Fig. 11.27. Columnar cooling joints in one large block broken off the top of cooling Unzen dome. Unzen volcano was active between 1991 and 1993 (Kyushu, Japan)

◀ Fig. 11.28. Erosional remnants of valley-filling pyroclastic block flows deposited in November 1994 in Buyong valley at the foot of Merapi dome (Java, Indonesia). See also Figs. 9.15, 11.30

laterally expanding surges, or are derived by fall-out from turbulent ash clouds that rise and expand from moving flows to spread for kilometers beyond the valleys. Pyroclastic block flows can be quite hazardous despite their small volume, because of their great velocity, high temperature, and especially the surges generated in transit.

Origin of Pyroclastic Block Flows

Pyroclastic flows are gravity-driven mass flows. Pyroclastic block flows are usually not formed by ash fountains or collapse of rising eruption columns. Instead they are mostly generated when highly viscous magma slowly rises above the rim of a crater, cools from the outside, and starts to collapse or crumble away. Many other factors have been invoked since the classic studies of Lacroix (188) for determining the driving mechanism. These include lateral expansions (*explosions dirigée*) from a lava dome, explosive degassing of dome-interior magma by decompression following major collapse, explosive degassing of blocks within the outflow in transit, and rapid expansion of heated air ingested in the front of a flow.

Escher (93) described deposits of nuées ardentes at Merapi as *a type consisting of avalanches of lava blocks, which are already hard but still hot, which break into numerous pieces and ash*

when racing down the slope and thus form a glowing cloud.

Taylor (374) in his famous analysis of the 1951 Mount Lamington eruption (Papua New Guinea) emphasized the importance of gravity as an energy source for the pyroclastic flows at Mt. Lamington, whether from vertical eruption columns or from collapsing domes growing in a crater.

A viscous lava mass that rises above a crater for tens to hundreds of meters can become unstable, partly collapse, and flow down the flanks of the volcano as a gravity mass flow. In addition, while the viscous mass of a dome is expanding due to the pressure of the continuously rising magma, gases that had been under pressure start to expand, especially once fractures form in the slowly rising hot mass. This gas expansion can accelerate the widening of the cracks and also contributes to the fragmentation of the blocks as they tumble down the slopes and break-up into smaller particles.

Neumann van Padang (249) described small collapses of lava, which he observed during November to December 1930 on Merapi, translated from the German: *The new addition (of magma) led to the fragmentation of the lava crust from beneath, which then tumbled down in blocks. During this down-slope flow the glowing lava blocks broke into pieces, the enclosed gases were freed and formed vapor clouds, which, together with the ash, also generated during the explosion and the old ash that had been thrown up into the air along the volcano's slope formed descending eruption clouds.*

Ash particles also form when the viscous mass of a rising lava dome starts to fracture – as seen in many video films taken between 1991 and 1993 at Unzen volcano in Kyushu (Japan) (392). These fractures propagate during rapid destabilization from the surface downward. Large amounts of ash are also constantly generated in the pyroclastic block flows when the blocks bounce off the ground and collide with each other.

Fragmentation in pyroclastic block flows descending the slopes of a volcano can become strongly accelerated or reactivated when the mass flow suddenly descends a steep slope, such as the scarp of a waterfall in a valley (Fig. 11.29). Huge ash clouds may rise from such critical breaks in slope, as at Unzen volcano (247). The death of some 70 people along the valley of Kalibuyong on the southwestern slope of Merapi in 1994 may also have been due to hot blasts generated when fragmentation in a hot block-and-ash flow was intensified as it raced over the scarp of a major cascade, probably aided by expansion of overridden and compressed air.

Surges, Blasts, Umbrella Clouds and Debris Jets

When the breakthrough in our recognition and understanding of pyroclastic flows was established in the early 1960s, chiefly by RL Smith and his associates, a subsidiary but equally fascinating group of deposits – that of *surges* – was identified. Their analysis proved even more difficult than that of pyroclastic flows. The existence of hot low-density currents accompanying nuées ardentes has been known for more than 100 years since St. Pierre was destroyed. However, characteristics of surge deposits have only been recognized during the last three decades. There is an entire family of deposits to which the name surge – or modifiers of it – apply, such as *ground surge, hot surge, dry surge, veneer deposit, ignimbrite overbank deposit, ash cloud hurricane etc.* Surges are associated with pyroclastic flows of various types (236 354). They are especially known as a particular form of high speed transport of pyroclastic particles in eruptions, caused, or strongly influenced by water-magma interaction (Chap. 12), where they have been called *base surges*.

Many ignimbrite and block-and-ash flow deposits are underlain, overlain, or laterally accompanied by relatively thin (a few cm to dm thick) lensoid ash to lapilli layers. The range of spectacular structures of surge deposits immediately evokes dynamic processes, high velocity, and therefore destructive potential. The most common characteristics of this diverse group of deposits include:

- well-defined bedding and cross-bedding or massive structure (Figs. 11.8, 11.19, 11.31)
- good to moderate sorting
- well-rounded pumice particles
- alternation of well-sorted and poorly sorted commonly lensoid layers
- lack of strong dependence of thickness on morphology, although slight thickening in valleys
- local lateral grading into massive flow deposits
- restriction to the proximal area around volcanoes (rarely more than 8 km away from vent)
- small total volume

These sedimentological characteristics of surge deposits cannot be explained by the simple models of fallout or lateral mass flow in valleys. There are several principal modes of occurrence of surge deposits, including on the top of a flow unit (95); at the base of an ignimbrite; as the proximal facies of fast pyroclastic flows (409); or as the marginal facies of ignimbrites (314). A significant aspect is that the pressurized flows can expand above the topography or valleys and surmount significant heights, thus affecting large areas. The low-density pyroclastic currents may expand straight ahead of valley-restricted pyroclastic block flows, as described in detail by Neumann van Padang (249) from the 1930 eruption of Merapi volcano.

The occurrence of surges during the transitional phase from the Plinian fallout to pyroclastic flow stage of eruptions has been studied in more detail (430). A phase of crater enlargement commonly occurs at the beginning of a Plinian eruption column collapse, thus generating pyroclastic flows. This is accompanied by a series of pulsating shock waves containing relatively small masses of particles and occurring in time intervals of a few seconds to about one minute.

The term *blast* was applied as early as 1904 (146) to the devastating pyroclastic currents that wrought havoc to St. Pierre in 1902. Gorshkov (123) described what he termed giant directed blasts associated with sector collapse from Bezy-

▼ Fig. 11.29. Model of pyroclastic block flows generated during collapse of a lava dome. Surges issue from the front of such hot block flows. Clouds of ash rise at the point where the hot blocks become increasingly fragmented when tumbling down former steep gradients, such as a waterfall, by which process they liberate gas explosively (317)

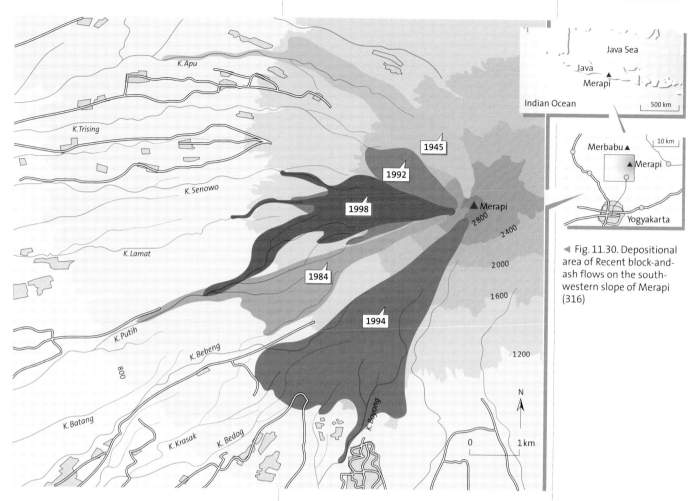

◀ Fig. 11.30. Depositional area of Recent block-and-ash flows on the south-western slope of Merapi (316)

▼ Fig. 11.31. Surge deposits representing the proximal facies of pyroclastic flows whose flow units are shown in Fig. 11.17, about 500 m north of Laacher See crater. Transport direction from right to left (*black arrow*). Note erosional unconformity at the stoss side (*arrow*) and almost vertical sigmoidal bedding generated by high velocity surges. Laacher See area (Eifel, Germany)

mianny volcano in Kamchatka. The most thoroughly studied blast is that generated during the initial phase of the Mt. St. Helens eruption on 18 May, 1980. The velocity of this pyroclastic current was so large (90–110 ms⁻¹) that it overtook the preceding rock avalanche that had been triggered by collapse of the northern sector of the volcano. The blast devastated an area of some 600 km², the millions of trees aligned along the curved path of the blast providing one of the most impressive sights following a major volcanic eruption (Fig. 10.20; Chap. 10).

A most interesting transformation from lateral to vertical motion took place in the Mt. St. Helens blast when it lost much of its bedload, heated ambient air as it moved, became buoyant, and lifted off the ground (356). A *giant umbrella cloud* formed as a result. In effect, giant umbrella clouds are commonly generated from the surface of pyroclastic flows s.l. and rise much higher than many eruption columns that develop convectively from gas thrusts at the vent. Many widespread ash layers are deposited from these *co-ignimbrite clouds*. Other recent examples of giant umbrella

▲ Fig. 11.32. Well-sorted, lobe-forming deposits generated during the explosive opening eruption of Arenal volcano (Costa Rica) in July 1968. The deposits are interpreted as having been transported by volcanic jets during Vulcanian eruptions. Note poor vesicularity and angularity of most clasts and small amount of red hydrothermally oxidized lithics

clouds rising to great heights from the surface of hot pyroclastic density currents include the 1956 eruption of Bezymianny, Pinatubo 1991 and Redoubt 1989 (Fig. 10.3). The sources of the gases that carry the fines upward are either internal or ingested air or vaporized water or glacier ice on the ground. The facies changes in the Mt. St. Helens deposit, from more massive near the volcano to stratified and cross-stratified distally, suggest that the depositional mechanisms of pyroclastic flows and surges are transitional.

On Martinique in 1902, hot block flows descending from the collapsing dome of Montagne Pelée were restricted to the valley of Rivière Blanche, while the complete devastation of the town of St. Pierre was due to the expansion of hot low-density currents. This ash-poor blast destroyed the city, and sunk and ignited ships in the harbor, penetrating every nook and cranny in ships anchoring even several hundred meters offshore, with many people dying by inhaling the hot ash.

In July 1968 Arenal volcano in Costa Rica came to life following several centuries of dormancy. Today it is one of the most active volcanoes in Central America and has become a major tourist attraction. The bulk of the 1968 Arenal lapilli deposits were formerly interpreted as laid down by pyroclastic flows. However, they are not valley-confined but form lobate sheets that roughly mantle the topography. Moreover, the sheets are almost as long as they are wide, reflecting an eruptive system that rapidly expanded laterally after leaving the vent. Grain size sorting in these deposits is generally excellent, with a dominant grain size mode between 1 and 5 cm (Fig. 11.32). Ash is nearly absent. These characteristics are incompatible with transport by a ground-hugging flow in which the bed-load was concentrated

in, or confined to, a valley. An alternative interpretation is deposition by *volcanic* jets that proceed at shallow angles from a vent, the material spreading along the ground for some short distance. Similar jets are not uncommon, especially in eruptions triggered by magma-water contact. They illustrate that our understanding of fluid dynamics of volcanic particle systems always lags behind the wide spectrum of transport processes in nature, many of which cannot be classified neatly into the flow, fall or surge pigeonholes.

The debate on the origin of surges *sensu lato* and blasts might reflect mere academic quibbles and pencil sharpening of little interest to many readers of this book. Yet, the disastrous effects of many volcanic eruptions result from these fast, energetic, and strongly diluted blasts, and not from the actual pyroclastic flows (because the latter are commonly restricted to the valleys). For example, the havoc caused by surges during eruptions such as those of Vesuvius in A. D. 79, during the destruction of St. Pierre in 1902, up to the event on June 3, 1991, causing the immediate death of 43 people (including the famous volcano photographers Maurice and Katia Krafft) at Unzen volcano necessitates a thorough and penetrating analysis of their mode of origin. Effective hazard mitigation must build upon a good understanding of transport processes of volcanic particle flows (Chap. 13).

The Eruption of Laacher See Volcano 12 900 Years Ago

Laacher See is a water-filled crater bowl about 2.5 km in diameter and 65 m deep. It lies in a beautiful setting surrounded by lush forests in the western part of Germany, famous for a twelfth Century Benedictine Abbey. The origin of this large hole and of the conical hills forming its rim and dotting the area between the lake and the Rhine River (10 km to the east) was hotly debated during the birth of the science of volcanology between the late eighteenth and early nineteenth century. About 40 years before the German poet and scientist Goethe visited Laacher See, Collini had put forward an interpretation about its origins. Collini's ideas (1777) were quite similar to present-day thinking, but contrary to the opinions of Goethe, translated from the German: *that the Laacher See had formed from a very important volcano, which here had destroyed itself and became extinct.*

This interpretation was also accepted by other scientists at that time, such as Hamilton (138), the father of volcanology. This is the more remarkable because the heated debate between Neptunists, Volcanists and Plutonists had not yet reached its peak in the 1770s (Chap. 7). Goethe recognized the logic of the arguments of the Volcanists and Plutonists, but was unable to change his view of the nature of volcanoes, not even after the death of A.G. Werner, the guru of the Neptunists, in 1817 (Chap. 1).

Debates such as these characterize all sciences, because science develops by the antagonism between different interpretations and hypotheses. New ideas commonly have a difficult path to recognition and acceptance, until they themselves become canonized and are subsequently defended just as vigorously as the older interpretations that they replaced. These early debates about the true nature of the Laacher See volcano played a significant role in the history of volcanology. Even until recently, opinions have diverged about the origin of the lake basin and the actual crater, from which the pumice masses erupted.

The eruption of Laacher See volcano in late spring 12 900 years ago was the most powerful late Quaternary volcanic eruption in central Europe. It is a classic example of a complex Plinian eruption, whose mechanism changed drastically during the short duration of its evolution. The more than 6 km³ magma that was erupted over only a few days, dwarfs that erupted from all of the more than 300 scoria cones, maars, and lava flows of the remainder of the Quaternary Eifel volcanic fields put together. The Laacher See tephra forms the premier time horizon of the very late Pleistocene in Europe. The total volume of magma erupted is higher than that of similar Plinian eruptions, such as that of Vesuvius (A.D. 79), Mount St. Helens (1980), or El Chichón (1982), and slightly larger than that of the 1991 eruption of Pinatubo. The well-exposed tephra deposits of Laacher See volcano are textbook examples of major Plinian deposits. Their generation was governed by a highly complex interplay of intrinsic and extrinsic factors whose relative roles changed repeatedly during the course of the brief eruption (Figs. 11.33–11.43). Characteristics include:

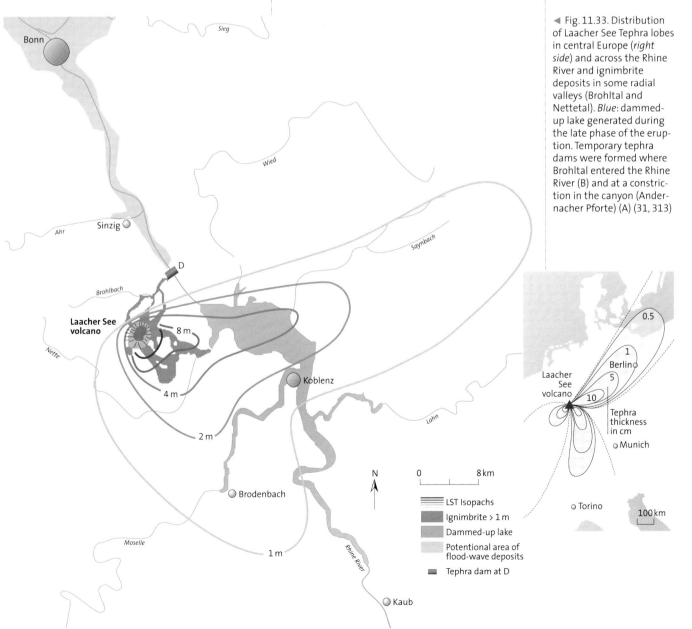

◀ Fig. 11.33. Distribution of Laacher See Tephra lobes in central Europe (*right side*) and across the Rhine River and ignimbrite deposits in some radial valleys (Brohltal and Nettetal). *Blue*: dammed-up lake generated during the late phase of the eruption. Temporary tephra dams were formed where Brohltal entered the Rhine River (B) and at a constriction in the canyon (Andernacher Pforte) (A) (31, 313)

- an extremely compositionally zoned and layered magma column evacuated from the top down. The upper part was almost free of visible crystals, rich in volatiles and therefore of low viscosity. The base of the erupted part of the magma column, located about 7 km below the Earth's surface, was poor in gas, crystal-rich (up to 50 vol%) and therefore highly viscous. Magma at greater depth was probably unable to erupt pyroclastically

- several feedback cycles in which external forcing (influx of water, conduit collapse) became do-

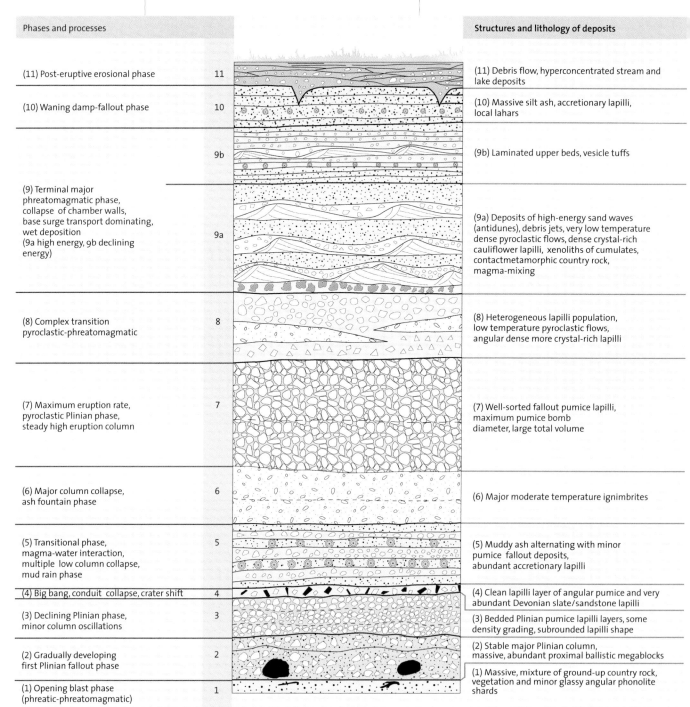

Phases and processes

(11) Post-eruptive erosional phase — 11

(10) Waning damp-fallout phase — 10

— 9b

(9) Terminal major phreatomagmatic phase, collapse of chamber walls, base surge transport dominating, wet deposition (9a high energy, 9b declining energy) — 9a

(8) Complex transition pyroclastic-phreatomagmatic — 8

(7) Maximum eruption rate, pyroclastic Plinian phase, steady high eruption column — 7

(6) Major column collapse, ash fountain phase — 6

(5) Transitional phase, magma-water interaction, multiple low column collapse, mud rain phase — 5

(4) Big bang, conduit collapse, crater shift — 4

(3) Declining Plinian phase, minor column oscillations — 3

(2) Gradually developing first Plinian fallout phase — 2

(1) Opening blast phase (phreatic-phreatomagmatic) — 1

Structures and lithology of deposits

(11) Debris flow, hyperconcentrated stream and lake deposits

(10) Massive silt ash, accretionary lapilli, local lahars

(9b) Laminated upper beds, vesicle tuffs

(9a) Deposits of high-energy sand waves (antidunes), debris jets, very low temperature dense pyroclastic flows, dense crystal-rich cauliflower lapilli, xenoliths of cumulates, contactmetamorphic country rock, magma-mixing

(8) Heterogeneous lapilli population, low temperature pyroclastic flows, angular dense more crystal-rich lapilli

(7) Well-sorted fallout pumice lapilli, maximum pumice bomb diameter, large total volume

(6) Major moderate temperature ignimbrites

(5) Muddy ash alternating with minor pumice fallout deposits, abundant accretionary lapilli

(4) Clean lapilli layer of angular pumice and very abundant Devonian slate/sandstone lapilli

(3) Bedded Plinian pumice lapilli layers, some density grading, subrounded lapilli shape

(2) Stable major Plinian column, massive, abundant proximal ballistic megablocks

(1) Massive, mixture of ground-up country rock, vegetation and minor glassy angular phonolite shards

▲ Fig. 11.34. Schematic stratigraphic column of Laacher See Tephra. Changes in eruptive processes (left side of diagram) are documented by changes in structures and textures (graphic representation and text on right side). See text for discussion. Many major Plinian eruptions show similar evolutionary stages.

minant when internal forcing mechanisms (magmatic volatile pressure) decreased

magma–groundwater interaction was especially powerful during initial and terminal stages,

likely triggering of the eruption by influx of hotter mafic magma into the resident cooler evolving magma reservoir

opening blast phase, and initial fine-grained hydroclastic ash layer

repeated enlargement and deepening of the conduit and collapse of the unstable crater walls, as reflected in the lithic component stratigraphy

lateral migration of the crater about halfway through the eruption and temporary simultaneous eruption from both vents as shown by two major overlapping depositional lobes

characteristic fall deposits (well-sorted lapilli layers, which mimic the morphology and whose thickness and grain-size decrease with distance from source in a semi-logarithmic manner)

fallout lobes with contrasting directional axes reflecting variable wind fields and dependence of height of eruption column on degree of magma-water interaction

massive, poorly-sorted low to very low T pyroclastic flow deposits (ignimbrites) restricted to valleys (Fig. 10.13)

ignimbrite overbank facies

• several types of surge deposits (Fig. 11.31)

• debris jet deposits

• flood wave deposits resulting from the instantaneous emptying of a temporary tephra-dammed lake formed during the eruption at a narrow of the Rhine river canyon (Fig. 11.33)

• a waning phase reflected in fine-grained ash accumulation

• thick lahar and fluvial deposits generated immediately after the eruption

• major climatic impact due to the sulfur-rich magma

Chemical and Volcanological Stratigraphic Framework

The deposits of the Laacher See eruption (LSE) have been subdivided into lower (LLST), middle (MLST) and upper (ULST) deposits as well as syn-eruptively reworked (SRLST) and post-eruptively reworked (PRLST) deposits (Fig. 11.34). These subdivisions reflect the most significant changes during the eruption in the eruptive mechanism, the position of the vent, the composition of the erupted magma and that of the country rock fragments (basement stratigraphy), and interactions with the environment. In the proximal tephra apron up to about 5 km east of Laacher See basin, laterally and ballistically transported tephra dominates. The bulk of the dominantly

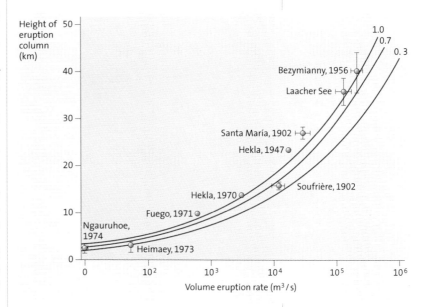

▲ Fig. 11.35. Height of observed eruption columns as a function of calculated volume eruption rates. The height of the Laacher See volcano eruption column has been inferred from the depositional fans. Curves denote different efficiencies of heat conversion (31, 424)

phonolitic magma was erupted in two phases characterized by major sustained and fluctuating high Plinian eruption columns. Conduit collapse, decreasing concentration of volatiles, possibly associated with three major compositional boundaries and water influx led to repeated collapse of the eruption column. The entire evolution of convecting to collapsing eruption columns, leading to generation of either dominantly fallout or flow deposits (Figs. 11.33, 11.35), well illustrate the boundary conditions postulated by Wilson et al. (425).

The Eruptive Center

The question of the location of the source crater of the widespread tephra deposits east of Laacher See may seem academic but had interesting economic implications when the first "energy-crisis" hit Europe in the early 1970s. A deep hole was sunk in 1975 about 8 km away from Laacher See basin in the hope to find a magmatic heat reservoir. The location of this drill site was based on a then widely accepted model that the tephra had erupted from several small suspected vents, one close to the drill site. However, this idea had never been based on convincing empirical or theoretical evidence. Only the lake basin is large enough to explain the high mass eruption rates required for Plinian eruption columns to develop. The drilling was an expensive flop and became known as the deep-freeze hole.

Opening Phreatomagmatic Phase

The highly explosive opening phase of the LSE was characterized by the interaction of rising magma and/or hot gases with groundwater as

▶ Fig. 11.36. The Laacher See crater depression just behind the scoria cones in the middle ground was the site of a large Plinian eruption 12 900 years ago, here shown by combining the landscape view with the eruption column of Mount Pinatubo of 15 June 1991. Pinatubo eruption column photograph: Harlow (Internet)

during the beginning of many other volcanic eruptions. The initial deposits consist, therefore, almost exclusively of country rock fragments and vegetation remains from downed, splintered, and defoliated trees. Minor glass shards are angular and nonvesicular reflecting fragmentation by thermal shock of the water-quenched rising magma. Initial blasts radiated through five morphological lows between older scoria cones surrounding Laacher See basin.

Initial Plinian Phase (LLST)

Once an initial conduit was reamed out, the magma was torn apart at depth by gas expansion and shearing, forming a sustained eruption column. The pressurized gas thrust eroded and wide-ned the crater during the beginning of this firs Plinian phase. Older basalt lava flows at the sur face were broken into huge blocks, some exceed ing 2 m in diameter, which were thrown to dis tances of more than 2 km from the crater (Fig 10.9). The eruption column probably rose to height of more than 20 km (Figs. 11.35, 11.36) Pumice lapilli and ash were transported by pre vailing winds at the tropopause for more than 1 000 km (Fig. 11.33) chiefly to the northeast where they form thin layers in lake sediments, bogs and swamps.

An initial continuous uprush was followed by an oscillating eruption column as reflected in a clear change from a lower massive to an upper layered part of LLST in which pumice lapilli are subrounded. Tubular vesicles in pumice dominate in this phase. Dense glassy lapilli scattered in the pumice lapilli layers vividly portray complex dynamic processes at the boundary of the rising magma and the Devonian slate envelope. Multiple sets of fractures and extreme degrees of fragmentation of slate and mixing with highly fractured crystals reflect quenching of the magma at the margin of the conduit, brittle fracture of both glass and surrounding rock and invasion of gasfluidized mixtures of both (Fig. 11.37).

First Major Hiatus

This first Plinian phase ended with a widespread impressive layer extremely rich in Devonian slate fragments (*Big Bang layer*) (Figs. 11.39, 11.40).

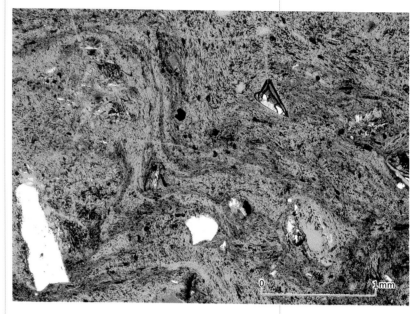

▼ Fig. 11.38. Change in relative height of the eruption columns during the evolution of the Laacher See Plinian eruption as deduced from the areal extent of tephra layers. Eruption columns were high (orange) during pyroclastic and low (blue) during phreatomagmatic eruptions. Intervals of oscillating column height are shown by small decrease in column length, column collapse by decrease to zero. The temporal succession of the main fallout tephra fans in central Europe is shown on the right. Oscillations between high eruption columns and northeast tephra fans to low eruption columns and southern tephra lobes reflect episodic impact of external water on the magma reservoir/conduit system (31 and unpubl)

The lithic fragments were probably generated when the walls around the deeply reamed out conduit became unstable, caved in and were thrown out in a major explosion, overpressure possibly being due to the collapsed and temporarily blocked conduit. The conduit instability itself may have been caused by decreasing gas pressure

▲ Fig. 11.37 Photomicrograph of dense glassy lapillus within LLST pumice lapilli deposit. The schlieren-like glass with collapsed vesicles alternates with dark finely ground-up slate and minute glass shards. The rock is interpreted to represent the marginal zone between the Devonian country rock and the rapidly ascending, vesiculating and fragmenting magma column. Complex textures are thought to reflect fragmentation of both slate and quenched phonolite melt, degassing and gasfluidized invasion of mixtures of both into the boundary zone. Large crystals are feldspar and titanite

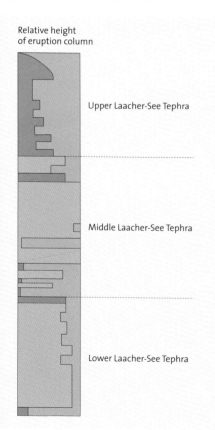

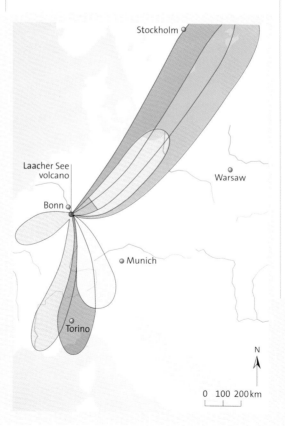

▶ Fig. 11.39. Medial facies of Laacher See Tephra. The base of LLST is hidden by slope debris. The fine-grained dark tuffs in the center (HBB) were deposited wet ending with minor overbank deposits of pyroclastic flows. Laacher See area (Germany)

following the rapid evacuation of the volatile-rich cupola of the upper magma column.

Mud Rain Phase (MLST A)

The dynamics of the entire eruption changed radically following the "big-bang" explosion and an additional crater was established in the northern part of the basin. Trace and major element concentrations across the boundary LLST–MLST show a clear and dramatic compositional gap (Fig. 3.17) in the eastern but more gradual changes in the southeastern fan reflecting eruption from two vents. Whether this could be the result of a relatively sharp compositional boundary in the magma column in the north, or the existence of linked but compositionally distinct magma reservoirs that were tapped successively is not clear.

A characteristic sequence of fine-grained tuff layers deposited wet as mud rain and small concentrically structured ash balls, *accretionary lapilli,*

includes minor Plinian pumice lapilli fallout layers (Figs. 11.39, 11.40). Such mud-rains and rain accompany many explosive eruptions because fine ash particles are excellent nuclei for atmospheric water vapor to condense around although much of the external water may have been supplied in the conduit. The oscillation between the contrasting layers may reflect the attempts of the rising vesiculating phonolite magma to reestablish a sustained high Plinian eruption column, episodically interacting with water having entered the collapsed and reamed-out conduit system. A sector graben formed during this phase (Fig. 11.41) radiating outward from the crater area with local subsidence exceeding 10 m. Subsidence died out prior to the main ignimbrite phase.

When the eruption column only reached the middle troposphere (that is when heat was consumed during influx of external water) ash was blown by lower altitude winds to the south, fallout tephra being found as far south as northern Italy.

Main Ignimbrite and Second Major Second Plinian Phase (MLST B)

Several pyroclastic flows were deposited at the end of this transitional phase, possibly generated by ash fountains reflecting a growing more sustained but still low eruption column. The pyroclastic flows were funneled through radial valleys (Fig. 11.33) and entered the Rhine River. Locally, the deposits are up to 60 m thick, representing the most easily accessible ignimbrites in central Europe. The unwelded relatively low T ignimbrites are composed of several flow units (Fig. 11.17). Minor fine ash was elutriated from the flows by hot rising gases to form convective ash columns that deposited thin ash layers on top of the flows and turbulently transported overbank surges which also covered higher ground.

The second and major Plinian phase began with an oscillating eruptive column phase as shown by more pronounced layering and evolved into a sustained column generating the main

▼ Fig. 11.40. Close-up of top of LLST characterized by abundant Devonian slate and sandstone fragments ("Big Bang" layer) and base of interval of mud rain tuffs. Note impact sags in green ash layer deposited wet

umice fallout beds with the largest proximal umice bombs, the eruption column possibly having reached >35 km. The change from tubular to roundish vesicles in pumice may largely reflect the widening in conduit diameter and decrease of shearing mechanisms in fragmenting the magma.

Waning Plinian Stage and Second Compositional Hiatus

The peak Plinian phase, largely confined to the northern vent area, ended with a characteristic layer composed of much denser, moderately phenocryst-rich, gray, more angular lapilli representing a major explosion. These clasts are interpreted to not only reflect water access to the magma reservoir (caused again by rapid evacuation) but also intersection of another compositional boundary in the magma column. The eruption columns decreased in height and became overloaded repeatedly generating smaller, less mobile and cooler pyroclastic flows lacking overbank facies. The interbedded small-volume fallout deposits represent increasingly compositionally mixed and heterogeneous pumice lapilli populations suggesting major recycling in the vent.

Late Phreatomagmatic Phase and Third Compositional Hiatus

The spectacular deposits of the terminal phreatomagmatic phase are composed of many rhythmic repetitions of (a) massive very low T pyroclastic flows, (b) medium-grained, cross-bedded, high energy dune deposits and (c) coarse-grained breccia layers, the latter two laid down by laterally moving base surges and debris jets (Fig. 12.23). The lapilli are dense, cauliflower-shaped, gray, extremely crystal-rich, and full of small chips of Devonian slate. The deposits reach proximal thicknesses of more than 30 m but were deposited mainly within 5 km of Laacher See. They are thus of much smaller volume than the preceding dominantly pyroclastic Plinian phases. All structural and textural criteria indicate a strong water influence most likely caused by massive ground water influx into the partially emptied and collapsed magma chamber (Chap. 12). Collapse of the walls of the magma reservoir is also indicated by the presence of abundant contact metamorphic country rocks and a host of different types of crystal cumulates some with vesicular glassy matrix indicating rapid eruption of *crystal mushes*. Quenched olivine-bearing basanite lapilli suggest influx of new hot magma into this more mafic part of the magma column, possibly having triggered the eruption.

Synvolcanic Lake Formation

A temporary *tephra dam* formed prior to the end of the eruption at a canyon-like narrow of the Rhine valley, where huge masses of pumice and ash had been washed together. A lake, at least 18 m deep, rose behind this natural dam (Fig. 11.33) (255). A clean pumice layer near the top of the stratigraphic column of the deposit in the lowland close to the Rhine is up to 30 cm thick and consists entirely of low-density ($<1\,g/cm^3$) pumice from the initial phase of the eruption (LLST). It is interpreted to represent light-weight pumice from the early stage of the eruption to have risen to the

◀ Fig. 11.41. Graben formed following eruption of Lower Laacher See Tephra. Subsidence of sector graben (growth fault) radiating from eruptive center took place rapidly over a few hours or days and terminated prior to eruption of Middle Laacher See Tephra as shown by horizontally bedded undeformed beds overlying the horst-and-graben morphology. The subsided block itself is faulted and the soft tephra deposits are strongly deformed along the margins of the graben. Wingertsberg (Laacher See area, Germany)

▶ Fig. 11.42. Laacher See Tephra reworked during drainage of dammed-up lake along Rhine River ca. 6 km east of Laacher See crater (Germany). The light-colored pumice lapilli are moderately well-rounded. The dark clasts are dominantly Devonian slate fragments and lesser amounts of pyroxene, amphibole and other phenocrysts. Tan-colored Rhine sand is locally admixed

surface of the lake to form a *pumice raft*. The floating pumice layer had covered the lake and was left behind in marginal flooded shallow water areas following drainage and finally rapid collapse of the unstable dam. The lake was emptied cata-strophically, *flood-wave deposits* being found as far north as Bonn, ca. 50 km downstream from the volcano. Next to the Rhine River, spectacularly cross-bedded deposits composed of rounded pumice, dark slate fragments and brown Rhine sand reflect the phase of reworking (Fig. 11.42).

Waning Phase

A brief phase of gullying (syneruptive reworking) that may represent not more than a few days or weeks was followed by an extended phase of low energy phreatomagmatic eruptions. These result-ed in fine-grained massive ash beds, some rich in accretionary lapilli, and with a large proportion of finely comminuted slate fragments (315). The deposits probably represent muffled eruption columns making their way through the collapsed conduit system. These widespread massive ash de-posits which overlie the spectacularly cross-bed-ded deposits resulting from rapid lake drainage superficially resemble loess layers. Similar silt-sized ash deposits are known from many volcanic areas, such as in Oregon or Japan. This phase may have lasted several months contrasting with the two to three (?) day-long main phase.

Posteruptive Reworking

The thick mantle of freshly deposited tephra had completely disrupted the drainage. Huge volume of tephra were rapidly reworked after the eruptio had terminated as is common following majo Plinian eruptions (Chap. 13, Pinatubo eruption). Numerous steep-sided erosional rills were proba bly caused by heavy rain cutting quickly through the friable ash covering the landscape by up to several tens of meters. The reworked deposits ar poorly sorted, horizontally bedded, dominantl coarse-grained, xenolith-rich and reach up to 20 m thick between Laacher See and the Rhin River. They are interbedded with local lake and lahar deposits.

Climatic Impact

The Laacher See eruption was once thought to b responsible for the younger Dryas, a 1,000-year long cold spell recognized in Europe and North America that interrupted the gradual warming during the Holocene. The age difference between the Laacher See eruption and the beginning of the younger Dryas is, however, about 200 years, too long for the eruption to have had an influence on this cooling. Despite this, the sulfur-rich magma composition, reflected in the beautiful blue, sul fur-bearing mineral haüyne (Fig. 11.43), had a major impact on climate for several years, at leas in the northern hemisphere (125, 313).

The Present

Today, CO_2-dominated gases are escaping from vents throughout the entire Eifel volcanic field, where in many places they are of economic importance. In addition, a strongly CO_2-bubbling area about 200 m long occurs along the east shore of the Laacher See lake (Fig. 4.22). The composition of these gases is magmatic and closely resembles those of Lake Nyos (119), a maar in Cameroon that erupted in 1985 (Chap. 13). The Nyos eruption is interpreted as due to the transfer of CO_2-charged deep waters to the surface by a landslide and rapid degassing of these waters, rather than due to a new influx of gases or magma. The area around the Laacher See basin is characterized by elevated microseismic activity (3). The question of whether the Laacher See volcano is extinct or merely sleeping will be discussed in Chapter 13.

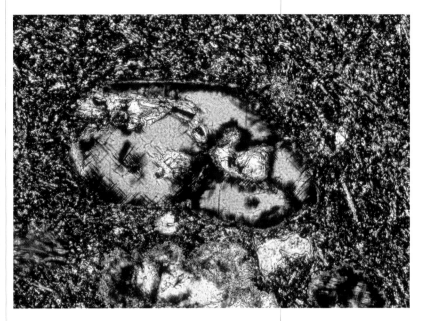

▲ Fig. 11.43. The blue color of haüyne, a characteristic phenocryst phase in the Laacher See phonolite, is due to a SO_4 group in the crystal structure. The phase was stabilized in the more oxidized upper part of the sulfur-rich magma column. Diameter of crystal 2 mm

Summary

Pyroclastic flows are some of the most mysterious and dangerous volcanic phenomena we know. They are also one of the most dramatic ways in which a volcano can erupt. They are fast, commonly very hot, and thus likely to destroy all life in their path. They are associated with ash-poor hot blasts, which can expand far beyond the valleys in which the main flows travel. It is frustrating and disconcerting that observers are unable to see what happens within pyroclastic flows because the bulk of the material travelling close to ground is completely hidden by billowing ash clouds that rapidly rise from the speeding flows. The character of small-volume, dense, pyroclastic block flows and their deposits had become recognized early in the twentieth century, based on the eruption of Montagne Pelée on Martinique in 1902. However, the Laacher See type (pumice-rich) and the huge, commonly strongly welded plateau ignimbrites (e.g., Yellowstone) were interpreted by most workers as lahar deposits or lava flows respectively until the 1960s.

Three basic types of pyroclastic flows/deposits can be distinguished:

1. The deposits of pyroclastic flows rich in vesiculated clasts are often termed ignimbrites. These pumice-rich flows are fed by ash fountains during highly explosive Plinian eruptions. The giant, commonly plateau-forming ignimbrite sheets are almost exclusively erupted from large calderas or volcano-tectonic basins that form in response to rapid evacuation of major magma volumes. Most of the very voluminous ignimbrites are mineralogically and chemically zoned and may mirror compositional gradients and layering in the magma reservoirs. The crustal roofs of the partially emptied magma reservoirs collapse during and after eruption of the ignimbrites.

2. The type of nuées ardentes erupted from Montagne Pelée in 1902 often form when parts of a slowly rising viscous dome of evolved magma collapses gravitationally and explosively. This generates a mixture of slightly vesicular hot lava blocks and ash, formed by attrition of the blocks when they tumble down the slopes of volcanoes. These valley-confined pyroclastic block flows are generally of small volume, are rarely more than 5 km long and their deposits can sometimes be confused with volcanic slope sediments deposited cold.

3. Pyroclastic flows are commonly preceded and/or accompanied by surges, or highly dilute pyroclastic density currents, which commonly leave only thin, commonly cross-bedded or planar bedded deposits. Surge deposits are a curious type of pyroclastic sediment whose fine grain size and thinness belie their often destructive character. Despite their low solids content, surges can be a major hazard beyond the valley-filling main pyroclastic flow deposits.

Frequently debated questions include the conditions under which pyroclastic flows form directly, without formation of an eruption column or from a crater by low ash fountains. Other disputed questions concern the amount of fine ash that can be elutriated from pyroclastic flows, the origin of hot surges at the beginning of many pyroclastic flows, and transport and depositional mechanisms in general.

No eruptions of very voluminous ash flows issuing from caldera ring faults have ever been observed with the exception of Tambora volcano erupting in 1815 (334). Fortunately, they are much rarer than the small ash, pumice, or block flows. The devastation that would be caused by such huge eruptions in densely populated areas would result in an unprecedented catastrophe.

The eruption of Laacher See volcano about 12 900 years ago is in many respects a classic Plinian eruption, representing the successive evacuation of a strongly compositionally zoned and layered magma volume. This eruption was also influenced by interaction with environmental factors, especially several episodes of groundwater intrusion into the partially emptied and collapsed magma reservoir. Major volumes of tephra resulting from ash flows and fallout were washed together at a constriction of the Rhine valley and led to temporary damming of the Rhine River near the end of the eruption. When the unstable dam collapsed, the up to 18-m-deep lake was rapidly discharged in the form of flood waves. The high eruption columns entered the stratosphere and significantly impacted the climate.

The reconstruction of the eruption of Laacher See volcano and the analysis of the observed eruption of Mount St. Helens (Chap. 10) have made it clear that external water influences the activity of volcanoes in many, and sometimes decisive ways. The processes during the interaction of magma and water and the volcanoes and deposits generated in this manner are the subject of the next chapter.

Fire and Water

The poet and scientist Goethe was a very gifted observer who made several important discoveries in anatomy and botany. Nevertheless, although he had the chance to personally observe evidence for higher temperatures below the Earth's surface, having visited Vesuvius and also Etna in Sicily, he still sided with A.G. Werner in the heated debate between the Neptunists and the Volcanists during the late eighteenth century (Chap. 1). However, in order to clarify his mind about the true nature of volcanoes and the source of the fire in the Earth, Goethe repeatedly visited an area of young scoria cones near Cheb in the west of the Czech Republic. Outcrops of the Late Pleistocene nephelinitic scoria cone Železná Hůrka are well-preserved, some 20 km south of the city of Cheb (Fig. 12.1). These show a textbook unconformity between strongly water-influenced eruption products, forming the outer initial tephra ring, overlain by partly welded lava spatter and scoria representing the hotter later stage of the eruption. It took almost 200 years, however, before scientists were able to correctly interpret the eruptive and fragmentation mechanisms based on this type of lithological evidence.

The notion that external (i.e., nonmagmatic) water can play an important role during volcanic eruptions has been with us ever since Alexander von Humboldt published his famous essay on volcanoes in 1823 (see below). The French scientist Buffon even assumed a connection between eruptions of Etna volcano and seawater, and thought one could extinguish volcanoes if one would seal them against seawater. However, these views about the role of water in volcano evolution were not followed up. Volcanologists traditionally assumed the major explosion or eruption processes to be due to oversaturation of a magma with a particular gas phase, especially H_2O, formation and growth of bubbles and buildup of pressure within them until they ruptured (Chap. 4). For example, maars, characteristic and common types of small volcanoes surrounded by low rims of lithic-dominated ejecta, were widely interpreted until the 1970s as gas explosion funnels caused by liberation of CO_2 from alkaline mafic magmas. The failure to recognize the error in this interpretation well illustrates how some poorly-grounded hypotheses can prevail for many decades. In this particular case, scientists had correctly interpreted

that CO_2 can escape from rising magma far below the Earth's surface – as proven by CO_2 inclusions in crystals of salt deposits traversed by basaltic dikes – and also observed that maar deposits are unusually rich in rock fragments. However, the time was not ripe to look at volcanic deposits in more detail and with a broader scope and employing a wider spectrum of observational tools and models.

> And so he [God] began to pit water against fire in the stomach of the volcano. Ivan, the volcano, had to swallow the waves that stormed within his overheated stomach. He swallowed and swallowed and swallowed for many years, until it became too much for him to bear and he exploded with a tremendous roar.
>
> *I. Gantschev, Iwan der Vulkan, Salzburg, 1980*
> *Translated from the German*

The Discovery

The discovery – or rediscovery – of the ubiquitous interaction between magma and external water followed a classical course. When scientists began to take a closer look at volcaniclastic deposits as well as actual eruptions, they soon realized that the gas-explosion hypothesis was no longer tenable. The trigger was the analysis of the products of a notorious explosive volcano, Taal volcano in the Philippines, probably the subject of Alexander von Humboldt's remarks (163) translated from the German: *In Central America (Guatemala) and on the Philip-*

▲ Fig. 12.1. Crater unconformity of the late Quaternary melilite-nephelinite scoria cone Železná Hůrka (Czech Republic). The older, bedded, colorful deposits dipping to the right are very rich in rock fragments and apparently formed by interaction of rising magma and groundwater. The dark, coarse-grained, slightly agglutinated scoria deposits dip left into the center of the crater

▲ Fig. 12.2. Vertical eruption column and basal base surges during an eruption of volcano Capelinhos in shallow water off the coast of the island of Fayal (Azores) in 1957. The deposits formed during this eruption are shown in Fig. 13.11 (Porter Studio)

pines, many natives formally distinguish between water and fire volcanoes, volcanes de agua y de fuego. With the first name, they characterize mountains, from which subterranean waters erupt, accompanied by strong earthquakes and loud noises.

Taal Volcano has grown in a large lake and the style of its 1965 eruption and the products of the eruption are exemplary (236). This eruption was an eye-opener to volcanologists. Field and laboratory analysis of deposits in several parts of the world subsequently provided a host of diagnostic criteria. Gradually the scientific community came to realize and accept that buoyantly rising magma or its advancing gas phase encountering groundwater, shallow water or water-rich sediments or water entering collapsed conduit systems appeared to constitute a second major mechanism

for causing explosive volcanic eruptions. Indeed, there is hardly a volcanic eruption that has not been influenced in some ways by magma–water contact. That such processes must occur abundantly is not surprising since there is plenty of water in the upper part of the Earth's crust, especially in highly fragmented and often very porous volcanic edifices–apart from the obvious situation where volcanoes grow in the sea or in lakes (Fig. 12.2). It is thus easy to understand that the initial, often very fine-grained products of a large explosive eruption have been generated by magma–water interaction and erupted in a manner significantly different from hot pyroclastic eruptions. These highly explosive water-influenced eruptions commonly precede, and possibly trigger, subsequent explosive pyroclastic eruptions. Magma–water interactions and the resulting deposits and volcanic landforms are the topic of the present chapter. The analytical work in the field was followed by experimental studies, comparison with vapor explosions known and feared in industry and theoretical models. Work is continuing on all these fronts with a refinement of criteria and experimental and theoretical efforts to better understand the basic mechanisms during encounter of magma and water.

The groundwork for recognizing volcanic eruptions due to, or influenced by, water–magma interactions was laid in the late 1960s to mid-1970s. Several robust criteria were recognized as diagnostic. Some of them are listed below as an example for how scientists assemble several different types of evidence from the deposits themselves as well as observed eruptions in order to reconstruct processes in the volcano-magma system prior to and during an eruption and deposition. This long list also illustrates the necessity to support changes in paradigm by particularly painstaking assembly of many types of proofs.

The Particles

- Glassy, angular and poorly vesicular to non-vesicular particles, evidence for quenching by water and thermal shock but incompatible with fragmentation by magmatic gas expansion

- Cauliflower- to potato-shaped, dense, commonly bread-crusted bombs rich in non-metamorphosed rock fragments indicating mixing of magma and fragmented country rock just prior to eruption

- Abundance or dominance of lithic fragments (Figs. 12.3, 12.4, 12.6, 12.7) evidence for fragmentation within the rock envelope around

◀ Fig. 12.3. Water-rich tephra, deformed plastically by impact of two pieces of basaltic scoria. Tuff ring on Linosa island between Sicily and Africa

▼ Fig. 12.4. Phreatomagmatic deposits of Hoyo Negro crater, erupted in July 1949 on the island of La Palma (Canary Islands). Pine trees in the lower part are still standing but were stripped off their branches, while some pine trees (*arrow*) have been transported out of the crater. Deposits consist almost entirely of older country rock, fragmented when rising magma interacted with groundwater

► Fig. 12.5. Thinly bedded and strongly cemented phreatomagmatic tuffs deposited on a soil horizon (massive brown layer in lower part of photograph) (*arrow*). Trees rooted in the paleosol are shown by vertical molds and indicated that the tephra was deposited very rapidly (see also Figs. 9.24–9.26). Eppelsberg scoria cone (Eifel, Germany)

▲ Fig. 12.6. Poorly sorted deposits of late Pleistocene Totenmaar (the maar type locality) consisting to more than 90% of Devonian slate and sandstone fragments (Eifel, Germany)

● Impact of country rock blocks into ductile plastically deformable sediments (Fig. 12.3) and clumping together of small particles during deposition reflecting wet to water-rich deposition of particles. Unburned wood, tree trunks (Fig 12.4) or their molds (Fig. 12.5).

The Eruptive System

● Short-lived but repetitive eruptive pulses, indicating numerous interruptions of rising magma by contact with water resulting in sudden but frequent explosive emptying of the upper conduit. Deposits are well-bedded
● Small eruptive volumes
● Low white steam-rich envelopes around black central stubby tephra jets with finger-like trails behind ballistically ejected clasts (Fig. 12.8)
● Characteristic eruptive environments such as fissures breaking through aquifers or during specific stages of otherwise pyroclastic eruptions
● High energy of explosive eruptions. The enormous overpressures calculated from the size of blocks in some deposits cannot be explained by the decompression of magmatic gases
● Characteristic volcanic landforms

Eruptions, governed or triggered by magma–water contact, are variously called *phreatomagmatic, hydromagmatic, hydrovolcanic* or *hydroclastic*. If no new juvenile lava fragments are present in such deposits, they are called *phreatic*. The style of phreatomagmatic eruptions ranges widely from relatively weak interactions, e.g., when low vis-

the conduit as an integral process arising from magma-water contact
● Abundance of small particles reflecting particularly intense fragmentation of magma (Fig. 12.5) and/or low permeability sediments confining groundwater.

The Depositional System

● Low temperature during deposition
● Poor sorting and lenticular or cross-bedding indicating lateral turbulent ground transport (Chap. 11). Transport of particle populations in fast, steam-rich ground clouds, so-called *base surges* (Fig. 12.2) resembling ring clouds observed during explosions of nuclear bombs

◄ Fig. 12.7. Black basaltic lapilli deposits weathered yellow at the top (where people are standing), overlain by thick phreatomagmatic deposits, containing clay fragments up to 2 m in diameter (*arrows*). The eruption became increasingly pyroclastic with time, as shown by the increase in black basanitic lapilli towards the top. The massive character of the lower 3.5 m of the phreatomagmatic deposit is due to abundance of finely disseminated clay and cohesive flow resulting from the interaction of rising magma, water and Tertiary clay deposits. Plaidter Hummerich (Eifel, Germany)

cosity lava flows enter water, to extremely energetic explosions, in which deposits are generated that consist of more than 90% fragmented country rock.

Rapid Cooling

For the field volcanologist, much of the evidence from which to deduce fragmentation, eruptive and transport mechanisms lies in the nature of the particles and the structure of the deposits. This type of analysis has been foremost in recognizing volcanic eruptions in which external water played a major role. The typical features of particles in hydroclastic deposits are their glassy nature, angularity and low vesicularity, all plausibly explained by quenching of a magma/lava upon contact with water (Fig. 12.9).

I discussed in the preceding chapters how magma is fragmented in pyroclastic eruptions, a process in which the expansion of gases and the contact with air has a cooling effect. The analysis of deposits of some pyroclastic flows and lava fountains shows that particles can be deposited in quite a fluid and hot state when they are erupted rapidly, i.e., close to the liquidus temperature of magma (between about 750 and 1 200 °C). Cooling thus occurs during the mixing of hot particles and gas with ambient cold air mainly in the convective part of eruption columns and, much less efficiently, during rapid flow along the ground. Water, on the other hand, can cool much more efficiently than air because it has a higher heat capacity and absorbs much heat of evaporation when it is heated above its boiling point.

It was shown in Chapter 3 that much heat is consumed during partial melting, while heat of crystallization is liberated from magma during the reverse process. When lava is quenched during contact with water, it cools so fast that no crystals, especially complex silicates, can grow in the melt. The melt is cooled much below its solidus temperature, and thus no heat of crystallization can be liberated. The result is an *undercooled melt* (Fig. 12.10). Glass can form well only below the solidus temperature at the *transformation temperature* (T_g), which is about 700 °C for basaltic melts (289). In industry, a substance is called a glass when its viscosity exceeds ca. $10^{11.5}$ Pa s. The lower the initial viscosity of magma, because of its chemical composition or the high concentration of H_2O, the more difficult it is to quench the melt to a glass. This is because crystal nuclei form very quickly at low viscosities and the growth rates of crystals are also very large. Hence, highly viscous rhyolitic melts form glass easily (ash, pumice, obsidian lava flows), while rapidly formed small *quench crystals* are found even in relatively quickly cooled basaltic particles.

The volume of glass is lower than that of a melt by a factor of about 5–10%. This is the reason why the thermal shock that occurs during quenching generates stresses in the glass that are released by fracturing. Angular blocky particles with sharp edges are the result. When the melt

▲ Fig. 12.9. Two rhyolitic breadcrust bombs 30 cm in diameter from partly phreatomagmatically generated Late Quaternary ignimbrite KP-1. Kutcharo caldera (Hokkaido, Japan)

forms thin threads or layers, it can granulate by such volume contraction. If only the surface of a lava flow comes in contact with water, a glassy skin or crust forms. Because glass conducts heat very poorly, further cooling of the fluid interior of a lava flow is retarded. A subaquatic lava flow will thus cool quickly in the beginning, then increasingly more slowly; the rind grows, simply put, proportionally to the square root of the cooling time, following the formula

$$D = K\sqrt{t}$$

where D is the thickness of the glassy crust, K is a constant and t is time.

Because cooling rates decrease toward the interior of a lava flow, increasingly larger crystals can form. Glassy crusts generally do not exceed about 1 cm in thickness (Fig. 12.11).

This is not the only effect of quenching. On the one hand, quenching inhibits degassing of lava; on the other hand, heat is transferred very quickly to the water. Due to the very high temperature of basaltic magmas (1100–1200 °C), water is heated at the point of contact greatly above its boiling temperature, even at large water depths where the boiling temperature is higher because of the higher hydrostatic pressure (Fig. 12.12). However, even in lava flows that flow from land into the sea and whose advance below water has been observed directly (237), visible vapor is often not generated. This is probably due to the fact that an

◀ Fig. 12.8. Eruption column of the almost exclusively phreatic eruption of the Usu volcanic complex (Hokkaido) in April 2000. Photo courtesy Hokkaido Regional Development Bureau

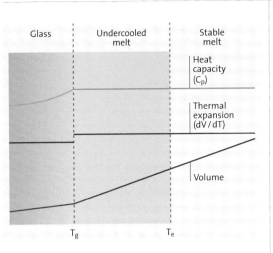

▲ Fig. 12.10. Changes in the physical properties of a magma that undergoes a change from a stable to an undercooled melt and to a glass at temperature T_e. Heat capacity (C_p) and thermal expansion (dV/dT) only change at a lower temperature called Tg (44)

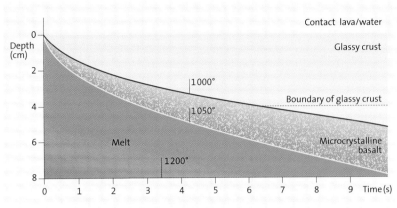

▲ Fig. 12.11. Formation of a glassy crust and microcrystalline zone (1 000 – 1 050 °C) at the margin of quenched lava. The thickness of the glassy crust and the microcrystalline zone increases with time until the melt has completely solidified. The lower part of the initial glassy crust is tempered by the addition of heat from below and becomes microcrystalline (227)

extremely thin vapor film develops at the boundary layer between a hot lava surface and water. This film is a good insolator, while the water vapor condenses immediately to water. This heated water rises convectively and is replaced by cold water.

Tephra particles in the Surtsey deposits and similar initial tephra cones of other volcanic islands are relatively rich in vesicles. This shows that degassing of the magma had already started in the upper few tens of meters below the sea surface, significantly contributing to fragmentation of the magma. The respective roles played by steam explosions and granulation by thermal shock are still uncertain. The properties of eruption columns in phreatomagmatic eruptions have recently been discussed in more detail (179).

The High Explosivity of Magma – Water Interactions

Fuel-coolant interaction (FCI) is the term for a process that in industry can have disastrous results. In FCIs, strong explosions are caused by the interaction of hot fluids (fuel), e.g., molten metals, and a cold fluid (coolant), e.g., water. Such a situation can arise during the meltdown of a nuclear reactor, which is the reason why scientists have studied this process in detail. In order to generate a sufficiently large surface for efficient heat exchange, the hot melt must be fragmented completely in milliseconds to generate a steam explosion. Fragmentation of melt is thought to be caused by the collapse of vapor bubbles, grown at the fuel-coolant contact, and their injection into the melt. Several scientists (63, 87) have invoked this mechanism during the eruption of Helgafjell volcano on the Island of Heimaey, off the coast of Iceland. They cited FCI as

a possible trigger of explosive hydroclastic eruptions for the first time in the volcanological literature. A recent summary is given in (445).

Explosive phreatomagmatic magma – water interactions are generated when rising magma encounters groundwater or surface water, in other words, when the heat, i.e., the thermal energy of a magma is rapidly transported into the fluid encountered. The kinetic energy of explosions resulting from this interaction are much higher than in dry magmatic eruptions, other conditions being equal. One important boundary condition is the depth at which magma encounters water, because the sudden change of water with a density

◄ Fig. 12.12. Density change of water (g/cm³) depending on pressure and temperature. The *dashed lines* represent three different geotherms (418)

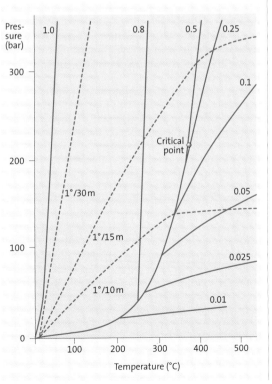

▶ Fig. 12.13. Change in volume ratio of steam ($V_{H_2O}^{gas}$) to fluid water with decreasing water depth

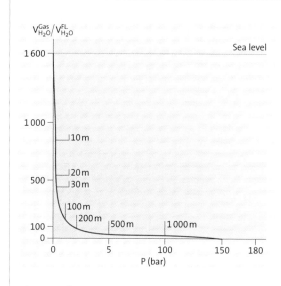

▼ Fig. 12.14. Dependence of form and structure of so-called monogenetic volcanoes on the ratio of magma interacting with external water. The largest mechanical energy, during which maars and tuff rings are formed, is reached at a ratio of about 0.1–0.3 (325)

of $1\,g/cm^3$ to vapor with a density of ca. $0.001\,g/cm^3$ only holds at the surface of the Earth. With increasing pressure and temperature, the density differences become very small. At the *critical point*, reached at ca. 350°C and 220 bar for fresh water (Fig. 12.12), the volume difference between water and vapor is negligible. In this case, liquid and steam cannot be clearly distinguished from each other and the state of the system is called a *super-critical fluid*. When water is heated to 100°C at a pressure of about 1 kb (ca. 3 km below the surface), its volume increases only by a factor of 6. At the Earth's surface, steam has a volume that is about 2000 times larger than liquid water (Fig. 12.13). Because the volume ratio of steam to water increases with decreasing pressure and therefore depth, explosions are most effective in the upper about 300 m, especially in the uppermost 100 m of the Earth's crust.

At very low pressure, heat transferred from magma to water along the boundary surface is limited to the formation of isolating vapor films the *Leidenfrost phenomenon*. These isolating films can collapse by seismic shock waves or spontaneous local collapse of the vapor film. In other words, the thermal energy of a magma is then transferred into seismic energy. This collapse can induce fragmentation of the magma and the country rock by increasing the critical strain, resulting in brittle fracture.

When lava flows of low viscosity enter water, explosive reactions are rare. Only when lava and water come together in a more or less enclosed space, e.g., in a dike, a sill, in a lava tube or in the crevasses of a spinose aa-type lava flow, magma and water vapor cannot escape and will thus mix with water. The seismic energy generated during these processes can then trigger the explosive reaction. When the increasing volume is not released by valves (such as in a pressure cooker), which in nature are fractures or pores in the rock roof above a magma column, an enormous excess pressure with high explosive potential can develop. Hence, the other boundary condition for a steam explosion is a partially closed system, at least temporarily.

This boundary condition is manifested very clearly in rare explosions in active geyser fields. Geysers develop where groundwater is heated by deeper magmatic heat sources. A water column below an open vent can become heated at its base to much more than 100°C, but it will not boil because the hydrostatic pressure at that depth is too high. When the heated water expands and rises convectively, the evaporation curve can be reached and the water is thrown out explosively, cools, falls back and a new cycle starts. If a vent becomes sealed completely by siliceous or calcareous precipitates, a process called *self-sealing*, high excess pressures can develop, which eventually blow out the temporary roof explosively.

The high explosive energy of vapor explosions may be due to the high ratio of the coefficient of the thermal expansion of H_2O to its total compressibility in the temperature and pressure field of the upper crust (324). At temperatures of 100°C and a pressure of 1 atmosphere, this ratio is about 15 bar/1°C. In other words, when the water becomes heated and its volume remains constant, the pressure rises by 15 bar for each degree of heating. A further boundary condition for the explosivity of some steam eruptions is the very fast intrusion of magma, such as in a dike (79).

The larger the specific surface area of the magma–water contact, the more effective the heat transfer is. In other words, a completely fragmented magma is much more explosive when contacting water than a smooth contact around a large magma body. An especially high and, therefore,

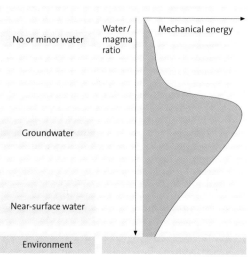

explosively efficient ratio of surface area to volume in tephra particles can be reached by simultaneous degassing. In other words, magmatic degassing at low pressure in the presence of water can lead to particularly explosive eruptions. The effect of a *rarefaction wave*, which migrates into the magma when the shock wave during an explosion migrates outward, by inducing nucleation of bubbles in the magma and inducing fragmenting is still poorly understood (e.g., 172).

Explosion experiments have been carried out with water and thermite, a highly exothermic mixture of Al_2O_3 and Fe, whose composition can be approximated to a silicate magma by addition of SiO_2. These show that the most effective explosive interaction (i.e., the transformation of thermal into mechanical energy) occurs at a magma/water mass ratio of about 0.1–0.3 (325, 429) (Fig. 12.14). At lower water pressure (i.e., at ratios of < 0.1), Strombolian (pyroclastic) explosions dominate. At higher ratios (>0.5), lava flows form. The presence of superheated water (i.e., over 100°C) appears to be an important prerequisite for the explosive transformation into steam (278). In this state, the chemical potential for evaporation is lower than that necessary to overcome the pressure of the surrounding fluid.

The interpretation of the origin of maars by steam explosions was initially based on detailed analyses of their deposits. These conclusions were unequivocally confirmed by the characteristics of deposits of recent maar-forming eruptions in Alaska in 1977 (174), as well as observations of actual eruptions and inspection of the resulting tephra deposits. Examples include Capelinhos volcano, off the coast of the island of Fayal (Azores, 1958; Fig. 12.2), Surtsey (Iceland, 1963), Miakejima (Japan, 1983, 2000–2002) (116) and Usu volcano (Hokkaido, Japan, 2000; Fig. 12.15). Pulsating, staccato-like explosions, interrupted by short periods of quiescence, are typical for all observed phreatomagmatic eruptions and result in well-bedded deposits.

Open Water Conditions

If we reduce the mythical elements, fire and water, to their real nature, the encounter of magma (lava) and water should be explosive. Yet most volcanoes on Earth erupt rather gently along mid-ocean ridges, albeit below a heavy lid of water, on average 2 500 m thick (Chap. 5). The simple physical reason for the absence of strong explosions during these eruptions is the load of the water column inhibiting both degassing of the magma as well as transformation of the water into voluminous low density steam.

The most common environment where the heat of magma encountering liquid water leads to sudden vaporization and strong explosions is the shallow water. In the introduction to this volume, I mentioned the origin of the volcanic island Surtsey (Iceland) in 1963 and the difficulty at that time to describe and analyze the processes during the encounter of water and magma. The attempt of growing subaqueous volcanoes to become islands has been observed repeatedly, especially along the Izu-Bonin Arc south of Japan. Apart from seismic shocks that may precede any visible activity, the first signs sometimes initially noted by fishermen are turbid and turbulent water, sulfurous odor and floating dead fish. It is the subsequent stage of intense and violent interaction of rising magma and water that is particularly hazardous. Booming detonations accompany the powerful and destructive blasts repeated in intervals of seconds to minutes, such activity lasting from weeks to months, depending on the magma supply rate to the surface. When the research vessel No. 5 *Kaiyo-maru* approached the emerging submarine volcano Myojinsho in 1952, blasts completely wrecked the boat taking all 31 people on board with it (Chap. 1). The power of such explosions appears to be especially strong when rising bulbous masses of very viscous magma crack, allowing water to deeply penetrate – as in the case of the Myojinsho dacite dome. Water entering the complex fractures has little free space to expand when suddenly heated much above its boiling temperature thus developing enormous overpressures when flashed to steam.

The initial stage at Surtsey when seawater was able to enter the submarine and early subaerial breached vents was also explosive, but observations from nearby were impossible. The eruption columns in the explosive stage showed the typical properties of phreatomagmatic eruptions: they were relatively low (a few tens to hundreds of meters high), they were black because of the high concentration of quenched tephra particles and became white when the water vapor condensed. The expulsion of larger blocks generated a typical cypress-like or cock's comb-type form at the edges of the stubby eruption columns. When a more or less closed tephra ring had formed around the vents at Surtsey in April 1964, keeping the influx of seawater to a minimum, basaltic magma erupted in the form of lava fountains, from which lava flows developed and advanced to the shore. This stabilized the ephemeral early tephra ring. Surtsey is still alive and by now colonized with many forms of life, excluding tourists who are not allowed to enter the pristine environment.

All volcanic ocean islands have gone through these stages (Chap. 6). Hyaloclastite cones form the initial shallow-water stages of volcanic islands and fringe many of them in places where magma has risen through the water-rich boundary zones between the submarine and subaerial edifice (Figs. 12.16 – 12.18). Apart from Surtsey, examples are the Azores, Madeira and the Canary Islands in the Atlantic, Koko Crater and Diamond Head on the eastern coast of the Island of Oahu (Hawaii) or the still active Anak Krakatau, the new volcano that started to grow in 1927 in the void left by the great eruption of Krakatau in August 1883.

The Initial and Terminal Phases of Eruptions

Rising dikes encountering groundwater is a common situation (Fig. 12.19). For example, phreatomagmatic processes have characterized the initial phase of many scoria cones in the Eifel (Chap. 7). A very large spectrum of deposits

▼ Fig. 12.16. Cross section through the wall of a tuff ring off the northwest coast of Madeira near Porto Moniz. The unconformity at the inner crater wall (arrows) indicates transport from right to left

◄ Fig. 12.17. Graham Island tuff ring, which formed between Sicily and Africa in June 1831 and had become completely eroded and disappeared below sea level in January 1832 (author's collection). See also Fig. 6.3

caused by magma–water contact can be distinguished, depending on the amount of water, porosity of the country rock, mass and ascent rate of the magma and so on.

The classic *maars* of the Eifel are a major tourist attraction and are visited during many geological field trips in Europe (Fig. 12.20). They are also known as the type locality for the most characteristic form of a volcano: a roughly circular crater, incised into the crust and sometimes filled with water. Maars are mostly a few hundred meters in diameter and are surrounded by a low

tephra ring (Fig. 12.21) (200). *Tuff cones* sit on top of the land surface and consist of a ring of pyroclastic, commonly palagonitized tephra (i.e., tuff), secondarily altered by interaction with groundwater, or when deposited warm and wet.

The tephra deposits of steep tuff cones with small craters and those of the low tephra rings which surround maar craters are distinguished from scoria cones by the absence or paucity of lava spatter, by the smaller grain-size and excellent bedding (Figs. 12.16, 12.17). Such tephra rings may consist of more than 90% of country rock

◄ Fig. 12.19. Schematic of the interaction of rising magma in a fissure intersecting an aquifer

▼ Fig. 12.18. Schematic cross section through tuff ring formed during near-surface contact of rising magma with water horizon. The crater wall migrates outward during the course of eruption (96)

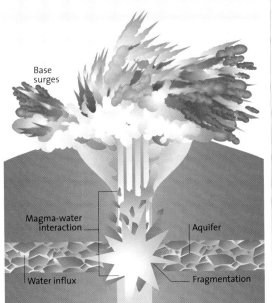

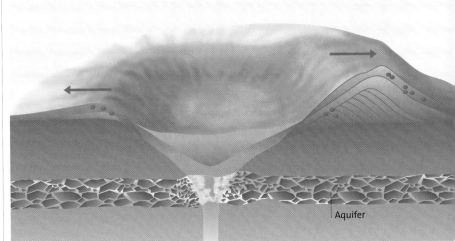

▶ Fig. 12.20. Late Quaternary Totenmaar near Daun (Eifel, Germany). See also Fig. 12.6

▼ Fig. 12.21. Holocene maar Marteles on Gran Canaria (Canary Islands). This maar has formed so recently that the crater filling continues to compact as shown by the episodically opening semi-circular ditches in the left part of the photograph (*arrows*). The maar formed when basanitic magmas rose below the axis of the major canyon Barranco de Guayadeque

agments, the diameter of large blocks in many
ses exceeding 1 m (Fig. 12.6). Next to scoria
ones, tephra rings and tuff cones are the most
mmon types of volcanoes on Earth. Lava flows
e commonly, but not always missing from these
mall volcanic edifices. The characteristics of the
posits and observed eruptions, as well as experi-
ental data indicate that the explosive encounter
f magma and water occurs at greater depth in
aars than in tuff rings.

Many large explosive Plinian eruptions start-
d phreatomagmatically, judging from the nature
f the initial deposits at the very base (Fig. 12.22).
his interpretation has led to the question whether
e contact between a rising magma column and
ater and the ensuing steam explosion and shock
ave could not represent an important trigger for
e decompression, bubble nucleation and sub-
quent explosive pyroclastic eruption of a mag-
a. In a single volcanic field, chemically similar,
ghly evolved magma erupts explosively in some
ses, but in others either does not erupt or sim-
ly oozes out as exogenous domes. The initial
hases of many highly explosive historic volcanic
uptions (such as the famous eruption of Mon-

▲ Fig. 12.22. Basal phreatomagmatic deposits of late Quaternary Plinian Laacher See eruption over-lying dark soil. The massive deposit consists dominant-ly of Devonian slate and sandstone fragments, older tuff particles and clay. Twigs and tree molds are common. Pumice lapilli in the upper part of the pho-tograph herald the begin-ning of the Plinian phase

◄ Fig. 12.23. Base surge deposits of the Upper Laacher See Tephra. Trans-port direction out of the crater (about 1.5 km to the left) to the right. Layers with dunes and antidunes are separated by coarse-grained layers that were deposited by low-angle debris jets, a combination of fall and flow

gne Pelée, 1902), are characterized by eruption
f water, mud flows and unstable dark eruption
olumns. The conclusion that pyroclastic erup-
ons of highly evolved magma are not only initi-
ed by phreatomagmatic eruptions, but can actu-
ly be triggered by magma-water interaction
02), was supported by the catastrophic eruption
f Mount St. Helens on 18 May 1980 (Chap. 10).

Magma–water encounter also characterizes
any late stages of large eruptions (Fig. 12.23).

This is evidenced by the abundance of cogenetic
plutonic nodules and contact-metamorphosed
country rocks reflecting collapse of the conduit
system and/or walls of a partially emptied high-
level magma reservoir with its rind of slowly
cooled and crystallized magma and adjacent thermally metamorphosed country rocks. It is the creation of a void and sudden pressure reduction
that may lead to wall collapse and inrush of
ground- and/or surface water into the still hot

▶ Fig. 12.24. Red oxidized slightly weathered pyroclastically generated massive scoria on top of scoria cone overlain by gray bedded phreatomagmatic deposits on the rim of Holocene Asososca maar in Managua (Nicaragua)

and partially magma-filled void. The final reaming out is thus largely governed by repetitive steam eruptions in which lithic fragments increasingly dominate as the liquid magma supply becomes exhausted.

Morphologically low areas in volcanically active regions of the world are particularly prone to phreatomagmatic eruptions. The densely populated Managua area (Nicaragua) is a typical example characterized by many very young maars and many older volcanic deposits are of phreatomagmatic origin (Fig. 12.24). Hydroclastic eruptions are a particular hazard to the capital, as demonstrated by a huge late Holocene phreatomagmatic eruption from nearby Masaya caldera generating widespread base surge deposits.

Lakes are a common companion of volcanoes, be it small crater lakes or large lakes in caldera basins. While the physical processes that characterize growth of new volcanoes in these lakes is identical to that in growing seamounts, the overall evolution of the volcanic system is the reverse. In other words, a volcanic edifice that may have started to grow hot and dry will go through especially prolonged and major phreatomagmatic stages once a lake has developed in its center. Such changes are common in long-lived caldera systems. For example, eruptions of large volume pyroclastic flows from late Quaternary Kutcharo caldera in Hokkaido (Japan) were accompanied by caldera formation. During a later powerful plinian eruption, the conduit appears to have

opened within the caldera lake that had accumulated inside the caldera basin. The great abundance of fine-grained initial ash deposits rich accretionary lapilli are clear testimony for massive magma–water encounter during the initial stage of eruption (Fig. 12.25). The magma had become pulverized much more intensely than during pyroclastic fragmentation. Moreover, the thick ash blanket covering the ground many meters thick was so wet that, when quickly loaded by subsequent thick pyroclastic flows, some of the ash became mobilized due to the high water pore pressure and squeezed up locally to form clastic dikes (Fig. 12.26). Even the pyroclastic flow deposits representing some 100 km³ of magma, contain more fine ash in the matrix than is common probably because the particles were wet and therefore less efficiently removed during flow. The terminal stage of the eruption (when magma supply had drastically dwindled) was also characterized by fine-grained ash blankets largely deposited the form of wet ash.

Phreatomagmatic eruptions repeatedly accompany a particular type of caldera collapse in a void created by lateral subterranean drainage magma. Such a process is probably much more common than appreciated up to now since withdrawal of magma preceding phreatomagmatic eruptions is difficult and mostly impossible recognize in older volcanoes. The generation phreatomagmatic eruptions during formation a caldera on top of Miyakejima volcano in Ju

◄▼ Fig. 12.25. Overview of main part of rhyolitic ignimbrite KP I erupted ca. 30 000 years B. P. from Kutcharo caldera (Hokkaido, Japan). Basal ca 5 m thick very fine-grained accretionary lapilli-rich deposits just below base of outcrop (see inset for close-up of accretionary lapilli). Three basal fine-grained flow units, each up to 3 m thick, are overlain by white up to 5 m thick coarse-grained flow units showing imbricate structures. Transport direction from left to right. Overlying fine-grained accretionary lapilli-bearing tuffs not exposed. The eruption probably took place through a caldera lake. Magma-water interaction was most violent and thorough during the initial and terminal stages but less influential during high magma discharge rates when the main pyroclastic flow pulses formed. Ca. 10 km northeast of Kutcharo caldera rim (Hokkaido, Japan)

◄ Fig. 12.26. Clastic dike cutting main white ignimbrite flow units shown in upper part of figure 12.25. The ashes were probably injected upwards when the water-saturated very fine-grained basal ashes were loaded shortly after the main pyroclastic flows had collapsed and compacted, possibly triggered by earthquakes. Photo courtesy Mari Sumita

▲ Fig. 12.27. Active volcano Poas in Costa Rica. The bedded deposits around the crater formed during repeated phreatic eruptions. Diameter of crater lake is about 50 m

2000 (Chap. 9) was interpreted as due to heating of groundwater by a small volume of magma intruded along fractures into a subsiding and most likely highly fractured rock column (116) (Fig. 9.44). This raises the question if the formation of some maars may be due to rock column subsidence into a void following lateral drainage of magma rather than interaction of an ascending dike with groundwater. This mechanism has the advantage that much of the fracturing of country rock occurs during rapid vertical subsidence aided by expansion of groundwater flashed to steam rapidly heated when magma invades the open framework breccia. For example, the formation of Marteles maar on Gran Canaria (Fig. 12.21) may be due to such a mechanism since a scoria cone and lava flow formed just prior to that of the maar less than 1 km away.

Phreatic Eruptions

Many volcanoes erupt steam and exclusively lithic nonjuvenile particles alternating with pyroclastic or lava eruptions (Figs. 12.27). Ash and steam eruptions can reflect very high level dome intrusions (*cryptodomes*) that get stuck a few tens or hundreds of meters below the surface, as is common in eruptions of Usu Volcano in Hokkaido (Japan) (Fig. 12.15). The disastrous eruptions of

Usu volcano from March 31 to August 2000 during which 300 houses were completely destroyed and some 500 partially damaged, are a recent example of the ubiquitous occurrence of steam and mud eruptions. Of the 65 (!) craters formed during the eruption all but three erupted exclusively country rock fragments, dominantly in the form of wet ash plastering the landscape with many layers of mud and excavating holes up to 50 m deep and 100 m in diameter (Figs. 12.8, 12.15).

Hydrothermal explosions (see above) representing a special type of phreatic eruption generated in geyser fields (see above) can excavate craters up to 1 km in diameter (244, 246). The famous sinter terraces and geysers of Tarawera in New Zealand were destroyed in 1886 by such phreatic explosions.

The analysis of particles ejected during an eruption harbor fundamental information whether or not new magma is involved in an explosive eruption. The failure to make a clear distinction between nonjuvenile fragments generated during purely phreatic or steam eruptions and particles of newly risen magma led to one of the largest volcano-hazard induced evacuations in history (Chap. 13).

Vulcanian eruptions are short (seconds to minutes) discrete explosions commonly associ-

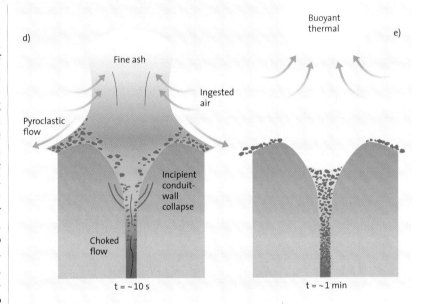

◄ Fig. 12.28. Model of explosive processes in and above a volcanic conduit during a Vulcanian eruption.

a) Conduit and crater before the eruption;

b) crater and conduit ca. 0.2 s after initiation of the eruption;

c) events about 2 s after initiation;

d) ca. 10 s after initiation;

e) ca. 1 min after initiation (242)

ted with abundant ballistic transport of lithic blocks or incandescent lava bombs and commonly stubby eruption columns composed largely of nonjuvenile fragments (242). They are typical for stratovolcanoes of intermediate composition. Shock waves are commonly observed indicating that the fluid is overpressurized at the vent with respect to the atmosphere (Fig. 12.28). They are named for the 1888–1890 eruptions at Vulcano in the Aeolian Islands. The eruptive products have many similarities with those of undisputed phreatomagmatic eruptions resulting in superheating of groundwater (302), the reason why these eruptions are discussed in this chapter. Other models include pressurization of a lava cap or lava dome by rapid microlite growth or closed-system degassing at depths of several km (242). Eruptions of volcanoes such as Sakurajima (Japan), Ngauruhoe (New Zealand) or Vulcano (Italy) are sufficiently distinct for several different mechanisms to operate and produce similar phenomenological characteristics.

Steam explosions may also result from lava flows or pyroclastic flows advancing over wet ground. Superheating of the sealed water – or captured ice blocks as in the pyroclastic flows generated in 1980 during the initial eruption of Mt. St. Helens – may result in spectacular craters, tephra rings or spiracles as in lava flows (Figs. 9.9, 10.18).

Rainwater

When I led a student field trip in 1976 to the Aeolian islands, we were initially unable to visit our main goal Stromboli volcano because of frequent rain showers. When the weather finally cleared, explosive activity in the famous crater had visibly increased, obviously caused by rainwater that had entered the conduit system. It is now undisputed based on observations in many volcanoes that

heavy influx of rain water can significantly increase fragmentation of hot lava especially in the outer rind of actively growing domes where large hot rock surfaces are exposed for some time. The major explosive eruption of Stromboli on April 5, 2003 following a few days of strong downpour may turn out to have been similarly triggered by magma-water interaction.

Fire and Ice

Table mountains or mesas are curious but impressive landforms. Most common mesas, such as those in New Mexico (the refuge of the Hopi Indians) are simply the remnants of hard rocks resisting erosion. These are commonly well-cemented sandstone, overlying softer rocks, such as shales. Hard, mostly basaltic lava flows or well-indurated welded ignimbrites may also form the hard cap to mesas or table mountains, such as in Oregon or

South Africa. These are not volcanoes proper. However, table mountains that are landmarks in Iceland or in some parts of British Columbia, represent a particular type of volcano that develops in a special environmental setting. In Iceland, they form a string, several hundred kilometers long, along the young volcanic seam that crosses the island from north to south (Figs. 6.31, 12.29). These table mountains with their flat tops and steep upper cliffs strongly contrast with their gently sloping neighbors, the shield volcanoes. Table mountains differ from the shield volcanoes by consisting largely of clastic volcanic rocks, chiefly complex breccias and pillow lavas.

Table mountains have long been recognized to represent *subglacial volcanoes*. They are found in volcanically active areas that were once covered by ice, or are presently still ice-covered, as in Antarctica. These subglacial volcanoes undergo similar stages in their evolution as the seamounts that evolve into ocean islands.

In the fall of 1996, many media agencies reported that a new volcano called Gialp had erupted in the center of the largest glacier of Europe, the 2-km-thick Vatna Jökull, which is almost 150 km long, covers 25% of Iceland and is underlain by half a dozen active volcanoes (131). One topic discussed daily was, when does the expected colossal flood occur? Such events release gigantic

volumes of water, up to half the discharge of th Amazon River, and are extremely attractive fo television. These glacier bursts, called *jökulhlaup* in Iceland, are floods of huge masses of rock and ice debris-laden glacial meltwater, formed by melt ing of the ice around the subglacial volcano, issu ing beneath the ice sheet. The gigantic flood sometimes carry blocks of ice and rocks man meters across for more than 10 km. Time and again, they destroy vast areas in south-eastern Ice land as in 1918 when Kattla volcano erupted, or i 1934 when large subglacial Grimsvötn volcan broke out. In 1996, however, there was a lon delay in jökulhlaup formation following the en of the eruption in mid-October, much to the frus tration of the media. On 5 November, the hug floods finally broke out of the glacier and deposit ed large masses of debris between the margin o the ice sheet and the coast, while transporting th fine-grained material into the North Sea. Lik many times in the past, several bridges and part of the main road between Reykjavik and settle ments on the sparsely populated eastern coast o Iceland were destroyed.

Table mountains were once thought to b basically the result of one long-lasting major sub glacial eruption. This was envisioned to star beneath ice, beginning by producing comple assemblages of pillow lavas and pillow breccia

▼ Fig. 12.29. Subglacial volcano Herdubreid (Iceland) rising to 1 682 m a.s.l. The lower poorly exposed deposits are dominantly sublacustrine turbidites and debris flow deposits. The steep upper cliffs formed during eruption in an ice cave when the volcano was covered by a Late Glacial thick ice sheet. The uppermost plateau is made up of sub-aerial lava flows

and finer-grained hyaloclastite, ending in subaerial lava flow eruptions when the volcanoes grew above the surface of the ice sheet. Many table mountains in Iceland are more complex, however, and consist of several cycles of volcanic activity and contrasting petrological characteristics (417; Fig. 12.30). The foot regions of table mountains, such as Herdubreid, the holy mountain of Iceland (Fig. 12.29) are largely made up of lacustrine sediments, including turbidites and debris flows. These were produced when the volcano itself grew in a huge subglacial lake or during periods when the glaciers were partly melted. The very complex mixture of scoria, breccia and pillows forming a chaotic facies in the steep cliffs of table mountains appears to form during warm to cold deposition in an ice cave as the volcano burns its way through the glacier. In Iceland, the steep upper parts of many table mountains may thus have formed during the last phase of extensive glaciation at the end of the ice age less than 10 000 years ago.

Summary

External water is easily the most important environmental factor influencing many different stages during the evolution of magma-volcanic systems. The water includes seawater, groundwater, lake water and even rain. It was only in the last 30 years, i.e. soon after birth of the volcanic island Surtsey, that evidence was mounting for the central role of external water in many highly explosive eruptions. A dramatic example of the explosive encounter of fire and water was the powerful opening phase of the famous eruption of Mount St. Helens (USA) on 18 May 1980. As to be expected with a new paradigm, these new ideas became only reluctantly accepted by the scientific community. The long-held view that maars owe their origin to magmatic CO_2 explosions was defended for a long time, but is now only of historical interest.

The insights that the encounter of magma and external water is not only common but commonly highly explosive represent one of the success stories of modern volcanology. As with other major changes in paradigm in volcanology during the last few decades – understanding of pyroclastic flows, volcano sector collapses and lateral blasts – prime evidence for ubiquitous magma-water encounter came from careful interpretation of field evidence. Particularly diagnostic criteria include the glassiness, angularity and poor vesicularity of juvenile clasts, abundance of lithics, well-developed bedding, common structures including lateral transport of wet cool particle systems and large block sizes. When these and other types of criteria are viewed in a broader context, characteristic scenarios generally can be inferred.

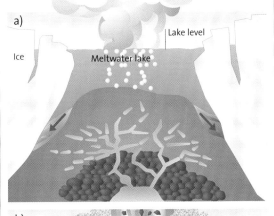

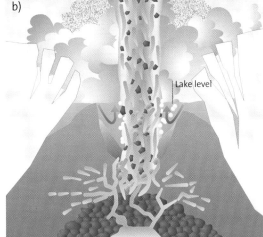

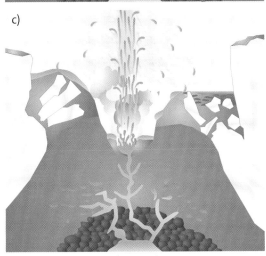

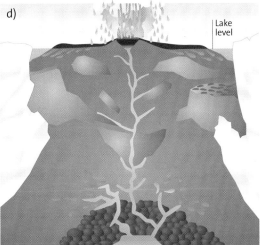

◄ Fig. 12.30. Development of a subglacial volcano (417):

a) Eruptions beneath ice cover. Formation of pillow lavas, dikes and a cover of granulated lava (hyaloclastites). Formation of a lake between margins of the ice sheet.

b) Increasing pyroclastic eruptions by degassing of magma, simultaneously quenched in water and formation of complex eruption columns.

c) Lava fountains and formation of breccias and partly welded agglutinates.

d) Subaerial stage with formation of lava fountains and lava sheets on the complex deposits, generated beneath ice and water.

Hyaloclastite, in part bedded, and with larger rock fragments and scoria

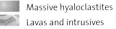

Massive hyaloclastites

Lavas and intrusives

Pillow lavas

Subaerial lava flow

In order to generate a phreatic/phreatomagmatic eruption, specific environmental conditions must be met, most importantly a water reservoir, be it shallow water, porous sediment or a water-filled fracture. The situation intuitively most easily to visualize is the growth of a volcanic edifice into shallow water when the volume ratio of steam to liquid water is large, a prerequisite for large explosive potential. Black stubby low eruption columns observed in many volcanic eruptions occurring in shallow water in the sea or in lakes are surrounded by white steam envelopes, cypressoid jets of blocks with trailing steam tails and basal ring-shaped clouds (base surges) in which material is transported laterally sometimes at great velocity. All oceanic volcanic islands have gone through this stage and many islands are fringed by cones that formed by magma-water interaction.

There are several other environments in which even greater explosive pressures can be generated as the encounter takes place underground. A common situation is the rise of dikes through shallow aquifers generating phreatic and phreatomagmatic explosions during the initial stage of an eruption. The very shallow emplacement of dikes rising from cryptodomes to within a few tens of meters of the surface can generate violent steam eruptions without the newly emplaced magma participating except for supplying the thermal energy and updoming of the crust. The late stage of eruptions when the conduit/magma reservoir system have partially collapsed allowing influx of groundwater to enter the subterranean hot void still partially filled with magma is also often characterized by phreatomagmatic eruptions. A somewhat different scenario is represented by the withdrawal of magma at depth, subsidence of a rock column and invasion of the fractured subsided mass by groundwater and commonly smaller amounts of magma that had not been drained or was newly supplied from depth. Even the rhythmic eruptions of Stromboli are at least in part due to interaction of rising magma and contact with ground/seawater.

In detail, the processes occurring during the contact between molten magma and water are highly complex and poorly understood. A central topic of research is the interaction and relative role of the three main fragmentation mechanisms at low pressure: flashing to steam of external water, magmatic degassing, and granulation of the magma by thermal shock. Similarly, the processes by which country rocks are fragmented in these types of eruptions is unclear. As shown by phreatic eruptions, rapid superheating of groundwater can lead to thorough fragmentation of country rock.

There are many models for the interaction between water and hot molten metals, developed to better understand possible reactions between a core melt-down of a nuclear reactor and cooling water. The idea is that the explosive fragmentation of a melt is caused by a layer of vapor that forms at the contact between the hot and cool liquids and which expands into the melt. It is not yet clear, to what degree these models can be fully applied to natural silicate magmas. Complications include the much higher viscosity of silicate magmas, the propensity for synchronous degassing of a magma when it interacts with water and especially the still poor understanding of the dynamics of vapor explosions (4).

The environmental impact of hydroclastic eruptions differs significantly from that of, say, Plinian and other pyroclastic eruptions. The volumes of hydroclastic eruptions are usually small and sedimentation occurs close to the source – commonly within about 5 km of the vent – because particles stick together in the wet cohesive transporting system. On the other hand, the high energy of base surges and of large ballistically transported blocks forms a major hazard proximal to hydroclastically erupting vents. Muddy sediments sometimes produced in copious amounts become quickly mobilized during an eruption and form destructive lahars. The sealing of the ground surface by hardened clay-rich muds is also more effective following such eruptions as contrasted with loose ash/lapilli deposits from pyroclastic eruptions that are eroded much more easily.

In previous chapters, I introduced fields in volcanology, in which the results of recent research into complex volcanic processes were the major topic. This basic research is the foundation from which more applied volcanology has developed, as discussed in the following chapters. The first of these (Chap. 13) is concerned with volcanic hazards on a local and sometimes global scale. The second (Chap. 14) is concerned with the impact of voluminous influx of volcanic gases into the stratosphere with clear global impacts following major Plinian sulfur-rich eruptions. As mentioned in the beginning of this chapter, the encounter of the two fundamental elements, fire and water does not only have disastrous effects. Geothermal energy and important magmatic ore deposits are the result of the nonexplosive interaction of fire and water. These and other benefits of volcanoes will be discussed in more detail in the last chapter (Chap. 15).

Volcanic Hazards, Volcanic Catastrophes, and Disaster Mitigation

Volcanic eruptions have occurred since the Earth formed about 4.6 billion years ago. Each year, about 60 of the roughly 550 active volcanoes on Earth erupt. The frequency, magnitude, and type of eruptions of volcanoes are unlikely to change in the foreseeable future. One in every six of the active volcanoes on Earth (5 % of all eruptions) has led to a loss of lives through its activity. About 260 000 people have died as a result of volcanic eruptions since A. D. 1600 (383) (Fig. 13.1). Cities and entire regions have been devastated. Disastrous volcanic eruptions are characterized by a rapid onset of their climactic phase and by a wide variety of eruptive behavior and effects: high and low temperatures, especially mass flows of different types (such as debris avalanches, pyroclastic flows and debris flows), but also including atmospheric transport of ash for hundreds of kilometers. During very powerful eruptions, huge masses of gases are injected into the stratosphere, forming aerosol veils that globally affect the climate and the ozone layer for years (such as following the Pinatubo eruption in 1991; Chap. 14).

Volcanic and other types of natural disasters are certain to increase in the future because the degree of vulnerability of societies is rapidly rising. The reasons are manifold. Volcanic soils are very fertile and therefore eminently suitable for land use. Population density in agriculturally cultivated areas around many volcanoes is high and constantly increasing. People settle higher and higher on some active volcanoes. Hence, the number of victims in such areas is likely to rise significantly in the future.

Major disruption is preprogrammed in areas where active volcanoes are in close proximity, especially where mass flows of various types are a major type of eruptive activity. Similarly, in developing countries, increasing investment in communication, power lines, pipelines, transportation lines, etc. have greatly increased vulnerability. The rising population density in sprawling urban areas and giant megacities close to active volcanoes are particularly acute cases of rising vulnerability. Other cases include local and regional air traffic across chains of active volcanoes. In addition, the complexities of energy, communication and supply networks are ever increasing.

Storms and subsequent flooding of large areas, caused by the altered radiation balance of the atmosphere, are increasing worldwide. Greenhouse gases have led to global warming and the warm air can absorb more water that is released in large storms. Increased precipitation combined with the massive ash supply following major volcanic eruptions will drastically increase the generation of lahars, one of the major long-term volcanic hazards.

While there were no volcanic eruptions with more than 1 000 fatalities during the 1970s, more than 25 000 people lost their lives during several devastating volcanic eruptions in the 1980s. If there had not been major scientific effort and well-organized disaster mitigation measures prior

> **Volcanoes are in fact indexes of danger, and the absence of them is the best security**
>
> *James D Dana, US Exploring Expedition, Philadelphia, 1849*

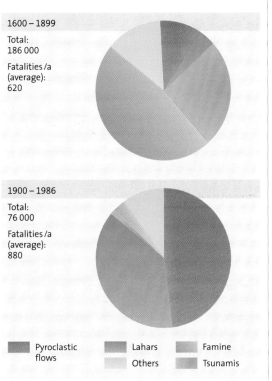

◄ Fig. 13.1. Main causes of deaths following volcanic eruptions in the time windows 1600 – 1899 and 1900 – 1986 (381)

1600 – 1899

Total: 186 000

Fatalities /a (average): 620

1900 – 1986

Total: 76 000

Fatalities /a (average): 880

Pyroclastic flows Lahars Famine

Others Tsunamis

to the 1991 eruption of Pinatubo (see below), perhaps as many as 10 000 people might have lost their lives. As a result of our increasing knowledge of the precursors to volcanic eruptions, each year thousands of people have to leave their homes for weeks or months, and some forever.

Terminology

The dictum in the Old Testament "Conquer the Earth" has dominated the attitude of western societies toward nature for centuries. Such may have been the background to the title of a book "Volcanoes declare war" by the famous volcanologist Jaggar in 1946. Our tendency to subdue nature has led to a *cul de sac*, as shown by the last quarter of the last century that saw increasing use of terms such as *natural catastrophes* or *natural disasters*.

Nowadays, an attitude is developing of how to *live with volcanoes*. In other words, society increasingly comes to accept often unpredictable and sometimes monstrous eruptions, to recognize the hazards when living close to potentially active volcanoes and to protect itself prudently to avoid a catastrophe.

In order to more fully appreciate the different aspects of the term *volcanic disaster* some terms need to be clarified to better understand the relationship between hazards, material effects, risks, coping and disaster.

- A *volcanic eruption* is a natural event that is the dynamic expression of processes in the interior of a living and sometimes dramatically active Earth
- Like many other natural processes, volcanic eruptions can present a volcanic hazard to people and property, a threatening event that is defined by a particular type, magnitude and probability of occurrence. A *volcanic hazard* takes many forms as discussed more fully below
- *Vulnerability* is defined as the degree to which the entire community is subject to impacts of hazards. This includes lives, property, essential environmental resources, economy and standards of living
- The term *volcanic risk* is applied to volcanic hazard in relation to life or property in a particular spot and at a particular time. *Risk* is then a combination of hazard features (which are specific to each volcano) and local valuable elements (including population, resources and infrastructure). People, property and natural resources are vulnerable to hazards when situated too close to a volcano, or, more specifically, to potential processes within a particular form of eruption. The term *risk* thus conceptualizes the combination of potential hazards with the value (in a broad sense) of elements potentially subject to the hazards

- *Disasters* result from the interaction of natural events with political, economic, social and technological processes. Disasters occur when people do not recognize the potential hazards of a volcano or particular types of volcanic eruptions, warnings by the volcano itself or by scientists, and hence do not protect themselves through mitigation and emergency response plans. In addition, political issues and insufficient communication between scientists, community leaders and emergency managers often lead to the worst types of volcanic disaster
- A volcanic disaster or catastrophe strikes when the scale or particular nature of a volcanic event exceeds the *capacity of a community to cope. The ability to cope* depends on the economic potential of a community or country and strength of private or public institutions capable to quickly and effectively respond to a hazardous natural event
- *Disaster mitigation* includes all activities toward reducing risk, either the hazards (via physical interventions, e.g. *sabo dams*) or vulnerability (via *hazard mapping, land use planning*). It also includes preparation of responsible administrative bodies and civil protection authorities as well as full information of the public and eventually of evacuation measures (Fig. 13.2). This is a formidable task since the spectrum of volcanic eruptions and therefore hazard types is large. When volcanoes have not erupted for hundreds or thousands of years, societies usually lose their sensitivity vis-a-vis hazards and do not develop appropriate disaster mitigation measures. The old adage often applies, the next disaster strikes when the memory of the last is lost. In many areas, people continue to live close to volcanoes that had a long history of disastrous eruptions. Some communities even expand increasingly closer to the main hazardous areas. Some volcanoes characterized by frequent explosive eruptions as well as high population density are listed below.

Volcanic Hazards

Sometimes the eruption magnitude (e.g. the volume of material erupted) is equated with the degree of hazard or risk, in other words, the larger the eruption, the greater the hazard or risk. This correlation commonly does not hold, however, other factors generally being more important.

Population Density

Whether a volcanic eruption leads to major loss in human lives or not depends on the population density in the proximity of a volcano and its state

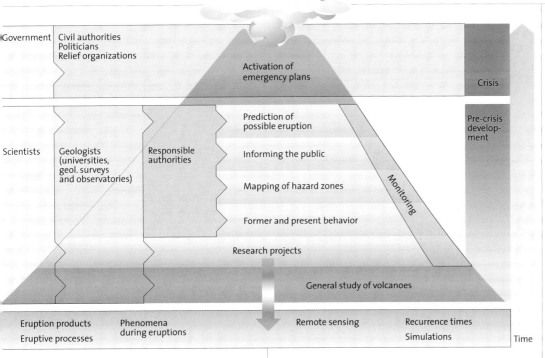

Government	Civil authorities Politicians Relief organizations	Activation of emergency plans	Crisis
Scientists	Geologists (universities, geol. surveys and observatories)	Responsible authorities	Prediction of possible eruption Informing the public Mapping of hazard zones Former and present behavior	Pre-crisis develop- ment

Monitoring

Research projects

General study of volcanoes

| Eruption products
Eruptive processes | Phenomena
during eruptions | Remote sensing | Recurrence times
Simulations | Time |

◄ Fig. 13.2. Research steps and disaster prevention measures required in the face of an active potentially erupting volcano (382)

of preparedness and organization. It does not generally depend on the volume of magma erupted or the type of eruptive processes.

For example, the eruption with the largest volume of magma erupted in the past century (Katmai, 13 km³, 1912) did not result in human victims because it occurred in an extremely remote area of Alaska (Fig. 11.5). It did, however, lead to large environmental impacts, both locally and globally with respect to climate. On the other hand, the magma volumes of some well-known and deadly volcanic eruptions in the last century were tiny. Eruptions resulting in more than 20 000 fatalities such as Mt Pelée (Martinique, 1902) (Fig. 13.3) and Nevado del Ruiz (Colombia, 1985) had eruptive magma volumes of much less than 1 km³.

The 8 May 1902 catastrophic eruption of Montagne Pelée was characterized by high velocity pyroclastic density currents, which destroyed the town of St. Pierre, with block-and-ash flows being channeled in the north. Similar eruptions took place during 20 and 26 May, 6 June and 30 August 1902. In contrast, only block-and-ash flows spawned by a growing and episodically collapsing lava dome were generated during the next major eruptive stages (1929–1932). Most were deposited in Rivière Blanche at the foot of a major collapse scarp.

During the famous eruption of Mount St. Helens in 1980 with less than 1 km³ magma, very few people perished, not only because of the timely and effective disaster prevention measures, but also because the area within about 50 km of Mount St. Helens was sparsely inhabited.

There are two main reasons for the fundamental contrast in the impact on society between these two eruptions. In the case of Montagne Pelée, the town of some 29 000 inhabitants was right in line of high-speed pyroclastic currents (Fig. 13.3). Had the town and smaller settlements in its neighborhood been built a few km north or south, the lethal effects of the volcanic currents would have been minimal. But then, major valleys on the slopes of volcanoes, commonly the zones of earlier sectorial flank collapse, are also ideal sites for settlements. The area of Katmai/Novarupta volcanoes, on the other hand, was uninhabited. Had the huge 10-km-wide and 20-km-long valley west of Novarupta crater been densely populated, the disaster would have been of a much larger magnitude than in the case of St. Pierre.

Life returned to St. Pierre after 1902 slowly, the present town having been reduced to a small quiet settlement in strong contrast to the thriving hub it represented in 1902. The question as to why the town had been resettled is legitimate especially since volcanic deposits came close to the town in 1930. Some 83 years after the catastrophic demise of St. Pierre, the town of Armero in Colombia was also annihilated with about 23 000 fatalities. The Colombian government then decided to make the area a national cemetery, moved also by the fact that the site of the town had been devastated twice in the preceding centuries by the same type of volcanic debris flow (see below). The hazards for the town of St. Pierre and, in fact, the entire valley, have not changed in the past 100 years. However, the methods for

▶ Fig. 13.3. St. Pierre (Martinique) prior to (a) and after (b) the disastrous eruption of May 8 1902. Commercial postcards

a)

b)

monitoring volcanic activity and recognizing important precursors early enough to enact timely evacuation have vastly improved during the past three decades. On the other hand, time spans between highly destructive volcanic events are generally significantly larger than a generation's lifetime. Thus, the fundamental benefits from maintaining at least part of the pre-disaster social structure and other benefits such as agriculture or favorable transport location override the fear of distant disasters.

Cities and megacities with high population densities and major growth rates within the reach of active volcanoes and thus places of major potential volcanic risk include:

- Tokyo, 100 km east-northeast of Mt. Fuji in Japan, a volcano that shows recent signs of awaking from its slumber since 1707
- Quito (Ecuador) some subburbs having been built on prehistoric and historic lahars from Cotopaxi volcano, 60 km to the southsoutheast, and close to the very active Guagua Pichincha volcano, last active in 2000 and Reventador last active in November 2002
- Mexico City (Mexico) in sight of the huge Popocatépetl volcano, which entered a new eruptive phase, starting about 1995
- Yogyakarta (Indonesia), at the foot of almost permanently active Merapi

- Naples (Italy) at the foot of Vesuvius, last active in 1944
- Seattle–Tacoma area (USA) in sight of presently quiet Mount Rainier some parts of the sprawling area having been built on Holocene lahars;
- Goma at the foot of very active Nyiragongo (Zaire) last erupted in 2001
- Manila (Philippines) close to Taal volcano, the last disastrous eruption having occurred in 1965
- Auckland (New Zealand) spreading across a Holocene volcanic field and
- Managua (Nicaragua) built on many young volcanoes, nearby active Masaya volcano having erupted violently by magma-water interaction in the late Holocene.

Pyroclastic Density Currents

All types of mass flows pose the greatest hazard to communities, because people tend to live in valleys that are the most convenient areas to settle since soil and water resources are nearby. Volcanic mass flows include pumice-rich pyroclastic flows, dense block-and-ash flows and lahars. While the first two are gas-inflated, lahars are debris-water mixtures. High velocity, hot, particle-poor blasts are especially dangerous because their pathways are not confined to valleys. The main loss of lives and the largest economic losses in the past century from volcanism resulted from pyroclastic flows, hot blasts and lahars. Examples include Mount Pelée (Martinique, 1902), Nevado del Ruiz (Colombia, 1985) and Pinatubo (Philippines, 1991).

Hot pyroclastic flows, which consist of gas and particles, at temperatures sometimes exceeding 800 °C, speed down the flanks of a volcano and may cover thousands of square kilometers (Chap. 11). They are characteristic for explosive eruptions of volcanoes above subduction zones (Figs. 13.3 – 13.5; Chaps. 8, 9). Pyroclastic pumice flows or block-and-ash flows are generated each year during many volcanic eruptions. People caught within their flow paths have little chance of surviving. Buildings, crops and forests are destroyed and warning periods are extremely short. On the other hand, the most likely pathways, i.e. the valleys, can be clearly delineated in hazard maps. Administrators responsible for land planning can effectively reduce the hazards when restricting settlements in valleys that are likely pathways for volcanic mass flows.

Pyroclastic flows and block-and-ash flows are commonly associated with surges, hot highly diluted turbulent currents that race ahead of, or spread laterally from, the denser valley-confined gravity flows. The destructive power of these currents unfolds when they expand above slopes and hills. On 3 June 1991, for example, 43 people were killed, including the world-famous volcano photographers Maurice and Katia Krafft, by hot surges, which expanded above the depositional area of hot block-and-ash flows at the foot of Unzen volcano (Japan) (Fig. 11.4). In May 1902, dilute pyroclastic density currents raced 8 km down the slopes of Montagne Pelée within a few minutes to reach the town of St. Pierre. Here, almost the entire population was killed (29 000 people). The actual pyroclastic block flows, however, were restricted to the valley of Rivière Blanche. Surges led to the death of about 2 000 people during the eruption of the remote El Chichón volcano in southern Mexico in April 1982. About 70 people died in 1994, when a pyroclastic block flow descended a valley on the slope of Merapi, the most active volcano of Indonesia, generating a lethal surge (Fig. 11.30). The top of volcanoes can collapse during or after large pyroclastic flow eruptions into the partially emptied magma chamber, to form calderas (Chap. 9). For-

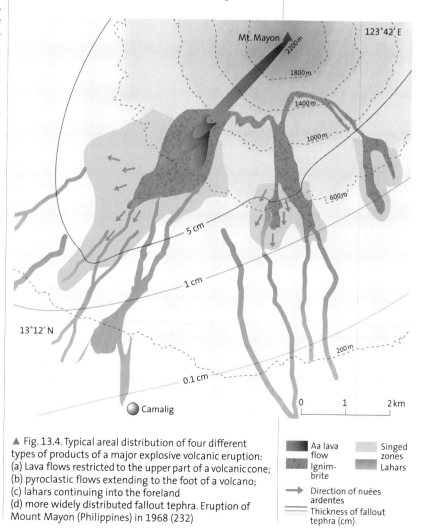

▲ Fig. 13.4. Typical areal distribution of four different types of products of a major explosive volcanic eruption: (a) Lava flows restricted to the upper part of a volcanic cone; (b) pyroclastic flows extending to the foot of a volcano; (c) lahars continuing into the foreland (d) more widely distributed fallout tephra. Eruption of Mount Mayon (Philippines) in 1968 (232)

Aa lava flow
Ignimbrite
Singed zones
Lahars
Direction of nuées ardentes
Thickness of fallout tephra (cm)

▶ Fig. 13.5. Surface of a pyroclastic block flow generated during a dome explosion of Augustine volcano (Alaska) in 1976

tunately, caldera-forming eruptions of gigantic scale have not happened during historic times. In densely populated areas, such an event would generate unimaginable and utter devastation.

Lahars

Lahars (volcanic debris and mud flows) are, next to pyroclastic flows, some of the most dangerous volcanic phenomena (207) (Fig. 13.4). They can flow for several kilometers, rarely also up to 300 km, move quickly with velocities that may exceed 100 km/h (!) – depending on particle concentration - and have a high destructive potential. About 10 % of all volcanic fatalities are caused by lahars. One can distinguish at least two types of lahars. Coarse-grained debris flows are of high viscosity, high matrix yield strengths (Chap. 4) and sediment concentrations are typically higher than 60 vol %. With increasing relative amounts of water, *hyper-concentrated flows* form, with intermediate viscosity, low matrix yield strengths and sediment concentrations between 20 and 60 vol %. In principle, both mass flow types can be highly destructive.

Many lahars form when pyroclastic flows s.l. enter riverbeds. The hazard potential is especially high when pyroclastic flows erupt on snow or glacier-clad volcanoes. A lahar formed in this manner during the relatively small eruption of the glaciated Nevado del Ruiz (Colombia) in November 1985. Hot surges and small pyroclastic flows and hot bombs melted the surface of the glacier. The debris-laden floods so generated mixed with the water of the river Lagunillas and incorporated large volumes of sediment before flowing to the town of Armero, more than 60 km away, killing about 23 000 people (147, 401, 402). Lahars are commonly associated with phreatic and phreatomagmatic eruptions because such wet deposits become easily mobilized on steep slopes.

In tropical areas, cloudbursts and cyclones (hurricanes) are common. These can wash freshly deposited ash and rocks from volcano slopes, to produce lahars, as during the eruption of Irazú volcano in Costa Rica over several years (8). In this way, lahars are commonly triggered for many years following a volcanic eruption. Lahars from the slopes of Pinatubo, which erupted on 15 June 1991, are expected to continue until at least the year 2005 (264). Lahars can also form directly during volcanic eruptions, for example in eruptions through crater lakes such as Ruapehu (New Zealand) and Kelut on Java (Indonesia) (see below).

Magma Composition and Tectonic Environment

The explosivity of an eruption is not only governed by the chemical composition of a magma and therefore its gas content, but also by the viscosity of a melt. Mixing of magma with ground- or seawater may also play a major role (see below). The higher the concentration of SiO_2 in a

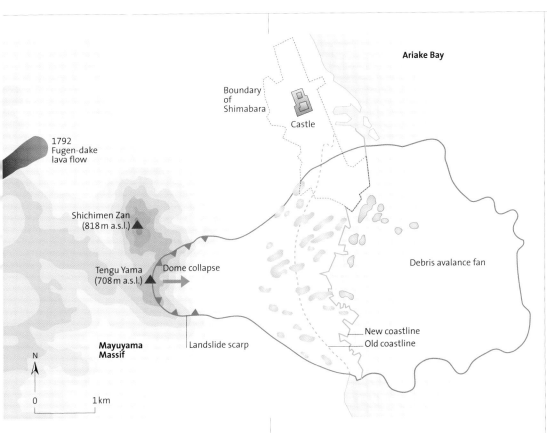

Ariake Bay

Boundary
of
Shimabara

Castle

1792
Fugen-dake
lava flow

Shichimen Zan
(818 m a.s.l.) ▲

Tengu Yama
(708 m a.s.l.) ▲   Dome collapse →

Debris avalance fan

**Mayuyama
Massif**   Landslide scarp

New coastline
Old coastline

N

0   1km

◄ Fig. 13.6. Source area (scar) of major debris avalanche and distribution of depositional fan generated in 1792 by collapse of a dome in Mayuyama Massif, south of the town of Shimabara (*dotted line*) (Unzen volcano, Kyushu, Japan). Most of the debris avalanche was deposited in the sea. The deposits pushed the old coastline (*dashed line*) into the sea and formed small hills and islands. The debris avalanche generated tsunamis when entering the sea. About 15 000 people died along the 120-km-long coastal strip around Ariake Bay, and 6 000 houses and 1 600 fishing boats were destroyed (329). See also Fig. 9.17

magma, the more viscous the magma and lower the gas diffusivity rates, the more explosive the eruption. Most active explosive volcanoes characterized by SiO_2-rich andesitic, dacitic and rhyolitic magma composition have grown above subduction zones, such as around the Pacific (Ring of Fire), in the Caribbean (Lesser Antilles) or in the Central (Aeolian Islands) and Eastern (Cyclades) Mediterranean. Many developing countries are thus strongly affected by volcanic eruptions, such as Latin America, the Caribbean, and the Southwestern Pacific, especially Indonesia and the Philippines.

The most active volcano on Earth, Kilauea on Hawaii is a major attraction to tourists and an ideal volcano for detailed scientific study, but the erupted lavas are neither explosive nor very dangerous (apart from the very rare phreatic and phreatomagmatic eruptions).

Sector Collapse, Debris Avalanches, Tsunamis and Environmental Factors

Dormant volcanoes can be deceptive, even for specialists. Under certain conditions, a volcanic edifice can suddenly collapse and produce extremely mobile gravity-controlled debris avalanches (Chap. 9). These accelerate downhill and can flow at more than 100 km/h over several tens of kilometers. Such collapses of entire volcano flanks, be it in connection with eruptions or after

extreme rainfalls, can be highly destructive. About 2 000 people died when the upper flank of Casitas volcano (Nicaragua) collapsed in 1998, following sustained cloud bursts generating rapid debris-laden floods.

Debris avalanches and debris flows when entering the sea, may generate *tsunamis* (e.g. during flank collapses on Hawaii or the Canary Islands). These can be highly destructive and represent a major hazard for the commonly densely populated coastal areas (Figs. 13.6, 13.7). Limestones found at 325 m a.s.l. on the Island of Lanai in the Hawaiian archipelago were interpreted (226) as the result of giant flood waves, generated during flank collapse of the southern flanks of the Island of Hawaii. This interpretation has recently been questioned, but the fact remains that volcano flanks collapse around the Hawaiian Islands with a periodicity of less than 100 000 years. These seem to occur mainly during the late stages of shield volcano growth (239). Submarine and subaerial flank collapses in the Canaries are even more frequent than in the Hawaiian Islands and also occur during very late stages of volcanic evolution (182, 308, 412) (Fig. 9.39). In 1792, the collapse of a parasitic lava dome of Unzen volcano (Japan) caused a tsunami resulting in about 15 000 fatalities (Fig. 13.6). The disastrous tsunamis when Krakatau volcano erupted in August 1883 (Fig. 13.8) may have been due to the entry of

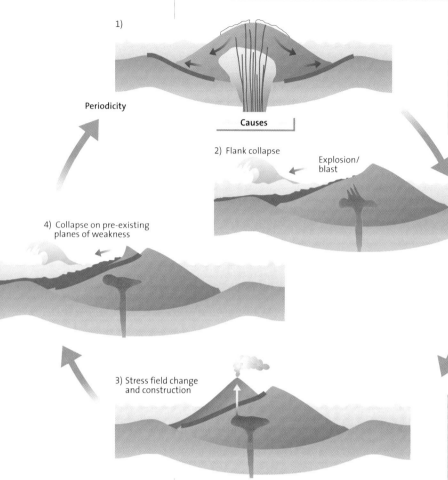

▲ Fig. 13.7. Sector collapse processes and cycles in oceanic islands
1) Sector collapse causes include: swelling due to intrusion of a new magma batch with or without eruption; hydro-thermal alteration of interior of volcano; repeated dike injection along rift zones; oversteepening of one flank of a volcano due to long-term sustained erosional attack, etc.
2) Flank collapse generating steam blast and tsunami and asymmetric unloading of magma reservoir
3) New volcano growing in the scar left by the collapse
4) Renewed collapse on pre-existing plane of weakness (412)

► Fig. 13.8. Time zones (minutes) of propagating tsunamis, generated when pyroclastic flows sourced in a collapsing eruption column impacted the sea during the eruption of Krakatau volcano in August 1883. The dark-colored coastal zones of Java and Sumatra became flooded by the tsunamis. About 36,000 people perished (337)

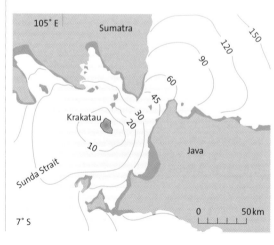

voluminous pyroclastic flows into the sea or a major phreatomagmatic (?) explosion (337). More than 36 000 people died when giant, more than 15-m-high flood waves devastated the entire coastal strip on the northern coast of Java and southern coast of Sumatra (Fig. 13.8). If the volcano had erupted on land, the zone of complete destruction would have been restricted to valleys around the volcano.

Volcanogenic tsunamis can be triggered by several causes or a combination of them. A large Plinian eruption destroyed parts of Santorini about 3 500 years ago, coinciding with the end of the Minoan culture. The eruption may have contributed to the decline of this culture. This eruption or the ensuing caldera collapse triggered large flood waves, which caused havoc on the Island of Crete in the eastern Mediterranean.

Environmental factors are commonly important in causing volcanic disasters. Examples are eruptions of glacier-clad volcanoes (e.g., Cotopaxi, Nevado del Ruiz, Grimsvötn and others), eruptions through lakes (e.g., the notorious and dangerous Taal volcano, Philippines), or the synchronous occurrence of a volcanic eruption and a taifun (hurricane) (e.g., Pinatubo on 15 June 1991).

The ubiquitous rivers that drain volcanic edifices pose a major hazard when blocked by volcanic landslide or massive tephra sedimentation. Uncontrolled collapse of such dams can devastate downstream areas for long distances. An example is a large lake accumulated behind a temporary tephra dam at a constriction in the Rhine river canyon during the late stages of the eruption of Laacher See volcano, about 12 900 years ago in the Eifel (Germany) (Fig. 11.33). This lake was up to at least 18 m deep but drained catastrophically when the unstable dam collapsed a few weeks or months after it had formed. This generated flood waves, whose deposits can be found as far north as Bonn, 50 km to the north and possibly even in the Netherlands (Chap. 11) (255). Pyroclastic flows blocked the Agatsuma River about 10 km downstream from Asama volcano in Japan during the famous 1783 eruption. More than 1 000 people were reported to have been killed by a lahar or flood waves when the dam collapsed 5 days after the beginning of the eruption.

Eruption Columns

Of the roughly 60 volcanoes erupting each year, about 10% are characterized by eruption columns that rise into the stratosphere (i.e., 8–16 km). As air traffic is increasing dramatically, so is the potential hazard from the interaction of aircraft with eruption columns and more distant ash clouds. During the past 2 decades, beginning with the unexpected eruption of Galungung volcano (Indonesia) in 1982, more than 80 modern jets were damaged when inadvertently flying through volcanic ash (49) and almost 10 large aircraft experienced in-flight loss of engine power and barely avoided a crash. In June 1991, about a dozen large airplanes experienced almost-catastrophes when suddenly flying through the eruption column rising above Pinatubo volcano. Thereafter, Manila airport was closed for several days, a precautionary manner, which is now routine in many airports, such as Quito during explosive eruptions of Guagua Pichincha in 2000 and Reventador in 2002, or Catania (Sicily) in July 2001 and October-November 2002 when Etna volcano was spectacularly on fire. The main cause of engine thrust loss is the accumulation and solidification of ash on the turbine nozzle guide vanes (Fig. 13.9).

A famous example is the near-crash of a new Boeing 747–400 with more than 300 passengers

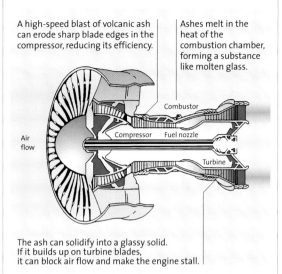

A high-speed blast of volcanic ash can erode sharp blade edges in the compressor, reducing its efficiency.

Ashes melt in the heat of the combustion chamber, forming a substance like molten glass.

Combustor

Compressor   Fuel nozzle

Air flow

Turbine

The ash can solidify into a glassy solid. If it builds up on turbine blades, it can block air flow and make the engine stall.

◀ Fig. 13.9. Jet engine cutaway showing areas of damage by volcanic ash (272)

▼ Fig. 13.10. Map showing flight path of KLM 867 and limits of detectable ash fall lobe from 15 December 1989 eruption of Redoubt volcano (Alaska). *Numbers* show main events:
1) plane starts to descend from 35 000 ft at 11:40;
2) plane encounters ash cloud at 25 000 ft;
3) power loss on all four engines after climbing to 27 900 ft;
4) engines 1 and 2 restarted at 17 200 ft;
5) engines 3 and 4 restarted at 13 300 ft;
6) plane lands at Anchorage at 12:25 (49)

aboard, more than 200 km northeast of erupting Redoubt volcano in Alaska on 15 December 1989 (272) (Fig. 13.10). When the jumbo started its descent into Anchorage, it encountered an ash cloud at an altitude of about 8,000 m at 11:46 h. The crew increased power and ascended by 1 000 m

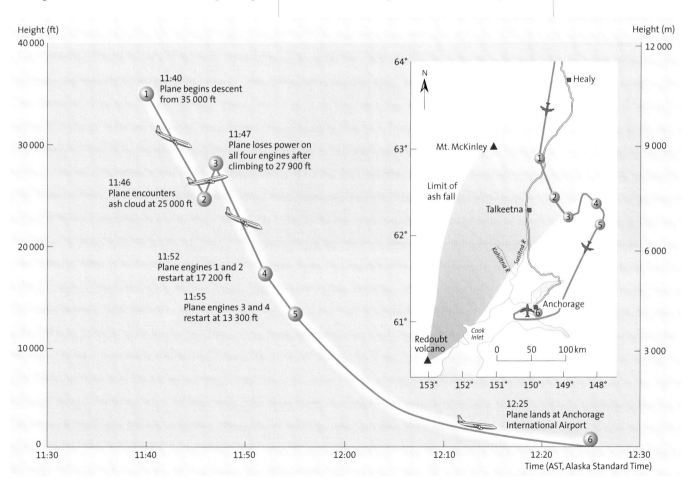

Height (ft)

40 000

30 000

20 000

10 000

0

11:30   11:40   11:50   12:00   12:10   12:20   12:30

Height (m)

12 000

9 000

6 000

3 000

11:40
Plane begins descent from 35 000 ft

11:47
Plane loses power on all four engines after climbing to 27 900 ft

11:46
Plane encounters ash cloud at 25 000 ft

11:52
Plane engines 1 and 2 restart at 17 200 ft

11:55
Plane engines 3 and 4 restart at 13 300 ft

12:25
Plane lands at Anchorage International Airport

Time (AST, Alaska Standard Time)

64°

N

Healy

Mt. McKinley ▲

63°

Limit of ash fall

Talkeetna

62°

Kahiltna R.

Susitna R.

Anchorage

61°

Cook Inlet

Redoubt volcano

0   50   100 km

153°   152°   151°   150°   149°   148°

◄ Fig. 13.11. House buried by ashes generated during a phreatomagmatic eruption of Capelinhos volcano off the coast of Fayal (Azores). See also Fig. 12.2

to 150 fatalities, remained unexplained until recently, and are now interpreted to be due entirely to suffocation by quietly liberated gases (191). The catastrophes of the Monoun and Nyos crater lakes (Cameroon) in 1984 and 1986, during which more than 1700 people died by a descending CO_2 cloud, were apparently not related to a volcanic eruption. Most likely, CO_2, dissolved at high pressure at depth to near-saturation level in the crater lake, was suddenly released. The causes for this sudden release are not entirely clear (177). CO_2-saturated deep waters were probably forced upward by an abrupt subaquatic landslide, became oversaturated at the lower pressure and discharged the liberated CO_2 in a dense cloud that flowed downhill and inundated the village.

The hazard of gas release for buildings is small, but economic losses such as the death of cattle can be significant. The early warning of an impending gas eruption is extremely short. Even the emission of gases (CO_2, SO_2, etc.) from fumaroles in active volcanoes can cause major damage to crops and are a significant health hazard (362). Detailed studies of such eruptions and the close monitoring of potential hazards are therefore necessary.

Plumes of SO_2 sourced in slowly degassing volcanoes can have disastrous effects on buildings and agriculture due to the formation of aggressive sulfuric acids as downwind from Masaya crater in Nicaragua, or Poas crater in Costa Rica. The unusually strong sulfur dioxide emission from Miyakejima (September 2000 to the present (Fig. 4.25)) has prevented a return of the more than 5000 inhabitants, evacuated since September 2000.

Ash Fallout

The rain of ash and ballistic transport of larger blocks is generally of immediate concern only for people who live near volcanoes, up to a few kilometers away. The damage may be significant, however, even at larger distances, when roofs collapse beneath the heavy load of rapidly accumulated ash (Fig. 13.11). Major damage by ballistically emplaced blocks to buildings occurred during explosive phreatic eruptions of Usu Volcano in 2000 (Fig. 13.12). Volcanic ash can also cause major damage in industry and agriculture. Machines and engines can be blocked by ash. Public transport and traffic can be strongly affected by major ash falls, as after the eruptions of Mount St. Helens and Pinatubo and recently at Etna (Sicily) during the major eruptions in July 2001

▲ Fig. 13.12. Blocks ballistically emplaced into wall of Kindergarten building during 2000 eruption of Usu volcano (Hokkaido, Japan). Distance to crater about 1 km

in order to climb out of the ash cloud. Within 1 min, all four engines decelerated below idle. Within 5 min the jumbo rapidly descended without power to 4000 m. At the last minute, the turbines, which had become clogged by molten and solidified ash, were restarted. The plane managed to land safely in Anchorage but damage amounted to more than US-$ 80 million.

Volcanic Gases

Volcanic gases accumulating in deep crater lakes and erupting episodically can spell disaster. Several mysterious eruptions during last century in the Dieng volcanic area (Java), which resulted in up

and the fall of 2002. At great distances, the ash can also be a health hazard, particularly if free silica is present. Silicosis, asthma and other respiratory problems may occur when particles are fine enough to be inhaled. In addition, acidic compounds, often associated with ash, can be a major environmental health hazard.

Dormant Volcanoes

Generally, the longer a volcano has been dormant, the larger and more explosive a new eruption is. Hence, the most hazardous volcanoes are commonly those in which eruptive phases are separated by hundreds or thousands of years of quiescence, as shown by the example of Pinatubo (1991; see below). A major cause of vulnerability is also the attitude of people who live in proximity to a long-term dormant volcano. It is thus commonly difficult to evacuate people from potentially affected areas. A recent example is the Soufrière Hills volcano on the island of Montserrat in the Caribbean. This volcano was not active since the island became colonized early in the seventeenth century. It started to erupt after a two-year-long phase of seismic unrest on 18 July 1995 and is still going strong in 2003 (357, 440). When volcanoes believed to be extinct erupt, catastrophes often follow, such as at Mount Lamington (Papua New Guinea, 1951) (374) or El Chichón (Mexico, 1982).

The Volcano Explosivity Index (VEI)

Newhall and Self (250) proposed a *volcano explosivity index*, based on subjective, qualitative descriptions plus some quantitative criteria (total volume of erupted products and height of eruption columns). The VEI has eight overlapping classes of increasing explosivity and can be applied to about 80% of the known, dominantly explosive volcanic eruptions. Using this system, the energy of large effusive eruptions with only minor tephra production (VEI = 0) is not properly assessed, however. On the other hand, these large effusive eruptions are generally not particularly hazardous to humans with the exception of the great Laki eruption in 1783 (Chap. 14).

In general, one can assume that highly explosive volcanic eruptions (VEI 3 to 6) have lethal effects. The periods of quiescence between these especially dangerous eruptions are very long, sometimes hundreds to tens of thousands of years (Fig. 13.13). Of the 21 most explosive volcanic eruptions (VEI 2 to 5) of the past 10 000 years, 17 erupted for the first time in recorded history (338). The hundreds of years of quiescence of so-called dormant, large, explosive volcanoes are thus quite deceptive.

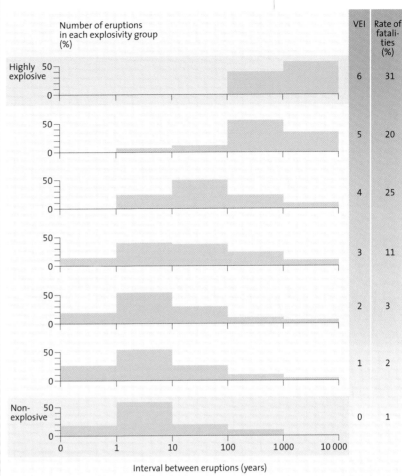

Can Volcanic Catastrophes be Avoided?
Diagnosis

A good doctor will take time with a new patient to find out about earlier health problems, including those of their family. This information is important for the subsequent in-depth diagnosis, which is later followed by suggestions for a therapy. By analogy, the careful analysis of the early history of a volcano is still the most important method to estimate the long-term probability of the occurrence of certain types of eruptions and their specific energy. Volcanoes commonly harbor an excellent historic and prehistoric event stratigraphy and chronology. This history can be reconstructed when the deposits are mapped, subdivided stratigraphically, structurally analyzed and dated. In addition, the chemical and mineralogical composition of the erupted lavas and pyroclastic deposits, and careful documentation of volcanological parameters (volumes, grain sizes, particle characteristics, eruptive and fragmentation mechanisms, etc.) provide important information of the earlier eruptive behavior of a volcano. This relies on the principle of uniformity, that is, a vol-

▲ Fig. 13.13. Frequency distribution of historic volcanic eruptions differing in explosivity. VEI = volcano explosivity index (250, 338)

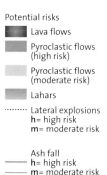

Potential risks

- ▨ Lava flows
- ▨ Pyroclastic flows (high risk)
- ▨ Pyroclastic flows (moderate risk)
- ▨ Lahars
- ⋯⋯⋯ Lateral explosions
  h= high risk
  m= moderate risk

- ⎯⎯ Ash fall
  h= high risk
  m= moderate risk

- ≡≡≡ Isopachs of fallout tephra (mm)

▲ ▶ Fig. 13.14. Prediction of possible types of volcanic emplacement modes during the imminent eruption of Nevado del Ruiz volcano, Colombia (*upper map*). *Lower map* shows the actual results of the eruption on 13 November 1985, consisting of a very minor tephra fallout fan and deadly lahars in the lower reaches of Rio Guali and Rio Lagunillas (256)

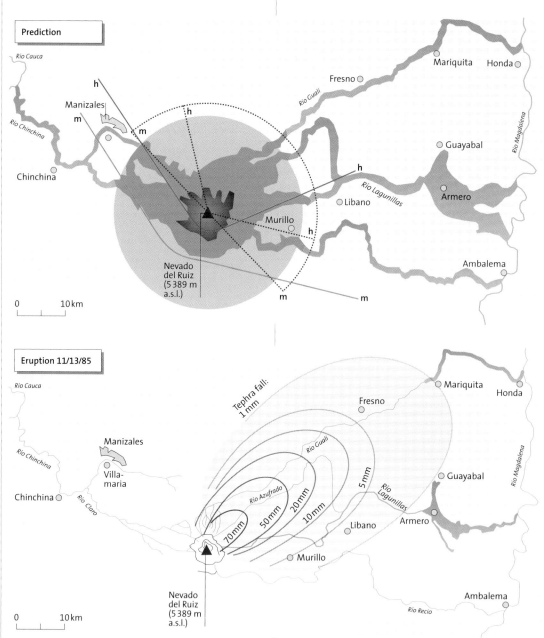

cano is likely to behave in the future as it has done in the past. From this data, hazard maps can be constructed, in which the distribution of the several types of eruptive products are shown (including lava flows, pyroclastic flows, tephra falls, lahars, volcanic debris avalanches and possibly tsunamis) (Fig. 13.14). The apparent probability that a volcano that erupted once historically, will erupt again is larger than in a volcano that appears to be dormant or extinct. Nevertheless, the periods of quiescence between eruptions in many volcanoes last much longer than the few hundreds or thousands of years of recorded human history. Large volcanoes that have erupted very frequently in the past include Kilauea volcano, Vesuvius (although the next event after the large eruption of 1944 is overdue), Hekla (Iceland), Etna (Sicily), Mayon (Philippines), Merapi (Java), Sakurajima (Japan), Komagatake (Japan), Arenal (Costa Rica) and Augustine volcano (Alaska). The best-known continuously active – but presently unstable – volcano is Stromboli in the Tyrrhenian Sea, with eruptive phases that last a few minutes to hours. What causes irregular intereruptive periods between different volcanoes is unknown and complex. The episodicity probably depends on factors such as the rate of magma generation in the source areas, the rates of ascent, he complex processes in magma chambers and he interaction with ground or surface water.

A volcanic eruption can last a few minutes, or aundreds of years, depending on how one defines t. The mean duration of volcanic eruptions on converging plate margins is about 65 days, in others (mainly intraplate volcanoes) 31 days (Fig. 13.15) 338). These are minimum values, however, because most of the energy is spent in the beginning of an eruption, while the slowly declining late phase is commonly not accounted for in the incomplete descriptions of older historic eruptions.

The prediction of volcanic eruptions, based on the statistical analysis of past events, is not exact enough to forecast eruptions precisely (e.g., to a particular date). Experience shows that people settle in volcanic areas in which eruptions can only be predicted with a probability of decades.

In addition to geological studies, the monitoring of volcanoes is a very important method to predict future eruptions, not only in time, but also the type of eruption and its vent location.

Forecast and Prediction

One of the premier goals of volcanological research has always been to improve the ability to predict eruptions. This is especially important in densely populated areas to enable timely evacuation. But even for volcanologists, in the face of the small number of scientists, expensive instruments and often remote setting of volcanoes, accurate timely predictions are essential to plan strategic measuring campaigns, prior to, during and after an eruption. If an eruption can only be predicted with a moderate probability, catastrophes cannot be avoided. Because periods of quiescence between eruptions in many volcanoes last much longer than a generation, people are more willing to accept risks rather than an evacuation. On the other hand, premature warnings and evacuations prior to events, which do not eventuate, have the paralyzing effect that nobody listens to the next warning, the cry-wolf syndrome.

Two classical examples, Guadeloupe and Hawaii, illustrate this dilemma. The Soufrière volcano on the island of Guadeloupe in the Caribbean started to show unrest in July 1975. The frequency of earthquakes increased so much that, when the volcano started to spew ash on 8 July 1976, catastrophic pyroclastic flow eruptions were anticipated. One expected events similar to those in 1902 on the neighboring islands of St. Vincent and Martinique (97). The daily number of earthquakes rose to 6000 in August, and scientists thought that the erupted ash not only contained pulverized older rocks but also particles of fresh lava. At the suggestion of scientists on hand the

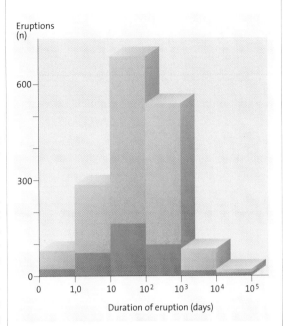

Eruptions (n)

Duration of eruption (days)

governor decreed evacuation. All 75000 people living close to the volcano were evacuated, the largest volcanism-related evacuation in history. The uncertain situation, which had been exacerbated because of political and economic factors, became even more complicated by development of fundamentally opposing views among the French scientists. Two groups of scientists formed and defended their opposing opinions in newspaper and television interviews. One group expected an explosive catastrophe, for which the other group saw no sign. In view of this rift, the French government asked a commission of international experts to convene in Paris. The panel agreed with the second group and saw no reason for concern. The volcano followed this suggestion, quieted down and has not erupted to this day. Should the signs for a larger eruption of Soufrière volcano be repeated in the near future, it would probably be difficult to convince the people evacuated in 1976 to leave their homes again.

A similar series of events began when a scientist of the Hawaiian Volcano Observatory dared to predict an imminent flank eruption to take place in 1976 on Mauna Loa, the largest volcano on Earth. He predicted it would occur at an altitude of 2800–3000 m asl. People prepared for the eruption (198), which occurred, as predicted in all detail–but only eight years later, in 1984 (199). The premature prediction of the flank eruption of Mauna Loa was based mainly on statistical or probabilistic analysis. For decades, summit eruptions of Mauna Loa had been followed by flank eruptions within a few years. Only after the summit eruption of 1975 was the volcano ill behaved;

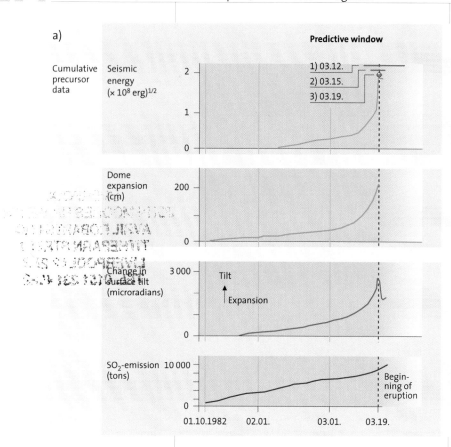

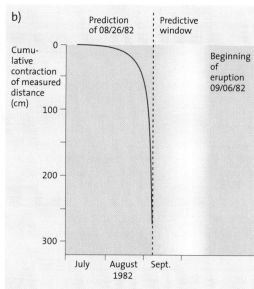

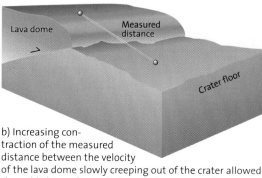

▲ Fig. 13.16. Prediction of two eruptions of Mount St. Helens in March and September 1982.
a) The increasing rate in change of four different parameters since early 1992 allowed to progressively decrease the predictive windows between 12 and 19 March and to predict the eruption precisely within hours

b) Increasing contraction of the measured distance between the velocity of the lava dome slowly creeping out of the crater allowed the publishing of a predictive window prior to the eruption on 6 September 1982 (370)

the expected flank eruption was significantly delayed.

Against the background of such experiences, with eruptions that did not occur or were not as soon as predicted, and the resulting political and social consequences, some scientists distinguish two types of prediction (370). A *forecast* is the general announcement that a volcano will probably erupt some time in the near future (months, years, decades), based either on an analysis of its former activity (like the forecast of the Mount St. Helens eruption in this century; Chap. 10) or on qualitative signs of unrest (drastically increasing number of seismic events, increased degassing, heating up, increased fumarole activity, changes in gas composition and rates in emission, etc.).

The term *prediction* is used for the relatively precise statement, concerning the probable vent location, the time of the eruption (at Mount St. Helens two to three weeks to a few hours prior to an eruption; Fig. 13.16) and the probable type of an eruption. Only if these predictions are made with utmost care and are validated by the volcanic events themselves, credibility can be achieved and successful mitigation measures realized. As our knowledge of the types of precursors inside a volcano improves, predictions will increasingly be based on deterministic and less on probabilistic models.

Monitoring

The four most important changes in a volcano, caused by the rise of a magma prior to an eruption, are (Fig. 13.17)
1. Volcanic earthquakes
2. Expansion of magma chambers (leading to surface deformation) (Figs. 1.10, 13.18, 13.19)
3. Increased gas flux
4. Heating

These physical and chemical processes and other less clear signs, such as changes in the electric, magnetic and gravity field of the Earth, are regu-

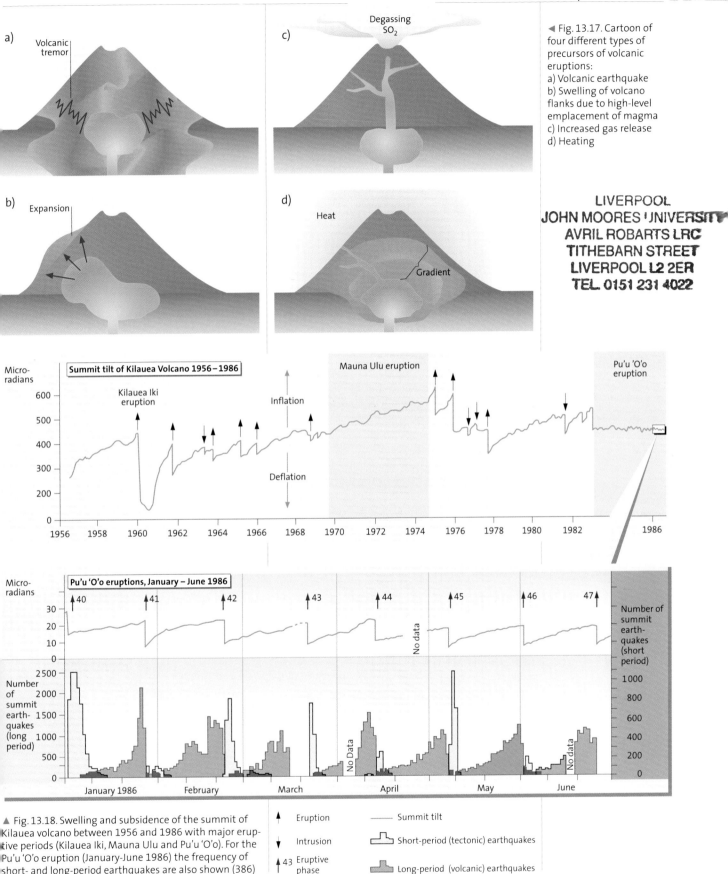

a)

Volcanic tremor

c)

Degassing SO₂

b)

Expansion

d)

Heat

Gradient

◄ Fig. 13.17. Cartoon of four different types of precursors of volcanic eruptions:
a) Volcanic earthquake
b) Swelling of volcano flanks due to high-level emplacement of magma
c) Increased gas release
d) Heating

Summit tilt of Kilauea Volcano 1956–1986

Micro-radians

Kilauea Iki eruption

Mauna Ulu eruption

Pu'u 'O'o eruption

Inflation

Deflation

600
500
400
300
200
0

1956 1958 1960 1962 1964 1966 1968 1970 1972 1974 1976 1978 1980 1982   1986

Pu'u 'O'o eruptions, January – June 1986

Micro-radians

40 41 42 43 44 45 46 47

No data

30
20
10
0

Number of summit earth-quakes (short period)

Number of summit earth-quakes (long period)

2500
2000
1500
1000
500
0

No Data

No data

January 1986   February   March   April   May   June

1000
800
600
400
200
0

▲ Fig. 13.18. Swelling and subsidence of the summit of Kilauea volcano between 1956 and 1986 with major eruptive periods (Kilauea Iki, Mauna Ulu and Pu'u 'O'o). For the Pu'u 'O'o eruption (January-June 1986) the frequency of short- and long-period earthquakes are also shown (386)

▲ Eruption

▼ Intrusion

▲ 43 Eruptive phase

——— Summit tilt

Short-period (tectonic) earthquakes

Long-period (volcanic) earthquakes

▶ Fig. 13.19. Collapsed house next to fault scarp with 10 m displacement, the result of domal uplift during the 2000 eruption of Usu volcano (Hokkaido, Japan)

larly monitored in several, but still lamentably few, volcanoes.

Most methods of volcano monitoring are still quite imprecise in detail such as if, when and how a volcano will erupt. A volume edited by Scarpa and Tilling (298) presents an excellent up-to-date assessment of presently available methods used for monitoring volcanoes. Most important are the analysis of seismic events and the deformation of the Earth's surface. These in many cases may allow short-term prediction of eruption onsets (days to hours).

Volcanic Earthquakes

Volcanic earthquakes can be defined as seismic events, which occur in, below and close to a volcano (generally within 10 km) or which are generated by volcanic processes (52, 218). Many documented large volcanic eruptions have shown increased seismic activity below and close to a volcano years, months, days or hours prior to an eruption. Even if earthquakes increase drastically in number only three days prior to an eruption, this is ample time for evacuation measures to be effected as demonstrated prior to the eruption of Usu Volcano in Hokkaido on 1 April 2000.

At the moment, about 200 volcanoes are seismically monitored, about a third of the around 550 volcanoes that have erupted in historic times. The number and quality of stations on each volcano are quite variable, however. The development of modern computers has allowed dramatic improvements in the speed and precision of locating earthquakes and determining the properties of seismic waves. With relatively close-spaced meas-

uring stations, the positions of earthquake hypocenters can be located with a precision of about 50–100 m horizontally and 100–130 m vertically (although deviations of 0.5 km horizontally and 1.0 km vertically are more typical). Many types of data can now be analyzed in real time. This permits a better assessment of the evolution of an eruption in progress and warning of specific events. With *real-time seismic amplitude measurements* (RSAM, 92) and *seismic spectral amplitude measurements* (SSAM, 360), important parameters can be measured automatically. Both systems register time series, which allow scientists to follow the evolution of volcanic processes prior to an eruption.

Classification and Origin of Volcanic Earthquakes

There are three types of seismic waves generated by earthquake sources in the Earth:
- Fast compressional or P-waves
- Slower shear or S-waves
- Surface waves, which are especially destructive

Shear waves do not propagate in fluids, allowing documentation of liquid magma bodies by their absence (Fig. 6.17). Earthquakes in volcanoes mostly occur less than 10 km deep, directly beneath an eruptive center. Sometimes there are exceptions to this. For example, the initial earthquake swarm prior to the eruption of Pinatubo occurred 5 km northwest of the volcano. Twelve days prior to the climactic eruption, the frequency of earthquakes decreased and strong earthquake activity started beneath where the caldera would form. During the large eruption of

Katmai volcano (1912), the distance between the collapse structure (Mount Katmai) and the actual Novarupta crater eruptive center amounted to 10 km. Earthquakes can thus occur all around an active volcano where stresses have built-up. These locations must not necessarily be exactly the spot where the magma rises and eventually erupts.

Many different types of seismic signals are generated beneath and within active volcanoes. Seismic waves can be generated by movements along fractures, opening of cracks or pathways, during fluid transport, by interaction of magma with the walls or fractures or vents and other fluid dynamic processes, as well as by explosive eruptions, pyroclastic flows, rock falls and collapse of entire segments of volcanoes. Today, four main types of seismic events are distinguished from each other (218; 224) (Figs. 13.20, 13.21):

- *high frequency earthquakes*, also called A-type earthquakes
- *low frequency earthquakes*, also called B-type earthquakes
- *explosion earthquakes*
- *volcanic tremor*

High frequency events are generated by brittle fracture in magma chamber roof rocks, due to stress caused by the increasing magmatic pressure, also by shear fractures in the country rock (adjacent to fractures or vents), and motions along faults, etc. They are distinguished from non-volcanic tectonic earthquakes in that they typically occur in swarms. High frequency or A-type events have a dominant frequency of 5–15 Hz. Because magmas can rise from great depth over large areas and for some time, high frequency earthquake swarms can last very long.

At depths of 1–3 km, significant changes occur. At first, the volatiles in a magma start to

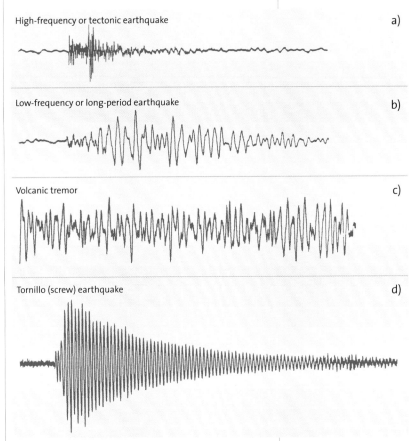

a) High-frequency or tectonic earthquake

b) Low-frequency or long-period earthquake

c) Volcanic tremor

d) Tornillo (screw) earthquake

form a free gas phase (bubbles). This changes the rheology and acoustic impedance of a magma and therefore its earthquake wave transmission and reflectivity behavior. For example, gas bubbles increase the viscosity and thus lower the acoustic velocity (218). A seismic event that is generated by resonance in a magma-filled space, is hence of low frequency. On the whole, the origin of most low

▲ Fig. 13.20. Typical waveforms of volcanic earthquakes from the area of active Redoubt volcano (Alaska) (218)

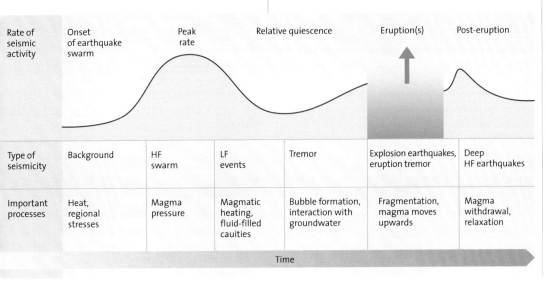

Rate of seismic activity	Onset of earthquake swarm		Peak rate	Relative quiescence		Eruption(s)	Post-eruption
Type of seismicity	Background	HF swarm	LF events	Tremor		Explosion earthquakes, eruption tremor	Deep HF earthquakes
Important processes	Heat, regional stresses	Magma pressure	Magmatic heating, fluid-filled cavities	Bubble formation, interaction with groundwater		Fragmentation, magma moves upwards	Magma withdrawal, relaxation

Time

◄ Fig. 13.21. Different types of volcanic earthquakes and processes beneath and around a volcanic edifice prior to and after volcanic eruptions. A similar evolution of earthquake types and distribution was observed and measured during the eruption of Mount Pinatubo (Philippines) (218). *HF* High frequency; *LF* low frequency

frequency earthquakes is related to fluids, not only the formation of bubbles and their collapse in magmas, but also the flow processes of rising magma and heated groundwater. Low frequency events commonly show P- but rarely S-waves and their frequency is in the range of 1–5 Hz.

The most typical volcanic earthquakes are characterized by relative constant amplitudes and frequencies, and are probably caused by turbulent motions in a rising magma column. They are called *volcanic tremor* or *harmonic tremor* (Fig. 13.20). Volcanic tremor is a continuous signal, which can last for minutes or days. The main frequency of tremor, 1–5 Hz, resembles those of the low frequency events. Harmonic tremor or spasmodic tremor are special cases of the general volcanic tremor. Harmonic tremor is of low frequency with a monotonous sinusoidal tremor with slowly changing amplitude, whereas spasmodic tremor is a high frequency, pulsating, irregular signal.

Tremor signals have been studied in greatest detail at Kilauea volcano. Most of these seismic events, up to several hundred per day, have their origin (hypocenter) at a depth of 2–4 km. Whether or not other earthquake sources at ca. 20–30 km depth beneath Kilauea are also caused by flow and pressure of the rising magma is uncertain. Likewise, it is not clear whether or not the rare earthquakes generated at 60 km beneath Hawaii really signal the beginning of magma ascent (Fig. 13.18).

Earthquakes, generated during explosive eruptions, are characterized by atmospheric shock waves, which can be registered on seismograms. Additional less common deep earthquakes, generated at 10–40 km depth, are commonly observed following large eruptive phases. These earthquake swarms probably reflect changes in the local stress field after partial emptying of a deep magma reservoir.

Prediction Based on Volcanic Earthquakes

In order to judge whether volcanic earthquakes herald a forthcoming eruption or not, it is important to monitor the background seismicity in a volcanic area for several years. Increased seismic unrest in active volcanoes can be caused by increased heat flow beneath a volcano; movement of groundwater, volcanic gases or magma; and motions of glaciers and rock falls or landslides. Only if one has learned to interpret seismic signals of these events with some assurance, can earthquakes that herald an eruption be interpreted in a plausible manner.

Well-documented case histories allow to postulate a general sequence of volcanic earthquakes swarms, which comprise some or all of the following components (218) (Fig. 13.21):

- background seismicity
- swarms of high frequency events
- relative quiescence after the peak of seismic activity (but before an eruption)
- low frequency events
- volcanic tremor
- eruption
- deep earthquakes following an eruption

Earthquakes of small and intermediate magnitudes (M < 5) occur in calderas or during sector collapse events. Volcanic earthquake swarms are distributed log-normally and generally last about 5 days. High frequency earthquake swarms last longest and do not always terminate in eruptions. On the other hand, low frequency swarms and volcanic tremor are shorter in duration and commonly occur directly prior to an eruption. Because intrusion and extrusion cause similar types of seismic signals, false alarms probably cannot be avoided, given the present state of knowledge.

Perhaps, an eruption could be described based on an empirical law, which describes fatigue of the material (FFM = failure forecast method) (401). Failure can mean a brittle fracture of rocks or fluid-saturated porous rocks, critical point of fluid pore pressure or the opening of fractures. This analysis is aided by using RSAM- and SSAM-systems (see above), which use the average absolute amplitudes of seismic signals from several seismometers.

Earthquakes in larger volcano structures, such as calderas, may not herald an impending eruption, or commonly begin long before an eventual eruption. For example, at Rabaul caldera in Papua New Guinea, strongly increased seismicity was recorded in the general caldera area from 1983 onward. A large eruption only occurred 13 years later, on 19 September 1994, being immediately preceded by a phase of intense local seismic unrest.

Volcanic Eruptions and Regional Tectonic Earthquakes

Earthquakes are classified, independently of volcanoes, depending on the depth of their hypocenter, into shallow (0–70 km), intermediate (70–300 km) and deep earthquakes (300–700 km). The deep events comprise less than 5 % and the shallow earthquakes more than 75 % of all seismic events. Intermediate and deep earthquakes occur dominantly along convergent plate margins, the zones where most large active volcanoes are located (Chap. 8, Fig. 8.1). It is worthwhile to look for causes common to the occurrence of volcanoes and deep earthquakes. Both types of events, volcanic eruptions and earthquakes, are expressions of sudden releases of energy in the Earth's crust.

and mantle. Large earthquakes may trigger or accompany volcanic eruptions or at least disturb the volcano-magma system including the hydrothermal system. Nevertheless, the precise temporal, areal or genetic connection between tectonic earthquakes and volcanic eruptions is poorly understood. Examples where earthquakes apparently triggered eruptions, are the eruption of Cordón Caulle, 48 h after the huge (M = 9.5) earthquake in Chile in 1960, small summit eruptions of Kilauea volcano shortly after the magnitude 7.1 Kalapana earthquake of 1975, and the 1707 Mt. Fuji eruption. A large M 7.3 earthquake in Landers (California) in June 1992 generated earthquake swarms in 17 volcanic areas in the western USA, at distances of up to 1250 km. Perhaps the seismic waves significantly influenced the volatile pressure in a magma in some as yet poorly understood manner (36, 157) (Chap. 4).

An areal and temporal relationship may exist between large fracture zones and volcanoes in Central and South America, an area characterized by shallow overthrusts (45). The volcanoes in these areas are generally inactive for a few years to decades prior to large earthquakes but begin to erupt a few months or years after such large earthquakes. Examples include Coseguina (1835) in Nicaragua and Santa Maria (1902) in Guatemala, the two largest volcanic eruptions in Central America in historic time.

Expansion of Magma Chambers

Beginning with the pioneering work of Eaton and Murata (90), measuring tilt on the roof of volcanoes is the premier method to detect an accumulation of magma in the interior of Kilauea volcano. Monitoring using tilt meters indicates whether a volcano swells or not. The expansion of the surface of a volcano can be measured very precisely with modern geodetic instruments. Expansion is an indication of a relatively slow rise of magma into a magma chamber or from this into the upper crust, which precedes many volcanic eruptions (Fig. 13.22). Tilt and distance measurements have also been used successfully in much more difficult terrain (steep slopes, covered with loose material or glaciers etc.), such as on Mount St. Helens (370) and Etna. A spectacular example of uplift and accompanying graben formation occurred during the 2000 eruptions of Usu Volcano in Hokkaido (225, Figs. 1.10, 13.19).

Degassing

The amount and/or composition of gases that are released from a volcano (Fig. 13.23), can change prior to an eruption. SO_2-emissions may increase months or years prior to volcanic eruptions. New

◀ Fig. 13.22. Measuring the degree of inflation of the surface of Kilauea volcano (Hawaii) in 1970 with laser geodimeter

undegassed magma appears to have risen into the upper crust releasing gases without immediately erupting lava or tephra. On the other hand, CO_2-emission can drastically decrease a few days prior to an eruption, such as prior to the Pinatubo eruption (see below). The ratio of F/Cl may be another useful indicator to monitor volcanic activity. The acid HF may impact strongly on the environment due to its excellent solubility in water and its rapid adsorption on ash particles or vegetation. Adsorption on vegetation is a serious problem to agriculture in areas that are subject to volcanic gases for long periods.

▼ Fig. 13.23. Late Werner Giggenbach, famous gas-geochemist, using the Giggenbach bottle to sample gases rising from a mineral spring in the Eifel (Germany)

▶ Fig. 13.24. Measuring temperature in a hot block that traveled with a pyroclastic block flow on the slopes of Augustine volcano in 1976 (Alaska)

Heating

Measuring temperature is a standard monitoring technique in active volcanoes (Fig. 13.24). A volcano is heated up only slowly, however, when new magma is emplaced into its interior. A rise in temperature can be measured directly in the surface soils or via infrared monitoring by aircraft or satellite. There are commonly indirect effects of ground heating, such as the increasing temperature in springs near a volcano, sudden melting of snow or change in the magnetic field.

Remote Sensing

The rapid development in remote sensing by satellites during the past two decades has greatly increased the potential to monitor volcanoes in the entire electromagnetic spectrum. Remote sensing is simply detection and measurement of electromagnetic radiation emitted from the surface of a volcano, by lava flows or by ash clouds rising above it (100, 296). The sensor can be fixed to the ground or be installed in a plane or satellite. The radiation can be reflected sunlight, as in a photographic image, heat as in the case of a thermal infrared receiver, or the reflected pulse from a radar system that bombards a volcano with microwave energy.

Remote sensing not only compliments ground observations and monitoring using conventional methods, but also offers entirely new methods such as detection of crustal deformation via synthetic aperture-radar. For most volcanoes not monitored on the ground, satellite-based remote sensing is the only possibility to rapidly acquire data on the precursors of eruptions and also the subsequent eruptions themselves.

Satellite-based volcano monitoring presently focuses mostly on eruptions in progress, including monitoring of ground deformation, thermal changes, and various parameters of eruptive columns. Since about 1980, measurement of SO_2 emissions of volcanoes has been carried out using the TOMS (total ozone mapping spectrometer) instrument on the Nimbus-7 and Meteor-3 satellites. This method is ideal to estimate SO_2 fluxes during an eruption, as well as estimating global mass fluxes of volcanic SO_2. The importance of sensors in the visible part of the electromagnetic spectrum is limited because of the common cloud cover of volcanoes. Radar satellites, by contrast, allow signal reception during any type of weather, but they cannot register thermal radiation. Multispectral sensors with high areal resolution are less suitable for repeated monitoring of volcanoes than sensors with a lower resolution.

Many new developments are currently under way, however, be it new satellites or, more importantly, the development of computer-based methods that are able to digest and interpret the huge amount of data. There is always some delay in monitoring actual eruptions. The other problem is the enormous cost. Like all monitoring techniques, possible precursors for an imminent volcanic eruption can only be estimated realistically when background behavior is known. In other words: a volcano must be observed over some time to recognize significant deviations from its "normal state".

Extinct Volcanoes?

One of the most common questions asked by laypeople and journalists centers around the problem whether or not a certain volcano can be regarded as extinct. Such questions are very difficult to answer. As discussed above, several apparently extinct volcanoes have caused catastrophes by their unexpected re-awakening. Volcanoes and volcano fields can be active for millions of years with relatively short periods of volcanic activity alternating with long periods of quiescence (hundreds, thousands or millions of years).

There are no precise or generally accepted definitions for the terms *active volcano*, *sleeping* or *dormant volcano* and *extinct volcano*. All volcanoes that have erupted within the past 10 000 years are commonly regarded as active (338). The recent (2002) adoption of this age limit by the Japan Meteorological Agency has resulted in the "increase" in number of active volcanoes in Japan from 86 to 109, previous definitions being based on an upper age limit of 2 000 years B. P.. Sometimes, all volcanoes that erupted since the boundary between the late Pleistocene and Holocene (in Central Europe the Holocene started about 11 000 years ago) are defined as active. The geological boundary between the last ice age and the Holocene was chosen as it is often easy to define. To use such definitions, however, one has to distinguish between different types of volcanoes; for some either definition would be misleading.

The most important criteria for regarding volcanoes as extinct or dormant, are the total lifetime of a volcanic complex or volcanic field and the overall frequency of individual eruptions. This requires very precise analysis to arrive at reasonable estimations of the probability of future eruptions. We can visualize periods of years, decades and possibly hundreds of years. But even geologists, who are familiar with geological dimensions (i.e., with the past 4.6 billion years) often resort to the perception of non-specialists and regard volcanoes that have not erupted for several generations as extinct.

The significance of this problem becomes very obvious with respect to long-term storage of nuclear waste. Billions of dollars have already been spent to prepare the storage area for all nuclear waste generated in the US at Yucca Mountain (a huge cave system dug into Tertiary ignimbrites in Nevada). The age of the youngest scoria cone, Lathrop Wells, 15 km south of the storage area, is about 0.07–0.1 million years. This has to be compared to a period of 10^4 to 10^5 years until the radioactivity of high-level waste has decayed. The question of whether or not volcanic eruptions may occur in the area of the deposit is therefore crucial to the final decision to house waste at Yucca Mountain (65).

Volcanic Hazard Mitigation

Volcanic eruptions cannot be influenced by man now or in the foreseeable future. They are governed by processes that start deep in the Earth and that may reach into the stratosphere. When a volcanic catastrophe occurs, the reasons are often rooted in society itself. There are, however, limited possibilities to reduce some of the effects of volcanic eruptions.

Most important is to define hazard zones, which are based on a very detailed analysis of the early history of a volcano, as well as an estimate of the possible scale and nature of eruptive processes at a particular volcano. Each volcano behaves differently, requiring careful analysis of the local factors.

The path of some lava flows can be influenced slightly by building barriers, unless they flow with high velocity and/or high effusion rates. This has been done on Etna volcano in Sicily several times, but the success has been limited (51). An ingenious method was attempted to slow a lava flow on the Icelandic island of Heimaey. The flow was about to close the entrance of this important fishing port. The Icelanders sprayed the advancing lava flow front with copious amounts of seawater so that it would cool and build itself a barrier, which it could not overcome (419). After months of water spraying, the flow stopped. It was not clear, however, whether the water stopped the flow progress or whether it was a drop in effusion rates as the eruption finished. In any event, the cost of rebuilding the town itself would have been several times the cost of spraying, including the large pumps ships brought in to assist.

Two principal types of *sabo dams*– a Japanese term literally translated meaning *sand protection*– are installed on the slopes of many volcanoes in Japan, a country characterized by steep mountains, high precipitation and many young and active volcanoes. Sabo dams are generally built at right angles to the drainage. Many thousands of such dams have been built on Japanese mountains, including practically all major volcanoes. They break the velocity and, therefore, the erosional energy of mass transport down steep drainage paths, and act as temporary sediment traps. They include massive concrete structures or more modern rock dams (Figs. 13.25 – 13.27), and some are augmented by video cameras or wires to monitor advancing lahars or pyroclastic flows. During the 2000 eruption of Usu volcano (Hokkaido, Japan), sabo dams were effective sediment traps for lahars (Fig. 13.27), preventing more extensive damage to buildings. Sabo dams at

▲ Fig. 13.25. Lateral dike protecting a village at the foot of Mt. Mayon (Philippines). Lahars commonly rush through this valley (see also Fig. 13.4)

▶ Fig. 13.26. Aerial photograph of Lake Toya, town of Toya and two of the more than 60 craters formed during the almost entirely phreatic eruptions in 2000. Several sabo dams on the outskirts next to the town proved very useful to retard the advance of lahars which destroyed several houses on the outskirts of Toya. Recent Showa Shinzan dome in background, main crater of Usu dome complex at right and site of enlargement of sabo dam (arrow) shown in Fig. 13.27. Photo courtesy Hokkaido Regional Development Bureau

Sakurajima volcano in Japan are also monitored by video cameras, or other optical devices and wire sensors devices. *Dikes* parallel to valleys have been constructed on several volcanoes in Japan, Indonesia and the Philippines to direct the course of block-and-ash flows and lahars. This can be effective at least for small-volume lahars. Dikes and sabo dams are most effective on volcanoes that erupt frequently, such as Mount Mayon

(Philippines), and Merapi (Java), Tokachi (Japan) or Sakurajima (Japan), where the channels used by volcanic mass flows are well-known and the individual eruption volumes relatively small. Sabo dams are not effective when the flow volumes are very large, e.g., when glaciers or snow melt. Moreover, sabo dams and dikes are very expensive to build.

Temporary lakes commonly form during an eruption by very rapid accumulation of large masses of tephra within river channels. These tephra dams are naturally very unstable and can collapse easily as shown by the examples of Asama and Laacher See volcanoes discussed earlier in this chapter. Countries that can afford the expenditure thus spend much money to drill tunnels through such natural dams, as was the case following the eruption of Mount St. Helens in May 1980. A very instructive example of how to control the effects of volcanic eruptions is Kelut volcano in Indonesia. This volcano ejected its crater lake during each eruption, which led to the formation of catastrophic lahars on its densely populated flanks. An eruption in 1919 killed more than 5 000 people. To prevent a further catastrophe, tunnels were dug to significantly lower the lake level. There were only seven fatalities during the next eruption in 1951. In the 1960s, a large water volume re-

...nained in the crater. Zen and Hadikusumo (441) ...oiced concern of another catastrophe during the ...next eruption, which they thought would occur ...n the next five years. These fears were sadly veri-...ied. Kelut erupted in the spring of 1966 with ...ahars killing some 200 people. New tunnels were ...lug and when the volcano erupted in 1990, no ...people perished by lahars, although 32 lives were ...ost when buildings collapsed as they became ...verloaded with tephra (Smithsonian Institution ...990).

Near-vent areas most prone to volcanic ...mpacts can be relatively easily delineated. These ...nclude notches in the upper part of a volcano ...nd major valleys down its flanks. Thus *mitigative* ...*measures* (such as evacuation and land-use re-...trictions) can be carried out in time to minimize ...osses. At greater distances from volcanoes, it is ...ossible to evacuate people early enough in well-...nonitored volcanoes. Warning periods of several ...nours, days or weeks in many cases can be ...chieved. A challenge with scientific monitoring is ...o provide important volcano status information ...nd predictions in a form understandable to non-...pecialists. Other formidable tasks are to convince ...ocal authorities of the threat and to distribute ...imely and accurate information in leaflets, print ...nd electronic media. Plans for evacuating people ...n case of imminent danger must be worked out ...n detail ahead of time. The communities must ...lso be informed of fundamental aspects of evac-...uation scenarios and on the mitigative actions ...hat they themselves can carry out. It is useful to ...arry out community disaster training as is done ...t regular intervals around the almost permanent-...y active Sakurajima volcano in Japan.

The issues of disseminating scientific warning ...nformation effectively came to a head following ...he catastrophic results of the eruption of Nevado ...del Ruiz in Colombia on 13 November 1985. ...Among the several contributing factors to this dis-...aster was a failure to effectively communicate war-...nings to administration and emergency manage-...ment authorities. Following several discussions, ...IAVCEI along with UNESCO produced a video, in ...which volcanic hazards and their impacts are ...explained and portrayed very drastically. Maurice ...and Katia Krafft (both of whom perished in June ...1991 at Unzen volcano) took a major role in ...preparing this video. It was ready in early 1991 ...when Pinatubo (Philippines) started to show ...signs of unrest in April 1991. This video was ...quickly distributed and shown with great success ...during education and warning programs in many ...villages surrounding Pinatubo. Newhall (pers. ...comm., 1991) estimates that up to 10 000 lives ...may have been saved because the willingness of

▲ Fig. 13.27. Enlargement of sabo dam by rock fill on the flanks of Usu volcano near Toya following the eruption in 2000

people to become evacuated rose significantly after having seen the video.

The main risk management problems en-countered during eruptions of Mount St. Helens (1980) and Pinatubo (1991) were (251):

- the time for monitoring, data analysis and dis-semination of warnings was very short (in each case 2–3 months prior to the large eruptions)
- the skeptical attitude of the local administration and other local authorities and also the general public. Purely scientific arguments were not very effective
- the unrealistic public expectation of how accu-rately volcanologists were able to predict an eruption. The public expected simple answers, not the discussion of different possible scena-rios
- the necessity to work on the main types of ha-zards close to a volcano meant that risks at lar-ger distances were neglected
- extreme pressure was exerted on volcanologists to provide accurate and timely warnings as well as to make decisions on predictions and evacua-tions. Along with emergency management offi-cials, volcanologists were extremely overworked during weeks prior to an imminent eruption. This had the potential to lead to mistakes or poor decisions in some cases
- the manpower, financial support, and monito-ring systems or equipment were often insuffi-cient or delayed until after the main phase of activity.

Lessons Learned From Two Large Volcanic Eruptions

Two recent major volcanic eruptions show dra-matically how catastrophes either develop (Neva-do del Ruiz), or can be avoided (Pinatubo).

Nevado del Ruiz
Precursors and Early Stages

Eruptions of Nevado del Ruiz had generated lahars with catastrophic results already in A.D. 1595 and 1845. The town of Armero was built 72 km east of the crater of Ruiz in the 1930s, on the deposits of the largest lahars of 1845. Steam eruptions of the volcano were noticed in late 1984. The activity increased slowly during the next few months. A stronger eruption on 11 September 1985 generated a lahar that flowed 27 km. Teams of volcanologists from several countries studied the volcano and reached the conclusion that a large eruption was very likely (147, 402). Lahars generated during such an eruption could flow as much as 100 km. The teams suggested:

- a hazard map for Ruiz (Fig. 13.14) to be published in the local press (done on 10 November 1985). The zones of maximum hazard on this map strongly resemble the actual distribution of lahars of 13 November 1985
- a delineation of safe areas, into which the population could be evacuated in case of a likely catastrophe
- permanent visual monitoring of the volcano
- constant communication with the areas in danger to relay warnings immediately after the onset of an eruption (20).

Course of the Eruption and Effects

Following weak explosions on the afternoon of 13 November 1985, a stronger eruption started in the evening. Hot surges started to melt the ice and snow caps on the summit of the volcano. Descending melt-waters eroded and incorporated loose debris along their path. Local radio stations asked the people to remain calm and stay indoors because of the possible ash fallout from the eruption. An order of evacuation made by the Red Cross never reached the town of Armero. The first

Nevado del Ruiz (Colombia) (Figs. 13.14, 13.28)

Location: Colombia

Elevation: 5 390 m a.s.l.

Glacial cover: glacier cap with an areal extent of 25 km² prior to November 1985

Characteristic activity: highly explosive pyroclastic flows, lahars, volcanic debris avalanches

Rock composition: andesitic-dacitic

Tectonic position: subduction zone

Date of eruption: 13 November 1985

Major historic eruptions: 1595, 1845

Recent eruptions: several smaller eruptions since 1985

Was the volcano known to be dangerous?   Yes

Were warnings published?   Yes

Were emergency plans ready?   No

Were there indications of an imminent eruption?   Yes

Was the catastrophe avoided?
    No. Approx. 23 000 fatalities (1985)

Damage: 5 092 houses, 50 schools, 2 hospitals, 58 factories and ca. 3 400 ha agricultural land were destroyed by lahars (total value exceeding US$ 1 billion)

Is the volcano now monitored?   Yes

lahars arrived in Armero one and a half hours after the strongest eruption. People in the town were fully unprepared and three quarters of the population of Armero did not survive the catastrophe (Fig. 13.28).

Nevado del Ruiz has remained active following the eruption of 13 November 1985 and is now monitored more closely. On several occasions since 1985, thousands of people were ordered to evacuate from areas near the volcano. However, despite the tragic example of Armero, most people did not want to leave their homes.

Lessons to be Learned

If the communication between scientists and authorities had functioned and if emergency plans had existed to provide timely warning to Armero after the beginning of the eruption, the people would

▶ Fig. 13.28. Debris flow deposit, about 1 m thick, laid down by the lahar that devastated the town of Armero following the eruption of Nevado del Ruiz (Colombia) on 13 November 1985. About 23 000 people were killed by the lahar. The area has been declared a national cemetery by the Colombian government

have had one hour to leave the dangerous zone. Although the town would have been destroyed, losses of live could have been minimized.

In areas at larger distance from well-monitored volcanoes, warnings can be issued up to several hours before impact, so that people can be brought into safety relatively easily. The lack of such monitoring at Nevado del Ruiz was the reason that a timely warning of the approaching lahars did not reach the town. It is also tragic that the administration of Armero did not take seriously the earlier general warning issued by geologists. On the other hand, the town should have never been built at this place, because of the previous inundation by even larger volcanic debris and mudflows in 1595 and 1845.

In hindsight, Voight (402) listed the main steps that would have helped to avoid the catastrophe of Nevado del Ruiz in November 1985:

- recognition and documentation of the hazard
- preparation of a priority list of the vulnerable elements, communities and areas
- open public discussions of the scientific risk assessments, despite social anxiety
- early planning of critical decisions, since time is the most important variable
- early testing of the warning system to check its reliability
- early assessment of possible technical problems of warning systems
- careful integration of different types of media for public announcements.

Pinatubo

Precursors and Early Stages

On 16 July 1990, an earthquake with a magnitude of 7.8 occurred about 100 km northeast of Pinatubo, along the large Philippine fault. This may have influenced the magma reservoir beneath Pinatubo at a depth of 9–11 km. In hindsight it became clear that there had been weak signs of re-awakening of the volcano for many years, and thus an imminent eruption. The signs included changes in the composition and temperature of the hot springs, increasing fumarolic activity, chlorine-rich, strongly acid fluids in geothermal wells (1988–1990), which, because they were not productive, had been terminated.

Two and a half weeks following local earthquakes, steam explosions on 2 April 1991 showed that Pinatubo had re-awakened. From late April to early June, the oldest deposits of Pinatubo were studied by a crash team, mainly from the US Geological Survey, and quickly dated. It became clear that larger eruptions had also occurred in the recent past, in which voluminous pyroclastic flows dominated. Past events were separated from each

Pinatubo (Philippines)

Pinatubo based largely on papers in the excellent volume by Newhall and Punongbayan (251) (Figs. 13.29, 13.30)

Location: Philippines, Luzon Peninsula, about 120 km northwest of Manila

Altitude: 1 745 m a.s.l. prior to, 1 485 m a.s.l. after the eruption

Volcanic history: active at least during the past 30,000 years. Non-eruptive intervals lasting from a few hundreds to several thousands of years

Characteristic activity: highly explosive, pyroclastic flows, lahars, debris avalanches

Rock composition: dacite, partly hybridized by mixing with basaltic magma

Tectonic position: part of a chain of compound volcanoes that form the Luzon volcanic arc, about 100 km east of the eastward dipping Manila subduction zone

Date of eruption: 15 June 1991

Earlier prehistoric eruptions: Unknown

Population on the volcano flanks: 30 000 people near the volcano and a further 500 000 in towns and villages on the ring plain that surrounds the volcano. The US Clark Air Base was 25 km east of the volcano and the US Subic Bay Navy Base about 40 km to the southwest

Post-1991 eruptions: minor intermittent eruptions

Was the volcano known as potentially dangerous? No

Were warnings published? Yes

Were emergency plans ready? Yes

Were there signs for an imminent eruption? Yes

Was the catastrophe avoided? Yes

Fatalities: Approx. 350

Damage (1991–1992): 8 300 houses destroyed completely, 73 400 partially; economic loss totaling more than US $ 700 million

How was the catastrophe avoided? Good communication between scientists and administration, emergency plans were ready and timely warning was possible. Strong logistic support by the American and later Philippine Air Force. Difficulties in communication were mostly due to the isolated position of the large volcanic complex and different administrative districts with differing political agendas and allegiances

Is the volcano now monitored? Yes

other by hundreds to thousands of years. The amount of SO_2 emitted rose from 500 t/d on 13 May to more than 5 000 t/d on 28 May. Together with permanent seismic activity, these data suggested that magma had risen beneath Pinatubo to such a height that its volatiles could be released as a free gas phase. In the preceding weeks, the rising SO_2 was probably dissolved in groundwater, so that a strong SO_2 emission may have reflected a vaporization of the hydrothermal system in the interior of the volcano. On 1 June, earthquake hypocenters

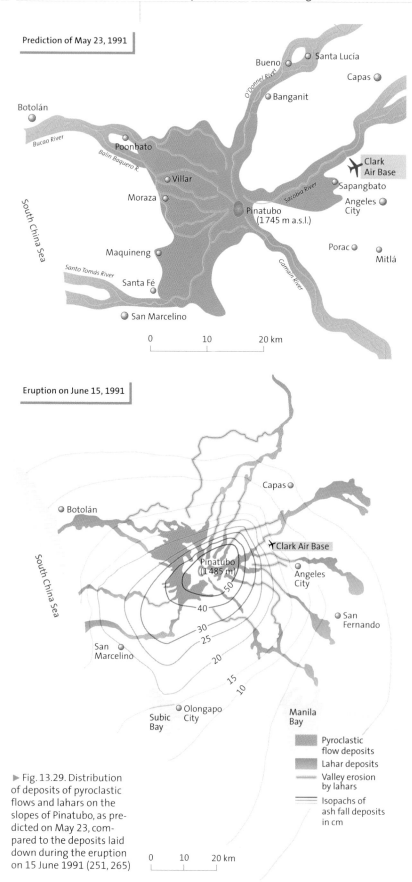

Prediction of May 23, 1991

Eruption on June 15, 1991

Pyroclastic
flow deposits

Lahar deposits

Valley erosion
by lahars

Isopachs of
ash fall deposits
in cm

▶ Fig. 13.29. Distribution of deposits of pyroclastic flows and lahars on the slopes of Pinatubo, as predicted on May 23, compared to the deposits laid down during the eruption on 15 June 1991 (251, 265)

were registered between the surface and a depth of about 5 km, about 1 km northwest of the summit. These earthquakes probably formed because the rising magma was opening a vent between the magma reservoir and the summit. At the same time, the SO_2 emission rate decreased from 1 800 t/d on 30 May to 1 300 t/d on 3 June and 600 t/d on June 5. On 3 June seismicity under the volcano increased, small ash eruptions occurred and a brief period of harmonic tremor was registered. In addition, the outer flanks of the volcano started to swell. On 5 June warning level 3 was announced with the possibility that a larger eruption with pyroclastic flows could occur within the next two weeks. On 6 and 7 June, the precursors intensified, steam clouds rose on June 7 to 8 km. Warning level 4 was announced with the possibility of a large eruption within the next 24 h; more people were evacuated. On 10 June, 14 500 people were evacuated from Clark Air Base. Following the first larger explosive eruption on 12 June the main radius for evacuation was extended to 30 km; the total number of evacuated people rose to about 60 000. Three days later, the largest volcanic eruption in the second half of the twentieth century occurred, topped only by the eruption of Novarupta volcano (Katmai) during which more magma (ca. 13 km³) was expelled.

Cause and Effect of the Eruption

After the start of the eruption on 15 June, the instruments stopped working at 13:42 h, because they had become saturated. Ash and pumice fragments, up to 4 cm in diameter fell at 14:30 h on Clark Air Base. Satellite data showed that the eruption column had reached the stratosphere at 15:40 h with a diameter of 400 km and a height of 35 km in the center and 25 km on its eastern margin. The main phase of the eruption began to decrease slowly after 3 h. The climactic eruption had come to an end at 22:30 h. The collapse of a caldera probably began at 16:30 h and lasted until 20:30 h. Sporadic small ash eruptions occurred until early September. A lava dome grew inside the caldera from July to October 1992.

Unfortunately, the eruption occurred when taifun Yunya had just reached the Luzon Peninsula on 15 June. The load of water-saturated ash falls caused many roofs to collapse, including hangars on Clark Air Base. Fallout pumice and ash fell simultaneously with ash flows, which had proceeded through the valleys. The tephra fall volume was estimated as 3.4 – 4.4 km³, that of the ignimbrites as 5 – 6 km³.

The ethnic group of the Aeta that lived on the forested slopes of Pinatubo was completely uprooted. They lived from wild fruit and animals and had their own specific religion. The two

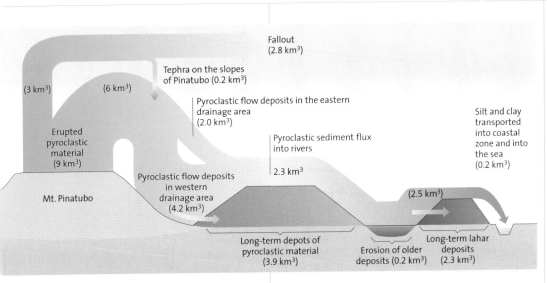

◄ Fig. 13.30. Volumes and depositional areas of different types of mass flow. These transported material during the eruption of Mount Pinatubo to the foreland and in part to the sea and were deposited for various periods at different temporary storage areas (264)

largest American military bases outside the U.S. (Clark Air Base and the Subic Bay Navy Station) were permanently closed. About 250 people were killed during the eruption and about 100 more in the subsequent lahars. The climactic eruption left an open crater system.

The unconsolidated ash flows reached a thickness of over 200 m in the valleys. The monsoon rain, which started in July, generated many additional lahars, many of them from rapid reworking of ash flows and therefore still with temperatures of more than 100 °C. These lahars, one after the other, covered villages, towns and fields and destroyed houses and bridges. Tens of thousands of people had to leave the area. At the end of the southwest monsoon in late October 1991, about 1 km³ of new lahar deposits covered an area of about 300 km², an area that prior to the eruption had been highly productive agriculturally. This reworking of ash in the form of lahars is predicted to last until at least about 2005. The enormous economic and social effects were caused especially by the continuous sedimentation in the creeks and rivers, which forced the lahars to spread broadly over the lowlands at the foot of the volcano (Fig. 13.30).

All Pinatubo lahars formed directly or indirectly from the strong monsoonal rains, episodically increased by tropical cyclones. During the eruption of 15 June 1991, several factors came together. These included a sudden and significant change of landscape by the ash-sealing of the pre-existing drainage systems. This generated an increase in surface runoff, as well as a huge masses of loose, easily erodable ash. The flowing water from heavy rains quickly took all loose sediments from the riverbeds, undercut their sides and changed them to sediment-rich mud- and sand-flows, often with more than 60 vol% sediment.

Lessons Learned

For the first time in history, an eruption of this magnitude was studied scientifically in detail. The number of new insights is remarkable. For example, the occurrence of long-period earthquakes prior to explosive eruptions; the early separation of a gas phase, deep in a magma reservoir; the sudden decrease of SO_2 emission shortly before the explosions; the interaction of the eruption column and the Earth's crust generating tremors with periods of 200–300 s that were registered globally; earthquakes with a magnitude of less than 5 were generated during caldera collapse. Morphological changes that normally take tens of thousands of years developed within a few days and years. The global climatic impact resulted in a flood of scientific studies on the effects of volcanic eruptions on atmospheric chemistry and climate forcing (Chap. 14).

Even though about one million people were in danger, only a few hundred lost their lives. By a multitude of means (informing the people, early preparation of evacuation, constant monitoring of the volcano with a variety of methods, establishment of safety zones, etc.) major losses in lives were avoided. However, through the destruction of settlements and agricultural land, tens of thousands of people lost their homes and their livelihoods. Moreover, the long-term effects (inundations, silting of rivers and the coast) had drastic economic consequences for the Philippines.

Pinatubo had been inactive prior to this eruption for 500–600 years. It was, therefore, difficult for scientists to convince the people that the mountain is a volcano that could erupt again. The most clear-cut precursors escalated only ten days prior to the eruption. A hazard map and safety system with five hazard levels was handed over to public administrations. It is a testimony to the high caliber and huge effort of these studies that predictions were clearly corroborated by the eruption and its deposits. Moreover, the IAVCEI-UNESCO-video, which shows dramatic images of volcanic hazards, was a very effective tool to motivate people to follow evacuation orders, etc. In addition, the eruption was not as large as some of the prehistoric eruptions in the same area. The small eruptions in the ten days prior to 15 June convinced even skeptics of the necessity for evacuation. The eruption was predicted relatively well and tens of thousands of people were brought to safety in time.

The eruption was probably triggered by intrusion of basaltic magma into a crystal-rich dacitic magma reservoir. During the eruption, more than $5\,km^3$ magma was erupted, but about $17\,Mt$ of SO_2 was injected into the stratosphere, along with $500-900\,Mt$ of H_2O, $3-16\,Mt$ of Cl and $42-234\,Mt$ of CO_2.

Harlow et al. (142) were able to predict the increasing eruptive potential based on a number of criteria including: changing earthquake hypocenters, increase of total seismic energy, change in the character of earthquake wave forms, increase of non-seismic precursor activities and an interpretation of the magmatic processes during the precursory interval. Seismicity increased at least two months prior to the eruption, although the exact beginning is unknown since seismometers were only installed following the first phreatic explosions. The ability to predict the increasing eruptive potential between late-April and mid-June on Pinatubo was based on several factors:

- the ability to set up a seismic network around the volcano relatively quickly
- the ability to analyze seismic data in real time and to relate these with geological observations and measurements of other volcanic phenomena
- the experience of interpreting seismic signals and precursory signals in other active volcanoes.

Hazard and Risk Assessment

Perhaps the most comprehensive and interdisciplinary broad hazard study with the ultimate aim to provide a robust risk assessment is presently realized at Mt. Fuji in Japan. The upper part of the volcano from $2\,300\,m$ a.s.l., the end of the road, to $3\,776\,m$ a.s.l. is climbed by about 150 000 people annually, while many more make it to the base station by car (Fig. 1.1). Local authorities estimate that some 20 million people visit the attractive foothill area of Mt. Fuji. This staggering number – even if on the high side – 15 % of the Japanese population – is due to the proximity of the 20-million metropolis of Tokyo, only 100 km away.

Trigger for the intensified risk assessment effort by consulting companies, universities, and geological surveys are the recent signs of unrest in the volcano mostly built of basalts during the past 10 000 years. The 2-mm isopach of the tephra fan produced during the famous Hoei eruption from 16–31 December 1707 reached to the center of present metropolitan Tokyo. A worst-case-scenario assumes a highly explosive eruption, such as in 1707, occurring during the rainy season and peak tourist population (Aramaki pers. comm.). The risk is estimated as roughly 20×10^9 (billion!) US\$ – one fourth of the damage resulting from the Kobe earthquake in 1995, the most expensive natural disaster of the past decades. The staggering costs are largely due to potential destruction of traffic and supply lines in metropolitan Tokyo.

Volcanic Eruptions and the Media

In this media age, volcanic eruptions can be welcome spectacular natural events. The fast moving ash clouds, or the impressive colorful lava flows and fire fountains can be presented in dramatic pictures that are quickly transmitted globally. For an impending volcanic eruption, media have also a very important function. With their help, people can be prepared for possible evacuation and warned in time to move. Radio and television can transmit alarms very quickly, while print media can disseminate detailed instructions about evacuation plans in some detail. When volcanic eruptions destroy cultivated areas and people are killed during eruptions, the dramatic television images serve to encourage other countries to help, even though such sensational pictures quickly lose their news value.

This is achieved by improving the communication skills of scientists when dealing with administrations, politicians, the media and people concerned. Moreover, efforts are increasing in many countries to cooperate with colleagues from social sciences, economists, etc. to more effectively assess crucial vulnerability factors, which can differ drastically between countries, different cultures, the local structure of communities, etc.

On the other hand, sensationalism in some media reporting does not correlate with the actual hazards. Such media reports and occasional media "grandstanding" of politicians and sometimes scientists makes it difficult for the non-informed to realistically estimate the actual hazard of an eruption, both in the short and the long term. Comparatively small eruptions with only minor economic consequences are blown out of proportion, especially at times when there is no other exciting news. The eruption of Ruapehu volcano in New Zealand in 1995/96 was a good example. The eruptions of Soufrière volcano on Montserrat (1995–present) have been reported in some countries with such sensationalism that the people on the island became deeply concerned about the image of their lovely island. Moreover, the long-term aftereffects of volcanic eruptions, whose economic impacts are commonly much larger than those of the eruption itself, are not an item that attracts much attention by the media. This can be bitter for the countries concerned when the amount of aid from other countries is tied to the news value.

Even in respected journals volcanologists are sometimes chided because their apparent inability to predict specific eruptions. This is clearly unjustified, because of the immense difficulties to predict type, locality and exact onset of particular eruptions, volcanoes and their roots representing extraordinary complex systems. It also does not recognize the efforts of highly qualified scientists, working at the site under most unfavorable logistic conditions.

It is the job of volcanologists to estimate volcanic hazards and therefore help to prevent catastrophes. Volcanic catastrophes, which range from thousands to tens of thousands of fatalities, to a short-term loss of livelihood, can only be effectively prevented or mitigated if the complex nature of people and society is taken into consideration. Volcanologists, who develop methods for predicting volcanic eruptions, or who are called upon to estimate if, when, where and how a particular volcano will erupt, therefore carry especially large responsibilities. Sometimes, as has occurred in some past events, they may get into extreme political trouble. This implies that volcanologists are bound to improve their methods of presenting the results of their work effectively to the general public.

Summary

The flanks and foothills around active volcanoes are endangered in many different ways. Particularly hazard-prone areas are volcano flanks and the valleys that radiate from them. The valleys are especially hazardous, because they channelize high-velocity pyroclastic flows and water-rich debris flows (lahars). Lahars can carry destruction for many tens of kilometers into the foreland (rarely >100 km) and can be generated many years to decades after a volcanic eruption. Over the last 200 years, most people were killed during volcanic eruptions by pyroclastic flows, surges, lahars and volcanogenic tsunamis.

Highly destructive flank collapses of volcanoes, producing huge debris avalanches and debris flows, are comparatively rare. Fallout of ash and coarser rock fragments causes the overloading of roofs. Collapsing of ash-loaded roofs can be highly and dangerously destructive relatively close downwind of a volcano. The entry of debris avalanches or large pyroclastic flows into the sea can generate destructive and lethal flood waves (tsunamis). These can also form as the result of major landslides on submarine volcano slopes. In many volcanic eruptions, several of these hazard phenomena can occur almost simultaneously. The interaction between rising magma and ground or surface water, as well as that between pyroclastic flows/surges with glaciers and snow, are especially dangerous. The hazards of fine ash clouds to air traffic have also become apparent in the past few decades by the more than a dozen almost-catastrophes, which occurred when large jets crossed ash clouds and their turbines became clogged. It is sheer luck that catastrophes have been avoided so far. In each case, the pilots of these aircraft were able to restart the engines after the planes had dropped by several thousand meters.

The greatest volcanic fatalities in the twentieth century resulted from relatively small eruptions in areas with high population density. The catastrophic effect was due to their sudden onset, which prevented a timely evacuation, or to political reasons, which also delayed response planning.

In general, it is most hazardous when an explosive volcano that has been at rest for hundreds to thousands of years, suddenly and unex-

▼ Fig. 13.31. Research vessel of the Philippine Institute of Volcanology. The notorious Taal volcano, situated in Taal Lake, is known for its disastrous eruptions

pectedly erupts again. This is because the risk potential at such volcanoes is commonly overlooked when people settle or expand settlements on its flank and in the foreland.

A volcanic eruption is the result of a large number of factors and reliable predictions can only be made in well-instrumented and well-monitored volcanoes. In order to predict the type of an eruption, its location, and if possible its onset, requires very careful analysis of the development of the history of a volcano. Monitoring methods should be tailored to the characteristic behavior of volcanoes prior to large eruptions. These can be used to document volcanic earthquakes, the swelling of the Earth's crust and increased gas emissions, especially SO_2. The rapid developments in remote sensing via satellites during the last few years have significantly increased the possibility to monitor volcanoes using the entire electromagnetic spectrum. For the overwhelming number of volcanoes, which cannot be monitored in detail on the ground, satellite-based remote sensing remains the only possibility to produce near-real-time data about precursors and possible subsequent eruptions. Large eruptions such as Mount St. Helens (1980) or Pinatubo (1991) were preceded by several months of precursors, but many volcanoes do not show precursors, or at least none that can be identified at present using currently existing instrumentation.

Seismology is at present the most important method to predict volcanic eruptions. A future goal of volcano seismology is to achieve a better understanding of dynamic processes in active magmatic systems. This includes the physical properties and the 3-D extent and evolution of the source regions of seismic events, both critical parameters to more precisely assess the eruptive behavior of a volcano, and therefore the timing of an eruption. Modern broadband seismometers and significantly improved methods for analyzing signals open the possibility to more realistically analyze the role of magmatic and hydrothermal fluids that generate seismic waves.

The areas potentially affected during volcanic eruptions can usually be relatively well predetermined, depending on how well a volcano and its deposits have been analyzed, allowing early mitigation planning and significant minimizing of damage. For areas at larger distances from volcanoes, warning times of up to several hours are possible for lahars and ash falls. People can thus be evacuated in time around well-monitored volcanoes, characterized by precursor activity.

Volcanic catastrophes have developed when people did not protect themselves early enough in the face of an impending volcanic eruption. One of the two largest volcanic catastrophes in the last century (Nevado del Ruiz, 1985) could have been avoided if the political administrative bodies had taken seriously the fear and determination of the people, or the warnings of scientists. In the city of St. Pierre, it would have been important to follow the wish of the people to leave the city. They had become frightened by the precursory eruptions of Montagne Pelée. Nevertheless, the traditional interpretation that it was the governor who forced or persuaded the people to stay in the city, in view of the impeding elections has been found to be erroneous. In the case of Armero, however, the administration should have taken the warnings of the geologists seriously. More importantly, the town should never have been built at this place. It had been inundated and destroyed by exactly the same type of volcanic debris flow twice in the preceding three centuries.

Many active volcanoes are in developing countries that commonly lack the scientific infrastructure and financial resources to study the prehistory of their volcanoes and to set up instrumental monitoring networks (Fig. 13.31). This problem is magnified by the pressure of growing populations that settle the fertile slopes of active volcanoes, such as in Indonesia, the Philippines and Latin America. This makes it increasingly difficult to prevent potentially hazardous areas from being occupied by people. The lack of authority of administrations often means they cannot prevent people from moving higher and higher on the flanks of highly hazardous volcanoes, such as Merapi in Indonesia or even Vesuvius in Italy, a first-world nation. To name another example: the airfield of the Boeing aircraft company in Seattle, a modern boom town in a technically highly advanced country, has been built on a 500-year-old volcanic mud-flow deposit from nearby Mount Rainier volcano. Clearly, societies today are much more vulnerable vis-à-vis volcanic and other natural hazards than in the past, particularly through the rapid growth of megacities and the increasingly intricate network of traffic, communication and pipelines and tight international networks and interdependency.

One effect of very large volcanic eruptions cannot be influenced or mitigated by people. That is the emission of volcanic gases into the stratosphere with global impacts on climate and the ozone layer, the topic of Chapter 14.

Volcanoes and Climate

Benjamin Franklin (1706–1790) was one of the most eminent men of his time. During his stay as ambassador of the United States in France, he noted that both the people of Europe and the US had become agitated by the coldest winter in memory (Fig. 14.1). A blue, cold and, as Franklin emphasized, dry fog, which could not be dissolved by the sun's rays, had started to cover the northern hemisphere in the summer of 1783. As far as we know, Franklin was the first scientist who related the blue fog and the dramatic deterioration of climate to volcanic eruptions (101). He was thinking of Hekla volcano in Iceland, but this time the most famous volcano of Iceland was not the source for the fog. Hekla was by the way interpreted as being the entrance to hell in medieval times and may have been on the mind of Jules Verne as the gate in Iceland through which his hero, Professor Dipenbrock started his journey to the center of the Earth at Snaefellsness volcano. In June/July 1783, about 12.5 km³ basaltic lava had erupted from a large fissure in Iceland, the Lakagir fissure, but only 0.7 km³ tephra was generated (332). Three quarters of the animals on Iceland (200 000 sheep, 30 000 horses, 10 000 cattle) died, apparently due to fluoride-poisoning (fluorosis) from the fluorine-rich gases and ash deposited on the pas-tures, as well as starvation because of the lack of feed. These effects probably lingered, due to the volcanic haze-induced deterioration of climate. About 10 000 people died in the resulting famine, one fifth of the popula-tion of Iceland. While Franklin wrote a scien-tific essay on the prob-lem, newspapers in the summer 1783 across Europe were full of speculations on volca-noes to be the source of the foul smelling air and depressing atmosphere. One newspaper in Germany even made up a story of an erupting volcano southeast of Frankfurt being the source of the oppressive weather. The flooding of central Europe in the summer of 1784 due to incessant rain was the worst in the past 1 000 years (122) most likely the aftereffects of the Laki eruption a year earlier.

The style of William S. Turner (1775–1851), a successful English landscape painter, was conven-tional until 1816. He had traveled a lot on the continent and the castles along the Rhine and Moselle have rarely been rendered so romantically as in his drawings and paintings. But when he

> I had a dream, which was not all a dream.
> The bright sun was extinguish'd, and the stars
> Did wander darkling in the eternal space,
> Rayless, and pathless, and the icy earth
> Swung blind and blackening in the moonless air
>
> *Lord Byron, Collected Poems, Oxford, 2000*

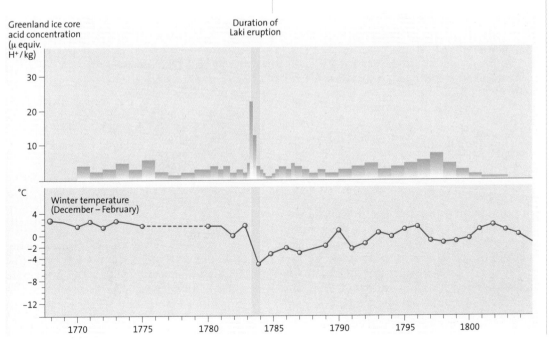

◀ Fig. 14.1. Change in average temperature in New England following the Laki eruption (1783) whose acid deposits have also been documented in ice cores from Greenland (331)

▲ Fig. 14.2. Early morning view from the Bay of Naples towards the city and volcano Vesuvius with the larger younger cone (*right*) having grown in the crater of the older volcanic edifice that had once collapsed (rim of older cone at the *left*). The smoke trail does not rise from the volcano but is emitted from smoke stacks along the shore

traveled in Europe in 1815/1816, Turner was so overwhelmed by the brilliant colors of the sunsets, that he changed his style radically, anticipating impressionism by more than half a century making him the most famous British painter of the 19th century. The cause of the lurid skies was the cataclysmic eruption of Tambora Volcano on 11 April 1815, on the Island of Sumbawa near Java. During this eruption, the uppermost kilometer of the volcano was destroyed, leaving a hole, 1.5 km deep and 5 km wide. About 50 km³ of magma were erupted in the largest eruption of historical time (334). The atmosphere was altered for several years from the mass of small aerosol particles, formed as a result of the eruption.

On Sumbawa and the neighboring island of Lombok, about 100 000 people lost their lives directly as a result of the eruption. Elsewhere, the events that caused a small revolution in painting, had drastic global effects. The climatic deterioration following the eruption triggered major social unrest in many countries. The lowest temperatures of the past two hundred years were measured in 1816 in Europe and the eastern coastal states of North America (the areas where regular temperature measurements had been recorded). The year 1816 became known as the *year without summer* (363). Lord Byron, the famous English poet and his guests spent the extremely wet and cold summer near Geneva, mainly inside, whiling away the time with gothic novels. During that summer, Mary Shelly developed the novel *Frankenstein*, and Lord Byron articulated the somber mood in his poem *Darkness*.

Similar colorful sunrises and sunsets were observed following many other large eruptions, famous examples being the huge eruption of Krakatau, between Java and Sumatra, in August 1883. These effects were described and interpreted in great detail, as in the famous report of the Royal Society of London (337).

The discussion of the impact of large volcanic eruptions on climate has assumed a new quality and importance during the last few years, because the composition of the atmosphere, increasingly tampered with by man, and its future development are of great concern (Fig. 14.2). This discussion has also become more rational, as more data have become available and theoretical models have become more sophisticated. Nevertheless, global warming and its immense societal effects will remain one of the premier long-term political issues. What is the role of volcanism in this debate?

The Scientific Revolution

Brilliant sunsets following large volcanic eruptions such as that of Krakatau in 1883 were commonly explained as due to ash or dust clouds increasing the *optical depth* of the atmosphere and therefore the scattering of sunlight. Deterioration of climate following major explosive eruptions were thus thought to be due to injection of ash clouds, consisting of glass shards and minerals, into the stratosphere. These particles were thought to absorb the incoming solar energy and radiate part of it back into space, therefore leading to a temperature decrease on the Earth's surface. This phase of scientific inquiry into the relationship between climate and volcanic eruptions peaked in the milestone work of Lamb (189). He postulated a relationship between large-volume volcanic eruptions and their injection of ashes into the atmosphere in historical times and observed or indirectly inferred climate deterioration. Today, this methodical approach has lost some of its relevance because of the scientific revolution following the eruption of El Chichón in 1982 (below). In addition, since the eruption of Mount St. Helens on 18 May 1980, evidence had been increasing that silicate ash particles, transported into the stratosphere, fall back to Earth relatively quickly (41).

The Chiapas region in southern Mexico has become notorious during the last few years mainly for its political explosivity. Shortly before the social unrest (and some consider that it triggered the unrest), the volcano El Chichón erupted on 29 March and on 4 April 1982. High-flying aircraft collected the aerosols after the eruption (Figs. 1.16, 14.3). Importantly, it was found that silicate particles only formed a minute fraction of the aerosols. The main mass consisted of drops of sulfuric acid, H_2SO_4. Simultaneously, the TOMS (total ozone mapping spectrometer) instrument on the Nimbus satellite platform in operation since 1980 for measuring ozone concentration (183), was able to measure SO_2 quantitatively in the atmosphere. About 13 megatons $(13 \times 10^{12} g)$ of SO_2 had been erupted from El Chichón, out of which about 20 megatons $(20 \times 10^{12} g)$ of sulfate (SO_4^{2-}) formed during the ensuing weeks (202).

The scientific revolution had, so to speak, occurred overnight. Ash particles were not responsible for the impressive sunsets and the short-term climate changes. Instead, volcanic gases were the culprits, mainly sulfur – injected into the stratosphere primarily in the form of SO_2.

Input of Volcanic Gases into the Atmosphere

The air directly above the Earth's surface is called the *troposphere*, being separated from the overlying *stratosphere* by a layer called the *tropopause* (Fig. 10.4). The tropopause is at about 7 – 10 km altitude at high latitudes, rising to about 15 – 18 km near the equator, the height also varying with the season at high latitudes. Hence, volcanoes erupting at high latitudes, such as in Iceland, Aleutians or Kamchatka, need only a relatively low energy eruption to inject their gas-particle mixture into the stratosphere, compared to those erupting near the equator. This is important because the stratosphere is largely dry. Aerosol clouds can thus reside in the stratosphere much longer, while they are quickly washed out by rain from the troposphere. Moreover, the troposphere is rather unstable because the temperature and therefore its density decreases from the Earth's surface to the tropopause, while the temperature increases again in the stratosphere. The magmatic volatiles H_2O and CO_2 and the halogens were discussed in Chapter 4, while sulfur is discussed here in more detail because of its significance for processes in the stratosphere. Magmatic CO_2 emissions are not thought to be important for the radiation balance

▼ Fig. 14.3. Migration of stratospheric SO_2 clouds emitted from El Chichón volcano on 5 April 1982 (202)

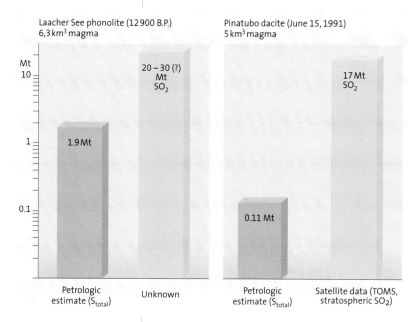

Laacher See phonolite (12 900 B.P.)
6,3 km³ magma

Pinatubo dacite (June 15, 1991)
5 km³ magma

Fig. 14.4. Comparison of stratospheric input of sulfur during the eruption of Laacher See volcano and that of Mount Pinatubo. The volume of magma erupted in both volcanoes is similar. Shown are the amounts of sulfur inferred by the petrological method and, for Pinatubo, also that measured by satellite (TOMS). The amount of sulfur emitted from Laacher See volcano was probably much higher (20–30 [?] megatons) based on a comparison with several recent eruptions (140)

because of the large CO_2 reservoir of the atmosphere (with the possible exception of large subaerial basaltic flood eruptions, Chap. 7).

Sulfur

Sulfur compounds, especially SO_2, are part of the cocktail of atmospheric trace gases. A significant fraction of the sulfur in the atmosphere is emitted from volcanoes and, in comparison with other magmatic gases, can be directly responsible for altering the radiation balance and indirectly for changing atmospheric chemistry (especially O_3). SO_2 is the primary volcanic gas that is measured regularly and quantitatively with instruments such as the COSPEC (correlation spectrometer), above volcanoes or open conduits (Figs. 4.21, 14.3–14.5). Since sulfur concentrations can be measured with COSPEC more precisely than other volcanic gases, there is a relatively large amount of data for S-concentrations. Remote sensing techniques such as the Total Ozone Mapping Spectrometer (TOMS) on the NIMBUS satellite platform have greatly increased our understanding of SO_2-mass fluxes from volcanoes since the 1980ies (213). Significant amounts of sulfur escape prior to eruptions from near-surface magma reservoirs, which can also be deduced from the contrasting sulfur solubilities in different types of magmas (48, 124, 233).

About 15% of the SO_2 in the troposphere is of volcanic origin. The bulk is emitted from small volcanoes or commonly from fumaroles. Gases above these sources only rise a few km and consist mainly of SO_2. Volcanoes, such as Arenal in Costa Rica (Fig. 8.15), emit about 200 t SO_2 (200×10^3 g/d =

0.2×10^6 g/d) daily into the atmosphere. Basaltic and trachytic to trachyandesitic magmas contain significantly more sulfur than highly silicic magmas (14), but the mass eruption rates of basaltic magmas are normally much too low to generate eruption columns that are able to rise into the stratosphere. In large eruptions of highly evolved magmas, however, huge masses of sulfur can be injected quickly into the stratosphere, such as in the case of Pinatubo (Philippines), 15 June 1991 about 17 megatons (17×10^{12} g) or Hudson Volcano (Chile), 12 August 1991, 2 megatons (2×10^{12} g) (Fig. 14.4).

The net sulfur emission depends on the eruption frequency and intensity. Large amounts of SO_2 are emitted from Etna volcano in Sicily, as much as 4×10^9 g/d on average between 1975 and 1987, which is about 8–9% of the estimated global volcanic sulfur output (18.6×10^{12} g/a) (6) (Fig 4.18). The average SO_2 emission of Kilauea eruptions between 1956 and 1985 is on the same order of magnitude, about 0.2 to >10×10^9 g/d. In Guatemala, a country with many active volcanoes, some of which are forming domes, about 80×10^6 g/d SO_2 is emitted from nonerupting magma bodies, possibly caused by magmas convecting in a reservoir (15).

The contrast in sulfur concentration in different magmas helps to explain why relatively small aerosol volumes were produced during the eruption of ca. 0.5 km³ magma of Mt. St. Helens on 18 May 1980 (113). By contrast, aerosol clouds were observed for over two years in the stratosphere following the smaller explosive eruption of El Chichón in Mexico in early 1982 (0.35 km³ magma) (334). The sulfur concentration in the El Chichón magma (ca. 0.6–0.9 wt%) was a hundred times higher than in the Mount St. Helens magmas (ca. 0.007 wt%) even high enough to cause the crystallization of the sulfate mineral anhydrite ($CaSO_4$). This mineral decayed during the eruption, adding its sulfur content to the rising gas cloud (83). The high sulfur input into the atmosphere appears to be mainly due to a sulfur-rich fluid phase, which was formed in a volatile (CO_2-H_2O-SO_2)-saturated magma. This also seems to have been the case for the extremely high sulfur input during the 1991 Pinatubo eruption (114).

The Sulfur Excess Problem

Only about 10% of the SO_2 masses emitted from the 1982 El Chichón and 1985 Nevado del Ruiz eruptions can be explained by degassing of the erupted magmas (202, 335). During the eruption of Redoubt (1989) and Pinatubo (1991), the discrepancy between emitted SO_2 volumes and that

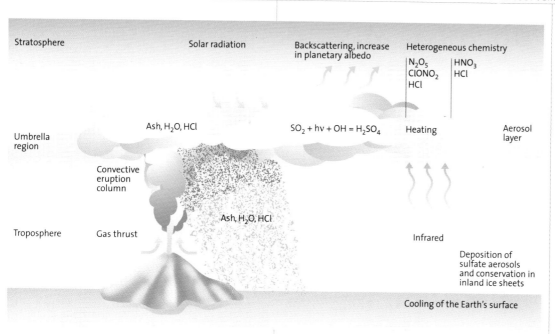

Stratosphere

Solar radiation

Backscattering, increase
in planetary albedo

Heterogeneous chemistry

N_2O_5 | HNO_3
$ClONO_2$ | HCl
HCl

Ash, H_2O, HCl

Umbrella
region

$SO_2 + hv + OH = H_2SO_4$

Heating

Aerosol
layer

Convective
eruption
column

Ash, H_2O, HCl

Troposphere

Gas thrust

Infrared

Deposition of
sulfate aerosols
and conservation in
inland ice sheets

Cooling of the Earth's surface

◄ Fig. 14.5. Gas input following large Plinian eruptions and formation of aerosols in the stratosphere. For discussion see text (213)

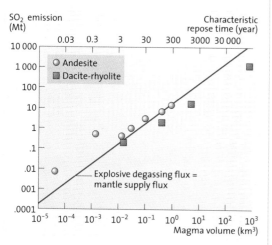

SO_2 emission
(Mt)

Characteristic
repose time (year)

○ Andesite
■ Dacite-rhyolite

Explosive degassing flux =
mantle supply flux

Magma volume (km³)

bubbles. The explosive degassing flux of SO_2 increases with increasing magma flux (except for some basaltic magma systems), the volumes in turn being correlated to characteristic repose times (410) (Fig. 14.6). By assuming

- a steady state balance between mantle supply and explosive loss of SO_2
- the historical record of explosive volcanism for the last 200 years
- a relationship between eruptive volumes and repose periods (Fig. 14.6) and
- a basaltic magma flux of 0.01 km³/a, the global mantle supply flux of SO_2 would amount to 2.7 Mt/a, roughly equivalent to the SO_2 flux from explosive volcanic eruptions.

Formation of the Aerosol Veil:
Gas-Particle Conversion

The photochemical reactions of SO_2 and H_2S with H_2O and OH^- over a few weeks in the stratosphere can lead to the formation of sulfuric acid droplets with a diameter of much less than 1 mm (Figs. 14.7, 14.8). Depending on the latitude and longitude of a volcano and the season of the year, volcanic aerosol layers, many several km thick, commonly form continuous veils at a height of 20–30 km. Following an eruption, they normally drift within a few weeks to months to the northern and southern polar areas, the main weather kitchens (Fig. 14.8). The concentration of volcanic aerosols in the atmosphere rises in the wake of large explosive eruptions to values that are more than one or two orders of magnitude higher than the normal stratospheric aerosol particle background.

◄ Fig. 14.6. Relationship between SO_2 emitted during explosive intermediate and silicic volcanic eruptions, volume of evolved magma and repose times. Line indicates steady state balance between the amount of sulfur introduced by prolonged degassing of basaltic magma at the base of a system during the repose period and the amount of SO_2/km³ of differentiated magma released by explosive eruptions (410)

of concentrations in the magma based on analyzed concentrations in melt inclusions in phenocrysts was even larger (114). SO_2 and other gases are apparently enriched in eruption columns because they form a separate gas phase that is released prior to and during an eruption.

Sulfur-rich, basaltic magma appears to be the major source for the apparent excess sulfur in the evolved magma and, in any case, they form the parent liquid for most highly differentiated magmas. It has been known for some time that hot basaltic magmas are often injected into differentiating magma columns prior to an eruption. These basalt injections appear to be a common trigger of an eruption (Chap. 4), while not erupting themselves or perhaps only as schlieren in mixed magmas. Volatiles such as SO_2 may be scavenged from the non-erupted magmas by rising CO_2 and H_2O

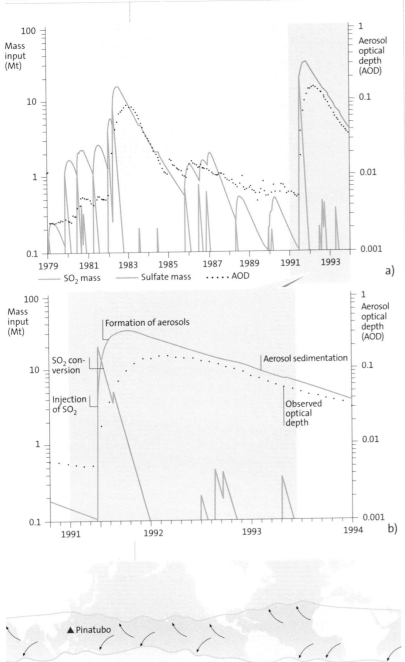

▲ Fig. 14.8. Migration of aerosol clouds that had formed following the eruption of Pinatubo volcano on 15 June 1991 towards the poles (213)

following eruptions. The aerosol droplets then slowly descend to Earth, whereby a particle with a diameter of 1.4 μm acquires a final sinking velocity of 15 m/day at a height of 30 km, to 70 m/day at a height of 10 km.

The Effects of Stratospheric Aerosol Veils on Global and Regional Temperatures and Ozone Depletion

Volcanic aerosols influence the temperature on the Earth's surface in a very complex manner (213, 389) (Figs. 14.9, 14.10). Initially, the volcanic aerosol veils lead to a heating of the stratosphere by absorbing solar energy in the tiny aerosol droplets, such as after the eruptions of Agung (March 1963), El Chichón (March to April 1982) and Pinatubo (1991), the latter producing the largest aerosol veil of the twentieth century (213, 281). This leads, simply speaking, to a reduction of solar radiation reaching the Earth's surface. In detail, however, the relationship between diameter, concentration and composition of aerosol droplets depends on the height of the aerosol cloud in the troposphere or stratosphere. The dominant effect of the aerosol layer is that the tiny aerosol droplets absorb not only solar – but also terrestrial radiation, i.e. infrared radiation reflected from the Earth's surface. This absorption of radiation heats the lower stratosphere. This *radiative forcing* results in massive changes in atmospheric circulation that in turn strongly affects tropospheric temperatures.

It is now known that regional effects on climate following large volcanic eruptions differ dramatically, depending on the migration (circulation) patterns of the air masses that are influenced by a volcanic eruption. Predictions of atmospheric models verified by observations have shown that the impact of the Pinatubo (and other sulfur-rich eruptions) varied greatly globally. For example, winter temperatures in North America, Europe and Siberia were higher than normal in 1992, but much lower than normal in Alaska, Greenland, the Middle East and China (281). These striking regional differences in atmospheric impact of large explosive eruptions severely limit the significance of paleo-temperature proxies – or help to focus the search for important proxies to areas prone to severe impacts, provided the present global circulation patterns resemble those of the past.

In any case, the regionalization and quantification of volcanic eruption impacts such as that

Following the eruption of El Chichón, the aerosols observed from the ground and from satellites initially formed two layers at altitudes of about 17 and 25 km. These coalesced in the fall of 1982 to form a single layer between 20 and 30 km altitude with a concentration maximum at about 23 km. Following the eruption, the aerosol layer migrated within three weeks around the globe (Fig. 14.3). Such large volcanic aerosol veils can remain in the stratosphere for about three years

following the Pinatubo eruption, is a major step forward (Figs. 14.09, 14.10). Nevertheless, global estimates of aerosol inputs are important for the entire radiation balance. Following the eruption of Pinatubo, the global temperature close to the Earth's surface decreased by about 0.5 °C from the middle of 1991 to the middle of 1993 (Fig. 14.9) (213). Global heating caused by anthropogenic greenhouse effects, also called *positive climate forcing*, decreased during this time because of the *negative climate forcing* caused by the Pinatubo eruption. Global warming resumed in 1994. Nevertheless, more subtle long-term effects, such as changes in patterns of ocean currents in smaller ocean basins, may last much longer.

The impact of large explosive volcanic eruptions on global climate was used as an analogue for modeling a global nuclear winter scenario following nuclear explosions (70, 388). The effect of the Pinatubo eruption further helped to clarify the impact of greenhouse gases on global warming. Figure 14.11 shows one of several recent predictive models for future temperature increases resulting from global warming. A novel aspect of these models is that they combine anthropogenic, solar as well as volcanic forcing.

A decrease in the atmospheric ozone (O_3) concentration attributed to volcanism was first measured following the eruption of Gunung Agung in 1963. The reason for this unexpected effect appears to lie in the surfaces of the H_2SO_4 aerosols that promote *heterogeneous chemical reactions*. These reactions change NO_x (e.g., N_2O_5) into nitric acid (HNO_3) and inactive anthro-

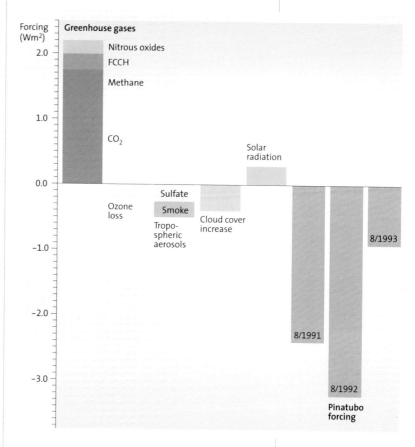

▲ Fig. 14.9. Positive forcing (increasing temperature) caused by greenhouse effect and negative forcing (decreasing temperature) between 1991 and 1993 due to the formation of aerosol layers in the stratosphere following the eruption of Pinatubo volcano (213)

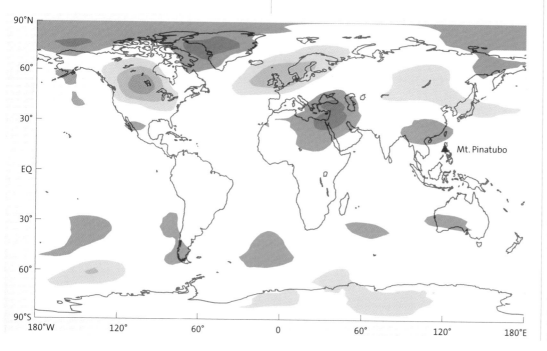

◀ Fig. 14.10. Regionalization of lower tropospheric temperature anomalies for the Northern Hemisphere winter (December 1991 to February 1992) after the 1991 Pinatubo eruption, a pattern similar for many tropical eruptions (281). Orange – yellow temperature scale indicates warming and blue cooling

▶ Fig. 14.11. Observed surface temperatures until the year 2000 (*black line*) and a simulation including both anthropogenic (greenhouse gases, sulfate aerosols and tropospheric and stratospheric ozone) and natural (solar and volcanic) forcings (*yellow*). Also shown are simulations for four different scenarios (A1 – B2, *blue, green and red lines*) and the combined total uncertainty for all scenarios (corresponding colored fields bordered by dashed lines). Up to about 2030, the simulations overlap considerably and indicate that the 5 to 95 percentiles of the forecast temperature distributions are 0.9 to 1.9 K relative to the pre-industrial (control) climate. After 2050, the different models diverge more strongly (364)

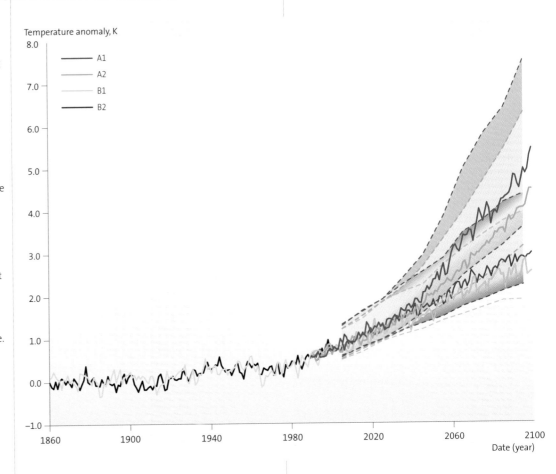

pogenic chlorine reservoirs in the stratosphere ($ClONO_2$, HCl) into reactive chlorine (Cl, ClO) that is most effective in breaking up the O_3 molecule (32). For example, stratospheric ozone concentrations between September 1991 and spring 1992 following the eruptions of the volcanoes Pinatubo and Hudson in June and August 1991, significantly decreased by about 15–20% at different altitudes (e.g., 159).

Which Volcanic Eruptions Load the Atmosphere Most?

Volatile elements are concentrated in more highly evolved magmas, because most of the minerals crystallizing earlier at higher temperature (olivine, pyroxene, plagioclase) do not contain volatile elements in their crystal lattices. The large concentrations of volatiles such as CO_2, N_2, SO_2, H_2S and probably large amounts of H_2O found in magmas generated above subduction zones are thought to be due to dehydration of subducted sediments, ocean crust or serpentinized mantle (Chap. 8). More highly evolved derivative magmas erupt mainly explosively, sometimes with very high mass eruption rates ($10^7 – 10^9$ kg s^{-1} during Plinian eruptions) from central vents or ring fissures.

One to two large eruptions per year, which contain abundant gases (H_2O, CO_2 and SO_2, during some eruptions more than 1 000 tons per day $= 10^6$ g/d) are important, because their gases are injected directly into the stratosphere.

Highly explosive eruptions, whose eruption columns reach the stratosphere, such as Pinatubo in June 1991, are not as rare as sometimes thought. About 50–60 volcanoes erupt each year and about 3 to 6 eruption columns reach the stratosphere each year (Fig. 10.4). This is probably an underestimate, because observations are often incomplete or imprecise from remote parts of the Earth, a problem that will become remedied by increasing satellite monitoring. Very large volcanic eruptions with magma volumes of between 15 and 100 km^3, such as that of Tambora (1815), are century events. Mega-eruptions with volumes more than 100 km^3 only occur every 10 000 – 100 000 years. Nevertheless, the few systematic studies show that the total budget of gases emitted into the atmosphere is also strongly governed by less spectacular eruptions or quietly degassing volcanoes. About two thirds of all volcanoes are located in the northern hemisphere (Figs. 14.12, 14.13), only 18% occur between 10° south and

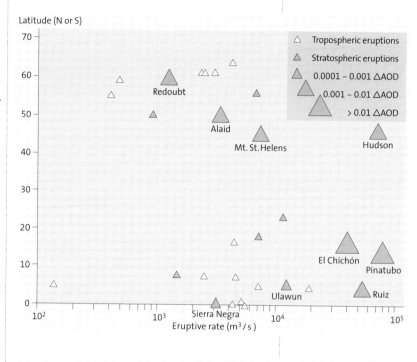

◀ Fig. 14.12. Distribution of volcanoes in different latitudes with contrasting eruptive rates, height of the eruption column (troposphere and stratosphere) and contrasting optical aerosol depth (AOD) (273)

the South Pole (338). The stratospheric injection of gas and ash thus mainly affects the climate of the northern hemisphere.

Mass eruption rates of basaltic magmas are normally much too small to generate eruption columns that rise into the stratosphere. Yet, large masses of aerosols can be generated after an eruption, during which much lava and gas but little tephra is produced. This is probably the case during the flood basalt eruptions (Chap. 7) and certainly was following the largest historic lava eruption along the Laki fissure in Iceland in 1783 (83, 223, 377). SO_2 can be concentrated in the stratosphere during such eruptions because heated air rises convectively in plumes above very hot and extensive lava flows.

Back *For* the Future

Analysis of climatic effects of older pre-industrial volcanic eruptions is fundamental if one wants to clearly separate the natural volcano signal from anthropogenic effects. Global and regional environmental changes of the geological past can often be reconstructed more precisely in geological deposits, ice sheets or tree rings, than by examining processes going on at present (Figs. 14.14, 14.15).

Natural archives, i.e. deposits of many types, contain different types of properties, also called *proxies*, that reflect the climate of earlier periods. Scientists attempt to recognize patterns and to quantify these in order to be able to project the future development of climate, at least how it would develop if not tampered with by man. The comparison between natural and anthropogenically influenced evolution may then lead to political commitments such as the *Montreal Protocol* (reduction of the global chlorofluorocarbon production), perhaps the most successful international agreement on environmental issues since the global ozone holes were found to be expanding.

Historical proxies that reflect former periods of climate deterioration include: catastrophic harvests, increased prices for grain, reduced tree ring widths (although tree rings sometimes increase after an eruption during periods of heavy rainfall), frost rings in trees, varve anomalies in lake or marine sediments, changes in coral growth rates or temperature sensitive element ratios in annual growth rings in corals and isotope changes in inland ice (Figs. 14.14, 14.15).

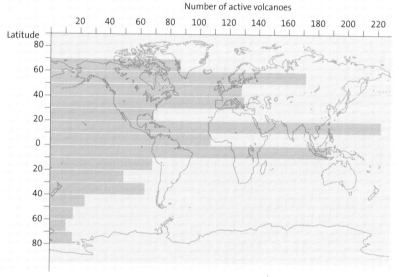

An elegant indirect method for reconstructing volcanogenic aerosol events, even of prehistoric eruptions to an age of about 50 000 years B.P., is the determination of the amount of acid enrichment in long ice cores, such as drilled in the Greenland ice sheet and in Antarctica (139). The aerosols sedimented out of the atmosphere following an eruption, or washed out by rain, are frozen in the inland ice and allow an almost complete reconstruction of past volcanic eruptions with hemispheric or even global effects.

Ice cores also contain other types of proxies such as stable isotope ratios and dust concentrations that are temperature or climate-dependant.

▲ Fig. 14.13. Number of active volcanoes depending on latitude (338)

Band of tree rings of reduced width

◄ Fig. 14.14. Cross section through a pine tree about 12 900 years old near Dättnau (Switzerland). The band of reduced thickness of annual growth rings (*arrow*) has been interpreted by Kaiser (168) as a result of climate deterioration following the eruption of Laacher See volcano. Photo courtesy F Kaiser

▼ Fig. 14.15. Temporal correlation of large historic eruptions (*DVI* dust veil index) with strong acid concentration in ice cores from Greenland and frost rings in trees (mainly bristle cone pine in California), generated by sudden influx of cold air in North America during the growth period of the trees (442, 443)

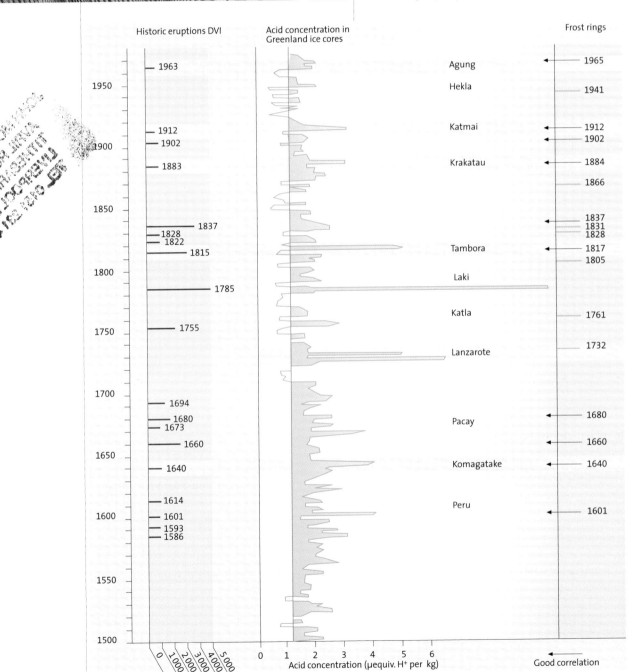

When ice core data are correlated to other types of volcanic eruption records, such as ^{14}C-dated deposits, historical accounts, tree ring information etc, the *acid peak stratigraphy* in ice cores shows a convincing record of large volcanic eruptions (provided the eruption did not occur during a phase of extremely reduced snowfall). Sulfite enrichment in ice cores has also been used to estimate the emitted HCl, HF and SO_2 masses (factor 2) of some historic eruptions (59). Acid peaks in ice cores occur about $1-3$ years after a large volcanic eruption, the time taken for aerosols to migrate to the poles and fall out (139). These can be correlated with temperatures deduced from oxygen isotope ratios. The minimum volumes of SO_2 emitted are determined directly from the deposits. In some cases, the volumes determined by petrologic and ice core methods agree with each other, but the petrologically reconstructed amounts of sulfur are generally much smaller than those determined by satellite, as discussed above (Fig. 14.4).

Another method is to determine the composition and abundance of magmatic volatiles (e.g., CO_2, H_2O, S, F and Cl) injected into the stratosphere in large amounts by analyzing glass inclusions in phenocrysts in tephra deposits, glass shards and pumice lapilli with instruments such as electron microprobe and ion probe. These allow in situ analysis of element concentrations and also isotopic ratios (83, 222, 332). Such analyses together with volume estimates of the entire tephra or lava erupted allow determination of the minimum mass of gases emitted during an eruption that was large enough to reach the stratosphere. This method has been used to estimate the sulfur emission during the eruption of a number of volcanoes, examples being Tambora (1815), Lagakir (1783), or eruptions as old as Laacher See volcano 12 900 years ago (140, 313).

The interpretation of volatile concentrations in melt inclusions in order to infer the flux of climate-relevant gases into the stratosphere is not straightforward, however. Glass inclusions in minerals often form after part of the gas phase has already been lost from a silicate melt by degassing. Similar concentrations of volatiles in melt inclusions and matrix glasses would indicate a relatively long coexistence of magma and a gas phase in the magma reservoir before an eruption. Larger differences indicate a rapid rise of the magma and significant degassing during eruptions. Hence, volatile compositions in quenched lavas and melt inclusions mirror the composition of the last equilibrium between melt with the rising gases and not that of the magma source or the parental magma.

The Chicken and the Egg

There is general agreement that long-term periodic global climate changes such as ice ages and interglacials are largely governed by *astronomic forcing*. Changes in the intensity and distribution of solar radiation can be clearly related to systematic variations in the orientation and position of the Earth in relation to the Sun and are called *orbital oscillations* of the Earth. When analyzing the kilometer-long drill cores in the ice sheets of Greenland and Antarctica, one of the surprising results was that climate changes also occur at very short intervals, sometimes within years to decades (7). For example, in the last Interglacial, i.e., the time between the last and penultimate glacial phase about 110 000 – 125 000 years ago, the climate fluctuated much more frequently and rapidly than previously thought. To explain these and other climate changes, additional mechanisms have been inferred, such as *plate tectonic forcing* expressed by rising mountain chains to divert flow of air masses and sporadic events of injecting large masses of greenhouse gases into the atmosphere.

Global periods of especially frequent explosive volcanic activity and simultaneous climatic changes appear correlated as indicated by temperature-sensitive proxies of marine microfauna (171). The link between volcanism and the microfauna record is through marine ash layers, resulting from particularly large Plinian eruptions. Many eruptions must generate more than $1 km^3$ of aerosols in the stratosphere to decrease the global temperature by $5-6\,°C$ over decades to sustain negative climate forcing (388). Even then, such a dirty atmosphere would be insufficient to completely suppress photosynthesis (266).

Whether or not single large eruptions or a series of many smaller eruptions can generate feedback mechanisms such that volcanic forcing overcomes astronomic forcing, is still a matter of debate. For example, one idea that has come up from time to time is that the isostatic loading and unloading of the lithosphere by accumulation and later melting of ice sheets would generate instabilities in the Earth's mantle. The unloading would lead to decompression of the mantle, generation of magmas by partial melting and therefore increased volcanism (167). An increase in volcanic eruptions, in turn, might lead to increased formation of aerosols, which would reinforce cooling until the astronomic forcing resumes its dominant role. On the other hand, fast climate changes accompanied by strong temperature decreases resulting in thick ice sheets may be able to generate volcanic eruptions, because loading of the lithosphere might destabilize the underlying

▶ Fig. 14.16. Increase in explosive volcanic eruptions could change the normal astronomical forcing of climate (butterfly as symbol for chaos) and lead to global warming. See text (after WJ McGuire, pers. comm.)

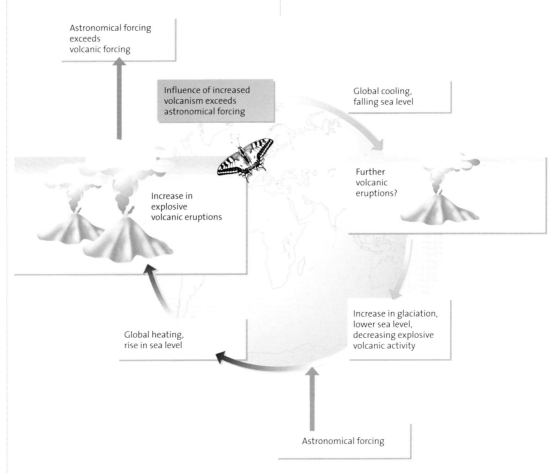

Are Mass Extinctions Due to Mega-eruptions?

The most famous, but not the largest, global mass extinction during Earth history, at the Cretaceous-Tertiary boundary (K/T-event) about 65 million years ago (popularly known as the *demise of the dinosaurs*) is causally related by some scientists to the rapid eruption of the Deccan Trap flood basalts. Extreme volumes of CO_2 (about $6-20 \times 10^{21}$ mol) were possibly liberated during the eruption of the Deccan basalt, which could have led to a global warming of about 1 °C (40, 276). This may have led to a destabilized atmosphere (217). Other authors, in contrast, postulate a cooling of the climate due to huge SO_2-emissions and hence massive production of sulfate aerosols in the stratosphere.

However, an alternative hypothesis for the K/T-mass extinction, that of a bolide impact, is much better founded at the moment. A major meteorite impact structure at Chicxulub in the Yucatan peninsula (Mexico) is most likely the *smoking gun* that evidences such a trigger mechanism for the K/T-boundary extinctions (9). A globally distributed dust layer about 3 mm thick on average is characterized by mostly altered glass spherules interpreted to have formed as condensation droplets from the impact vapor cloud, shocked quartz and an iridium anomaly, the original proxy for relating this layer to an extraterrestrial impact. It is instructive to visualize the huge mass and global distribution of this deposit that dwarfs any volcanic ash layer on record (Fig. 14.17).

Despite the impressive data set on this deposit reflecting more than 20 years of intense search on all continents and in all ocean basins, the actual mechanism that impacted the biosphere causing global mass extinctions is still poorly understood and highly controversial. Photosynthesis shutdown by a global cloud of fine dust as originally suggested by Alvarez et al. (9) and since invoked by many other workers appears unlikely since very little dust ($<10^{14}$ g) was of submicrometer–size (266). Shutdown of photosynthesis and global cooling are more likely to have been caused by the impact production of sulfate aerosols from the target rock, dominantly shallow water marine anhydrite deposits alternating with carbonates (e.g., 267).

asthenosphere and increase rates of partial melting (277) (Fig. 14.16).

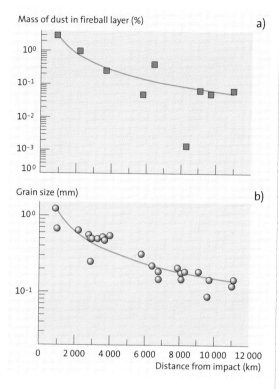

Mass of dust in fireball layer (%)   a)

Grain size (mm)   b)

Distance from impact (km)

◀ Fig. 14.17. a) Estimates of mass percentage of clastic debris (dust) in the Cretaceous-Tertiary fireball layer as function of distance from Chicxulub crater. *Solid line* is a power-law regression (266).

b) Relationship between shocked-quartz grain sizes in Cretaceous-Tertiary fireball layer and distance from Chicxulub crater. *Solid line* is power-law regression of maximum-grain-size data (266)

tor). Solid particles, i.e., volcanic ash or "dust", fall back to Earth after a few to at most 15 days because of their high density. Long-lived volcanic aerosols are condensates from the magmatic gas phase (especially SO_2). Gigantic amounts of sulfur are liberated within a few hours during large explosive eruptions of highly evolved magmas (about $15-20 \times 10^{12}$ g during the eruption of Pinatubo on 15 June 1991). Volcanic aerosols are generated by gas-particle conversion ($SO_2 + H_2 \rightarrow$ aerosol droplets) and, following large eruptions, are distributed over the entire stratosphere, spreading to the polar vortices where they influence the circulation patterns and therefore tropospheric temperatures. Such aerosol layers or veils may remain in the stratosphere for about 3 years, significantly influencing the stratospheric chemistry and the radiation balance. Large Plinian SO_2-rich eruptions can reduce global temperatures by about $0.2-0.5\,°C$ and perhaps even $>1\,°C$ for a few years. Moreover, volcanic aerosol clouds also have a major destructive effect on the stability of ozone in the stratosphere.

Many aspects of volcanic climate forcing are still poorly understood, however. For example: Which are the main chemical and physical processes that take place within an eruption column that may rise up to 40 km above the Earth's surface and how do they interact? What are the mass flux rates and transfer processes of various gases that are emitted from the Earth into the atmosphere? What is the duration of the more subtle climatic effects that are caused by a major eruption? Can significant thinning of the stratospheric ozone layer following large volcanic eruptions be recognized in the past?

In view of the rapid climate changes (over years), recognizable in ice cores, the question whether or not volcanic forcing influences global climate is back on the agenda. Can volcanic eruptions also force long-term changes in climate? What is the causal relationship between volcanic eruptions and past climate changes and mass extinctions? In other words, can volcanic eruptions trigger ice ages or cause global mass extinctions? Present consensus is that mass extinctions were probably caused by meteorite impacts. A major meteorite impact structure at Chicxulub (Mexico) is most likely the "smoking gun" that evidences such a trigger mechanism for the major K/T-boundary extinctions. However, debate continues as to whether periods of especially strong volcanic eruptions, in particular those of flood basalt volcanism, or other less spectacular causes could also cause the Cretaceous and older mass extinctions. The coincidence in time of the very brief periods of massive outpouring of flood

Summary

People have recognized for centuries that volcanic eruptions can significantly influence climate. Colorful sunsets and other atmospheric effects such as long-term haze, years of drastic cooling, poor harvests and other effects have been known to follow in the wake of some powerful eruptions. Until a few years ago, the mass of fine tephra particles, the "dust", has been thought to be responsible for such major volcano-induced atmospheric impacts. One of the more recent breakthroughs in volcanology is the recognition that the sometimes drastic impact of eruptions on the radiation and energy balance of the atmosphere is not due to the mass of solid particles injected into the atmosphere. Instead, satellite (TOMS) and radar (LIDAR) study of the stratosphere and volcanic aerosols, direct sampling of volcanic aerosols by aircraft or balloon, computer simulations, electrical conductivity of ice cores, and geochemical and mineralogical analysis of pyroclastic deposits all show that it is the mass of SO_2 injected into the stratosphere that is the main factor responsible for volcanic climate forcing.

The atmospheric effects of a volcanic eruption are primarily determined by the height of an eruption column, mass of magma/gas erupted and the amount of sulfur injected into the stratosphere. An eruption column reaches the stratosphere at mass eruption rates of more than $10^6\,\text{kg/s}^{-1}$ (high latitudes) or 10^7 (near the equa-

basalts during Earth history with mass extinctions is unlikely fortuitous.

Nevertheless, understanding the complex climate changes within the system Earth and separating natural events from anthropogenic forcing is very difficult. Clarification of climatic effects of pre-industrial volcanic eruptions is, however, very important for separating natural volcanic and other climate signals more clearly from the consequences of human activities that have resulted in the highest greenhouse gas concentrations during the past half million years. Only when the amount and effect of large natural (volcanic) emissions are better known (Fig. 14.15), can we estimate the additional impact of man on climate. These boundary values can then be used as the base for tolerable anthropogenic emissions. This all involves the interdisciplinary cooperation of volcanologists, atmospheric chemists, meteorologists and climatologists.

Moving away from these interesting speculations I will turn in the last chapter to some of the benefits of volcanism to man. I will not treat the widespread religious importance of volcanoes but their direct benefit to humans, which includes their beauty, their rocks, their subterranean heat and their supply of ingredients to the Earth's surface, all essential to life.

Man and Volcanoes: The Benefits

arly traces humans left in East Africa about 3 million years ago are footprints in volcanic tuffs (Fig. 15.1). It is as if man had walked through a thin layer of freshly poured-out concrete that subsequently hardened to the benefit of

▲ Fig. 15.1. Footprint of a Polynesian in ash, which had just been deposited wet and warm after the phreatomagmatic eruption of Kilauea volcano in 1790 (?). The footprints are well-preserved on the dry southern flanks of the volcano

posterity and anthropology. Seen in historical perspective, volcanoes are a threat to people, commonly only for very brief periods. Exceptions are places, such as the city of Kagoshima on the southern tip of the Japanese island of Kyushu, with its almost daily rain of ashes from the nearby almost permanently active volcano Sakurajima. The custom to dry clothes in the garden thus never developed in Kagoshima. In general, however, man has always benefited very much more from volcanoes than suffered from their eruptions. The undeniable benefits of volcanoes range from obsidian tools used in many early cultures, caves that can be dug easily in massive tuffs, fertile soils, attractive landscapes to geothermal energy. It is little known, for example, that a modern city such as San Francisco, receives the bulk of its electric energy from a young volcanic area, the Geysers, a couple of hours by car north of the Golden Gate metropolis.

The immense benefit man draws from volcanoes will be discussed below under five headings: heat from the Earth, volcanogenic ore deposits, volcanic soils, volcanic raw materials and volcanic landscapes.

Heat From the Interior of the Earth

Like passion, whose emblem it is, it can die
Susan Sontag, The Volcano Lover, New York, 1993

To exploit the heat in volcanic areas is nothing new to countries such as New Zealand, the Philippines, Iceland, Italy or Mexico (Fig. 15.2). The rising energy needs in modern society (even if more slowly than predicted only a few years ago), the decreasing supplies in valuable fossil gas and oil, the CO_2 emission of coal power plants, and the obvious problems of nuclear energy (especially the difficult problem of long-term storage of nuclear waste) all have spawned an increase of basic research into alternative forms of energy. The search for geothermal deposits and those metallic deposits that occur around magma chambers has thus been accelerated – and it is hoped that this energy can be utilized without losing sight of a sustainable environment.

Geothermal Energy

Temperatures increase from the surface of the Earth toward its interior. For example, it is getting warmer when one enters a mine, but not everywhere at the same rate. Mining can be carried out at depths greater than 3 000 m in southern Africa, an old continental shield where the heat flow is low. In areas of active tectonism or volcanism, however, the same temperature is reached much closer to the surface. In young rift zones as along the upper Rhine graben, where hot mantle material has risen from depth in the recent geologic past and increased the heat flow (i.e., the amount of heat rising per square meter), mines cannot operate at more than 800 m depth.

Even though unbelievably large amounts of energy are radiated from the surface of the Earth into outer space each day (ca. 10^{12} J/a) this heat cannot be used effectively as a source of energy. Nature has to focus heat before it can be exploited technically. Such a concentration is reached in the Earth in three steps and only if all three steps are combined in an optimal way, can we use the heat of the Earth effectively in power plants.

▲ Fig. 15.2. Geothermal power plant at Wairakei (New Zealand)

▼ Fig. 15.3. Schematic cross section through a geothermal system. Water circulating in porous rocks is heated and rises as hydrothermal water to the Earth's surface (hot springs, geysers) or is utilized in geothermal power plants (269)

The first step is speculative, but theoretically probable. As discussed in Chapter 2, we can assume that the Earth mantle convects. Hotter parts rise from greater depths (as in mantle plumes, Chap. 6), so fast that they do not lose their heat en route; colder lithosphere sinks along subduction zones into the interior of the Earth (Chap. 8).

The second step is the partial melting of convectively rising and thereby decompressing mantle rock. If sufficient magma has collected in some areas, it can detach, rise and may accumulate in magma reservoirs in the upper part of the Earth's crust. This concentration of heat (magmas with liquidus temperatures of 750–1 200 °C in a continental surrounding of 80–300 °C, 3–10 km beneath the surface), is in itself not focused enough. This is so because the conductive migration of heat in well-insulating dry rock is much too slow and too diffuse to be utilized technically.

The most efficient transport of heat is by fluids, such as magma or water. In the upper crust of the Earth, the mobile fluid is groundwater that circulates in pores and fractures. When heated above a magma chamber or a cooling, still hot, crystallized magmatic body, its density decreases, enabling the hot water to rise into the upper crust or to the Earth's surface (Fig. 15.3). This is apparent in many areas of young volcanism in the form of hot springs, or the often spectacularly erupting geysers. To be used in the form of energy, one also needs pressure. This condition is reached when the water, heated above its boiling point, cannot expand beneath a dense impermeable layer of rocks, such as clay-rich sediments. When drill holes are sunk through such an impermeable layer, the steam, which, depending on the temperature and pressure, is either dry (i.e., free of condensed water) or wet, can expand according to its temperature either directly in turbines, or through heat exchangers.

The utilized geothermal energy in power plants is thus dependent on a combination of several geological processes. This includes focusing of heat in the form of magma and transport into the upper part of the Earth's crust. The other necessary condition is plenty of water and permeable rocks, in which the water can circulate. In most current geothermal power plants, the heat source is a documented or assumed magma chamber in the upper crust. Well-known geothermal power plants are in areas of young volcanism, such as the Geysers in California (2 000 MW); Iceland (590 MW) – ca. 50 % of the country's energy consumption; Wairakei, New Zealand (250 MW) (Fig. 15.2) and Larderello, Italy (530 MW).

The larger and younger a magma chamber and the closer it is to the Earth's surface, the higher its geothermal potential. The presence of such young magma chambers must be inferred either indirectly through geological-petrological studies or via remote sensing, such as gravimetry. Examples are:
- negative gravity anomalies above SiO_2-rich and thus low-density magmas
- increased conductivity because of the presence of magma or hot rocks
- microseismicity around young magma chambers.

Because the solubility of minerals generally increases with temperature, chemical geothermo-

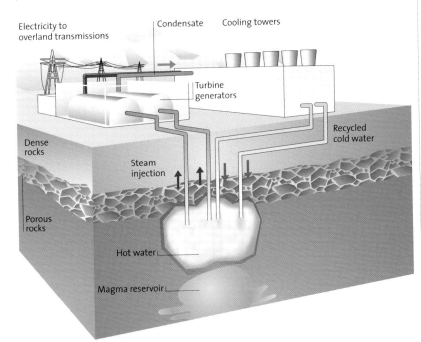

Electricity to overland transmissions

Condensate   Cooling towers

Turbine generators

Dense rocks

Steam injection

Recycled cold water

Porous rocks

Hot water

Magma reservoir

meters have been developed to infer the temperature at the origin of hydrothermal systems. A well-known geothermometer is the ratio of the elements Na/K/Ca with corrections for Mg and SiO$_2$. Light stable isotopes, during the last few years, were central to the discovery that a large part of the water in hydrothermal systems is of meteoric and not magmatic origin. In addition, high ^{3}He/^{4}He ratios indicate magmatic sources (mantle degassing).

Locally, one does not need young magma chambers to utilize the heat of the Earth. In thick, porous sediment series, with a regional or slightly elevated geothermal gradient, water can penetrate to greater depths and rise as heated hydrothermal water convectively in the pores of sediment or along fractures or faults to the surface.

Well-known heat anomalies in central Europe, where warm groundwater is used for space heating or for hot baths, are the Pannonian Basin in Hungary, Urach and Upper Rhine Valley in Germany, and the Paris Basin in France. Research into the geothermal potential of young magma chamber systems has been intensified in many coun-

tries. In addition, research has been carried-out where cold water is pumped into hot dry rock systems along artificial fractures, where it circulates, becomes heated, and rises back to the surface.

Geothermal energy will play a significant role in the near future only in areas with abundant young volcanism, apart from hot springs (Figs. 15.4, 15.5). Only when the costs for fossil fuel become too high, or if the CO$_2$ and SO$_2$ emissions of

▲ Fig. 15.4. Eggs boiled for tourists in fumaroles. Atosanupuri dome volcano (Kutcharo caldera, Hokkaido, Japan)

▼ Fig. 15.5. Sinter terraces in the geyser area of Mammoth Hot Springs (Yellowstone National Park, Wyoming, USA)

▶ Fig. 15.6. Former copper mine in submarine lavas of the Cretaceous Troodos ophiolite complex (Mitsero, Cyprus)

coal power plants become critical, or the hazards and problems of nuclear power plants or their acceptance in society have reached critical threshold values, will the exploitation of the natural heat supply of nature be utilized either in the form of hot water from depths or through the hot dry rock process.

Many magma chambers occur at shallow crustal depths (2–4 km) beneath mid-ocean ridges. The search for such magma chambers by seismic and other methods has been very difficult, however, and well-documented magma lenses have so far been found only in a few areas along the East Pacific Rise (Chap. 5). Their technical utilization is still a matter of science fiction, apart from the problem of transporting the energy over large distances or moving industry from the continents to the middle of the oceans. On the other hand, the enormous heat supply of mid-ocean ridge magma chambers has powered a fundamental recycling machinery during Earth history. Many important ore deposits on Earth were generated by streams of hot water continuously circulating through the young and hot ocean crust in the middle of the oceans.

Hot Water Valves on the Ocean Floor and the Formation of Ore Deposits

Volcanogenic ore deposits have been the fundamental economic base in many countries for thousands of years. Three characteristic types of deposits are known from present day mid-ocean ridges. The first type is rich in iron and manganese, the second only in manganese and the third is rich in sulfides and poor in manganese. The first, most common type, commonly forms deposits at the base of oceanic sediments above the volcanic crust. The third type is the most spectacular one, because it forms through the hot water chimneys, called *black smokers*, on mid-ocean ridges but also on active seamounts.

One of the most exciting and important discoveries in geology during the last few decades was the finding of black smokers on the East Pacific Rise in 1977 (206). A diving expedition of "Alvin" found a group of round chimneys about 10 m high and up to 0.4 m wide, about 1000 km north of the Galapagos Ridge. These not only consisted entirely of ore minerals, but were conduits, rapidly (2–3 ms^{-1}) emitting hot hydrothermal water into the cold seawater (206). At the exit of the hot water, dense clouds of dark sulfide crystals, ore minerals, especially iron, zinc and copper sulfides formed, the reason for calling the vents black smokers. The ore minerals rained out from the hot water fountains emitted from the chimneys formed small hills. The faunal associations around the chimneys, especially the *blood worms* (a new phylum in the animal kingdom), *giant clams* and *giant crabs* were part of a biological community that had developed completely anaerobic and outside photosynthesis. The assumption is that chemosynthetic bacteria oxidize H_2S emitted from the chimneys to elementary sulfur and sulfate. This oxidation generates energy that helps to incorporate CO_2 into organic compounds. The more complex organisms in turn feed on these bacteria.

The importance of this discovery is not minimized because it had been in principle predicted. In the early 1970s, some scientists noticed that the heat flow values measured along mid-ocean ridges were much lower than expected. They explained this discrepancy by postulating that cold seawater penetrated the permeable and porous volcanic crust to a depth of several km (196). The hypothesis was that seawater flowed downward into the crust over large areas, but resurfaced after being heated close to magma chambers in a focused manner. Apparently, the downwelling, initially cold seawater is not only heated when circulating close to hot magma pockets, but also becomes acid, and transforms into *hydrothermal solutions* up to about 350 °C hot when resurfacing at the seafloor. These hot and acid solutions corrode the rock through which they rise and become enriched in elements such as zinc and copper. At the orifice of the chimneys, the hot solutions are quickly cooled, become oversaturated and precipitate ore minerals, chiefly sulfides (Figs. 15.7, 15.8). In classical mining areas, such as Cyprus, an island that owes its name to the mining of copper (Latin: *cuprus*) during the time of the Phoenicians, these ore deposits occur in uplifted ocean crust, ophiolite complexes (Chap. 5, Fig. 15.6). These ore deposits are vivid expressions of the interaction of seawater that has penetrated the volcanic crust to a magmatic heat source and reappeared on the seafloor. The similarities between the processes postulated for present ocean crust-forming processes and the frozen effects in the fossil oceanic crust on Cyprus were impressive, although direct proofs were still missing.

Hydrothermal vents on the seafloor are thus places in which heated water exits. Massive sulfides are deposited first, whereby the hydrothermal solutions are impoverished in Cu, Ni, Cd, Zn, Hg, S, Se, Cr and U. Manganese crusts form subsequently by deposition from cooler oxidizing solutions, which only contain a few percent of the original hydrothermal solutions. Sediments enriched in iron and manganese form at relatively low temperatures.

The fundamental importance of black smokers goes much beyond explaining the formation of ore deposition. For several years, scientists had noted that elements supplied to the oceans by rivers and deposited in clastic and chemical sediments in the oceans were not in equilibrium. For example, rivers transport more Mg, S and K into the oceans than is fixed in the sediments. On the other hand, sediments contain more Ca, Mn, Cr

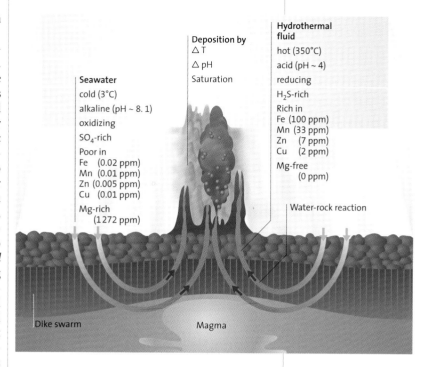

▲ Fig. 15.7. Scheme of hydrothermal circulation on a mid-ocean ridge (318)

▼ Fig. 15.8. White barite-rich smoker exiting at more than 300 °C from a vent in the Lau Basin. Photo courtesy Yves Fouquet

and other elements than delivered from the rivers. The flux of elements is thus not restricted to this system, but also includes the volcanic oceanic crust. Many different reactions occur in the black smokers at high temperatures, as predicted by some experiments. The temperatures of the hydrothermal waters, exiting at mid-ocean ridges (pressure at this depth 250 bar), cannot rise above about 390 °C, and just above magma reservoirs, which are about 2000 m deeper (pressure here ca. 450 bar), not above about 465 °C. At higher temperatures at these pressures, the liquid seawater would turn into a salt-poor vapor and a highly concentrated brine (25). According to these studies, the critical point for seawater is at 403–406 °C and 285–302 bar, appreciably above that of pure water (Chap. 12). The amount of heat removed in the black smokers from the underlying hot rock is so large that the lifetime of these chimneys is only about ten years. Indeed, active chimneys are surrounded for some distance by cooled and broken up chimneys, along with the skeletal remains of the bizarre faunas.

That many widely distributed sulfide ore deposits, which contain Fe, Cu, Zn, but also Pb, Au, Ag and other elements, form from hot watery, hydrothermal solutions, around magma chambers had been postulated for decades. Nevertheless, one had assumed for some time that the hot water was directly derived from the magma, whose slow crystallization into mostly water-free crystals resulted in a concentration of water in late solution, which would penetrate along fractures into the roof and there crystallize during cooling. According to present-day knowledge, the elements of such ore deposits are scavenged by leaching from the country rock traversed by heated-up external water or a mixture of magmatic and external water. This is another example of the encounter of the elements fire and water, triggering a chain of complex feedback mechanisms, similar to phreatomagmatic eruptions (Chap. 12).

Not all *hydrothermal ore deposits* can be explained by reactions between the heat of a magma and external water. The concentration of elements such as molybdenum, tungsten or tantalum in some continental ore deposits is much too large to be explained by leaching from the country rock. Such ore deposits are especially common around granite plutons, slowly crystallized magma bodies. The ores hence not only owe their heat energy to the magma, but also their element concentration and part of the water needed for transporting the elements.

Mining of the submarine massive sulfide concentrations, formed along mid-ocean ridges at intermediate and fast spreading rates (e.g., 284), in the near future is doubtful. Such an enterprise not only depends upon an upturn in world market prices for metals, but also on complex national jurisdictions. Up to which distance from a coast can the mining rights be extended? Apart from the political and technical issues there are also major ecological problems associated with a possible submarine mining endeavor.

Volcanic Soils

Tourists who land on the airport of Gando on Gran Canaria or the Aeropuerto Reina Sofía on Tenerife (the two largest of the entirely volcanic Canary Islands), directly proceed to Playa del Ingles or Playa de las Américas and only move between their bungalow and the beach, may think the idea that volcanic soils are very fertile is a myth, upon seeing the desert-like hinterlands. This is true in so far as nothing much can be seen of volcanic soils on these dry southern slopes. Nevertheless, when visiting the green northern side of these islands, where the trade wind clouds generally park around 800 m a.s.l., this view is drastically changed. A similar situation is encountered in Hawaii, where the contrasts between Kona and Hilo sides of the Big Island are even more staggering. The southern side of Kilauea volcano is a dry lava desert, to the enjoyment of geologists and tourists. 100-year-old lava flows look as if they were erupted yesterday.

There are plenty of volcanoes in Iceland, the entire small continent is volcanic. In fact, there is so much volcanic heat in Iceland that one can even grow bananas in geothermally heated greenhouses. On the other hand, there is a lot of rain in Iceland, much to the dismay of tourists. But the volcanic soils on this big island in the northern Atlantic are skeletal or raw, because it is simply too cold for soil-forming chemical and biological processes to proceed rapidly.

Volcanic deposits, especially ashes (weathering of lava flows takes of course much longer), have three properties to which they owe their exceptional quality as parent material for fertile soils. For one, ash particles are quite porous, especially larger lapilli. They hold moisture much longer than many other soils and release water slowly to the roots of the plants. These unique properties are frequently used to advantage in dry areas, such as the eastern Canary Islands. Basaltic lapilli, locally called *picón*, are spread on the fields, where they catch the moisture from the low clouds at night and release them peu à peu (Fig. 15.9). A second useful property of volcanic ashes is their enrichment in some elements, e.g., S, Mg, K, especially the high concentrations of some trace elements such as Se. This rich supply of nutrients can be an important boost for the growth of plants. The third factor is the glassy nature of the quickly cooled volcanic particles. Glass is thermodynamically unstable and alters relatively rapidly in the presence of water and in the right climate to crystalline and short-range-order phases, especially to clay minerals, the main mineral constituents of soils. Glass of lapilli or ash particles quickly becomes hydrated at its surface and is eventually converted to Al and Fe/Mg hydroxides. These secondary mineral surfaces and hydrated glass hold strongly onto water. All these factors, together with the humic acids of the dying plants, are the prerequisites for the Garden of Eden that is so fascinating in many volcanic areas (Fig. 15.10).

All these prerequisites: volcanic source material, warm climate and rain are present in many volcanically active areas, such as Indonesia, the Philippines or Latin America. But herein also lies a big problem. The high population pressure in developing countries forces people to move higher and higher on the fertile slopes of volcanoes, also the active and dangerous ones. Thus, many people have died during eruptions of volcanoes, such as Mount Mayon, Agung or Merapi, or have their properties destroyed in areas that actually should be closed to settlements.

▼ Fig. 15.9. Typical landscape on Lanzarote (Canary Islands). Pleistocene weathered scoria cone in the background. The half-moon-shaped walls made of pieces of basaltic lava protect the grapes against the wind. The black lapilli store the condensed water during the night when clouds pass over the island

▲ Fig. 15.10. Rice fields on the slopes of Merapi Volcano near Yogyakarta (Java, Indonesia)

▲ Fig. 15.11. Mural at entrance to road metal company in Masaya (Nicaragua) mining basaltic lapilli

▶ ▲ Fig. 15.12. Entrance to one of the many subterranean tunnels in the lava flow at Niedermendig (Laacher See area). For several hundred years, millstones were cut underground from this lava and were exported throughout Europe

▼ Fig. 15.13. Nephelinitic lava flow Hohenfels near Gerolstein (Eifel, Germany), a basalt popular with sculptors

▶ ▼ Fig. 15.14. Pumice mining at Nickenich (Laacher See area)

Volcanoes as Source for Raw Materials

Scoria cones and basaltic lava flows are mined in many areas for aggregates to serve various purposes, some highly specialized (Figs. 15.11 – 15.13).

Pumice is an ideal construction material because it can be mined very easily and, mixed with cement, can be turned into building blocks without much ado. These blocks are not very strong and cannot be used for high-rises or television towers. On the other hand, they insulate very well because of the high porosity of the pumice particles. They are thus sold e.g. in Germany as insulating light-weight building blocks. A major pumice industry has developed in the basin of Neuwied, east of Laacher See in Germany. However, much of the pumice cover formed about 13 000 years ago has been mined (Fig. 15.14).

On Merapi, the unconsolidated pyroclastic block flow deposits are utilized almost immediately after an eruption to build protective dams (sabo dams) next to the valleys that channel the block and ash flows.

Lithified tephra, *tuff*, forms by different processes. In the Eifel in Germany, e.g., the originally unconsolidated deposits of pyroclastic flows, deposited warm but not extremely hot became zeolitized with age, specifically their pores became filled by secondary minerals (zeolites), where soaked by groundwater. These rocks are still mined today and can be found throughout the country, for example in railway stations in Frankfurt, Hamburg and many other large cities.

Ignimbrite deposits have been used to advantage in many early cultures all over the world in places where caves could be dug easily without the

◀ Fig. 15.15. Bas relief at the wall of Hindu Temple under construction. The material is a poorly welded ignimbrite erupted from Batur Caldera (Bali). The ignimbrite is the main rock on which the famous stone sculpture industry on the island is based

◀ Fig. 15.16. Sculptures of "little gods" (Ojizosan) made of andesite lava. Such figures, many hundreds of years old, are common along roadsides in Japan. They are worshipped down to the present day. Omuroyama volcano (Izu Peninsula, Japan)

▼ Fig. 15.17. Slabs of surface layers of prehistoric pahoehoe lava flow decorating the entrance to a house. Restinga (El Hierro, Canary Islands)

walls collapsing. Caves at Göreme (Turkey), those carved by the Etruscans in Central Italy, or those dug by Indians in New Mexico are well-known examples. Moreover, welded ignimbrites that can be cut into blocks are used in many countries as building material. These often orange or reddish flamed rocks can be seen easily in many public and private houses in South America, Central America, on the Azores, Canaries, Italy, Armenia and many other countries. Volcanic rocks of many types have been a favorite raw material for sculptors over the centuries (Figs. 15.15, 15.16). Sometimes nature has already done the job by creating most attractive ornamental material, wrinkled surface slabs of fluid basalt lava being especially popular (Fig. 15.17).

The production of *cement* is a major economic problem in many developing countries because there is either not enough limestone in

▲ Fig. 15.18. Pieces of obsidian on the surface of a rhyolitic lava flow (Newberry crater, Oregon, USA). Obsidian was a popular raw material in many ancient and pre-historic cultures for a variety of tool and ornamental purposes. Even today obsidian is used as jewelry, for example in Mexico and Georgia

▼ Fig. 15.19. Illustration in Newberry Park of how Indians used obsidian from rhyolite lava flow (Oregon). See also Figs. 4.4, 9.12, 9.13

the country or the costs for buying oil for heating the furnaces to make cement are exorbitant. Zeolitized tuffs, when crushed, make an excellent cement with hydraulic potential under water. These cements have been used for building dikes in Holland or for the lower foundation of bridges across rivers. Such material is mined in many countries, such as Turkey and Indonesia.

Pavement made of fine-grained *blue basalt* has been popular in former times in many countries in Europe. The use of basalt has seen a recent renaissance e.g. in the restoration of churches, such as the Cologne Cathedral, because it is much less affected by noxious environmental gases, such as SO_2, than sandstone.

Obsidian is perhaps the best-known volcanic raw material used in many ancient cultures from Asia (Armenia) through the Mediterranean area to Latin America (Figs. 15.18, 15.19). The reason is that the glass is very homogeneous and commonly does not contain any phenocrysts. Thus, obsidian breaks with extremely sharp edges and can be used for many purposes, for example for cutting material, or as arrowheads. In some countries, surgeons even prefer obsidian knives to those made of steel.

The Attraction of Volcanoes and Volcanic Landscapes

Large earthquakes often result in major human tragedies as well as destruction of vital infrastructures and countless buildings. Fractures in old buildings may tell us about past earthquakes (Fig. 15.20) but remnants of these destructive powers of nature are commonly quickly removed and covered by new buildings. Obviously, areas devastated by earthquakes are not favorite tourist spots. The popular museum and information center in Kobe (Japan) built to be a reminder of the particularly destructive Kobe 7.3 earthquake on 5 January 1995 is an exception.

Spectacular volcanic eruptions, in contrast, appeal to all senses: they can be seen, felt and even smelled. Some continue for days, weeks or even longer and many draw huge crowds of sightseers. No doubt: volcanoes, active ones, quiet ones and extinct ones, are major magnets to tourists (Figs. 15.21–15.24). When lecturing to school children, I found that volcanoes only compete with dinosaurs as the most attractive topic. In learning about Earth materials and deep Earth processes, volcanoes and volcanic deposits are a powerful educational tool. And the entire topic of natural hazard mitigation and the case histories of volcanic catastrophes are superb teaching material to talk about natural hazards in general and how to cope with them.

The number of visitors in areas of young or active volcanism can be impressive. In the US, for example, the bestsellers Grand Canyon and Yosemite are followed closely in numbers of visitors by national parks in volcanically active areas. Yellowstone, where geysers have American dimensions (Figs. 7.19, 15.5), leads with more than 3 million visitors annually, followed closely by the Hawaiian National Park, where the most active volcano on Earth, Kilauea, attracts some 2.5 million visitors a year. Even the currently inactive Mount Rainier,

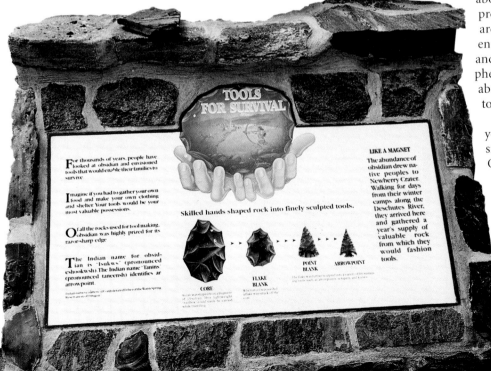

◀ Fig. 15.20. Church, destroyed during a large earthquake in 1956 on the rim of Santorini caldera (Greece)

he largest volcano of the conterminous United States, is visited by some 2 million people annually. Mount St. Helens, which awakened dramatically on 18 May 1980 and remained active until 1986, has been declared a National Volcanic Monument under the jurisdiction of the Forestry Service. It attracts an increasing number of visitors, 1 million alone in 1997. Beautiful Crater Lake in Oregon (Fig. 9.45), the deepest lake in the US cre-

ated by a huge eruption some 7 700 years ago, is another highly popular national park.

The easily accessible craters of active volcanoes Irazu and Poas (Fig. 12.27) in Costa Rica close to the capital of San Jose have been a must on tours in the country for decades. But even the relatively remote but more spectacularly active andesite volcano Arenal (Fig. 8.15) has developed into a major tourist attraction. The number of

▼ Fig. 15. 21. Pico de Teide (3 718 m a.s.l.), viewed from the bottom of Las Cañadas Caldera (ca. 2 000 m a.s.l.). Front of a highly viscous phonolitic lava flow broken up into blocks in the middle ground. Tenerife (Canary Islands)

▲ Fig. 15.22. Volcanoes Toliman (*left*, 3 158 m a.s.l.) and San Pedo (*right*, 3 020 m a.s.l.) on the rim of Atitlán caldera (1 562 m a.s.l.) (Guatemala)

bungalows and hotels around the volcano is growing by the year. The hot avalanches generated on 9 May 1998 and even more recently have further added to the attraction of this volcano. Revenues from tourism have now outpaced those from agriculture in Costa Rica, volcano attraction representing a significant share. In neighboring Nicaragua, the easily accessible Masaya lava lake and crater complex (Fig. 4.24) and the frequently erupting Cerro Negro scoria cone (Fig. 7.7) are fast becoming standard places to visit. Indeed, volcano tourism may become a significant economic commodity in Nicaragua.

In Europe, the most famous young volcanic areas are located in Sicily. The gifts of nature include the largest and most active European volcano Etna, and, in the nearby Aeolian Islands, the prominently active volcano Stromboli and the island of Vulcano, home of the ancient mythical god Vulcanus. Vesuvius on mainland Italy opposite the Aeolian Islands, once very active but quiet since 1944, attracts a huge number of tourists, not the least because of the once-buried cities of Pompeii and Herculaneum, where new discoveries are made even today despite centuries of excavations.

The unusually spectacular eruptions of Etna volcano in Sicily in 2001 and 2002 have resulted in a major increase in tourism, many people wanting to get a first-hand impression of Etna on fire. On the other hand, the sudden – and hopefully temporary – decline on December 26, 2002, of the famous intermittent eruptions of permanently active Stromboli that was almost continuously active for thousands of years, is a major blow to tourism in the Aeolian Islands. Even further north, in France and Germany, areas of young Quaternary and even Holocene volcanism are also drawcards for tourists. These volcano fields are also one of the cradles of volcanology, especially Central France, where milestone discoveries were made in the early part of the second half of the eighteenth century (Chap. 1).

Fire and ice are the central buzz words in the Icelandic tourism industry; this entirely volcanic land in the northern Atlantic attracts hundreds of thousands of tourists annually. Farther south, the Azores, Madeira and especially the Canary Islands with their spectacular volcanic scenery all owe much of their touristic appeal to their volcanic nature and activity.

◀ Fig. 15.23. Lake Toya, the most popular tourist area in Hokkaido (Japan). The dacitic domes have grown in the lake of Toya caldera

◀ Fig. 15.24. Tourist groups on the way to fumaroles in the crater area of Asahi-dake volcano (Daisetsu Mts., central Hokkaido, Japan). The crater area and the picturesque steam vents draw some 500 000 visitors annually

Craters have fascinated man for millenia, the early Greek philosopher Empedocles being said to have finished his live by jumping into one of the craters of Etna volcano, a dark side of volcano attraction that is still with us in some countries. In Europe, the craters of Stromboli, Etna, those of the logistically most easily accessible Vesuvius or Vulcano are prominent goals for people being moved by peering into the depth of the Earth. Bandama crater on Gran Canaria (Canary Islands) and of course Halemaumau on Kilauea volcano and the crater on active Oshima Island not far from Tokyo are examples in other parts of the world. Some craters such as Ngorongoro in east-

▶ Fig. 15.25. Famous painting by unknown artist picturing the explosive eruption of Asama Volcano in August 1783 (Honshu, Japan). Courtesy of Mr. Hiroo Misaizu

ern Africa or Mt. Suswa in Kenya are attractive not only for their wildlife but also the grandeur of the volcanic crater landscape.

Many lakes hosted in– or created by– volcanoes have few rivals in grandeur among lakes worldwide. Some of the most spectacular volcanic lakes have formed in calderas, maars or other types of craters. Examples include Crater Lake in Oregon (Fig. 9.45), Tianchi across the Chinese-Korean border (Fig. 9.51), Lake Atitlan in Guatemala (Fig. 15.22), lakes Ilopango and Coatepeque in El Salvador, Laguna de Apoyo in Nicaragua or the water-filled calderas and maars in Italy, France and Germany (Fig. 12.20). Caldera lakes are ubiquitous in Japan, the classic site being Hakone caldera close to Mt. Fuji. Caldera lakes abound in Hokkaido and include Lake Toya, the most popular tourist spot in Hokkaido (Fig. 15.23), bordered in the south by very active Usu volcano and, farther north, Akan, Kutcharo and Mashu caldera lakes. Examples on Honshu include lakes Towada and Ashino-ko. Other famous examples of caldera/crater-hosted lakes include Lake Toba in Sumatra, site of the world's largest Quaternary volcanic eruption, Lake Taal near Manila (Philippines) and lakes Taupo and Rotorua in New Zealand.

Lakes accumulated behind volcanically created dams – debris avalanches, lavas, and debris flows – have formed in many volcanic areas in the world. Lake Myvatn in northern Iceland is the classical example in Europe. The lake-dotted landscape south of very active Komagatake volcano in Hokkaido, where a major debris avalanche dammed up rivers and creeks in 1640, has developed into a major tourist attraction. Similarly, a basaltic lava flow spread from the flanks of Mt. Fuji in A. D. 864 caused the formation of several lakes in the Fujigoko, the five-lake area. The debris avalanches resulting from the spectacular collapse of the northern flank of Bandaisan dammed up Nagase river to create several lakes.

In many countries blessed with young volcanism, the most enjoyable benefit from– and direct impression of– volcanic heat are *hot springs*. The culture of hot springs is nowhere more developed than in Japan. Hot water pools can be found in many hotels throughout the country. Even more important for developing and maintaining social contacts are the public hot baths both outdoors and indoors, where neighborhood folks gather in the evening to relax from a hard days work and enjoy gossip. Bathing in hot springs is also part of the culture and a major tourist attraction in Iceland, Hungary (especially around Lake Balaton), Italy and other countries.

This chapter should not end without even a brief mention of the role volcanoes have played in the arts over the centuries. Rendition of the explosive energy and havoc brought about by destructive eruptions (Fig. 15.25) has been– and still is– one way to cope with disaster and to remind posterity of a past catastrophe. The 18th

century was a particularly rich period in which highly active Vesuvius came to fame in the works of many artists from several countries, some having made their home in Naples at the foot of the spectacular volcano (Fig. 15.26). It was the light effects and spectacular illuminations of lava fountains or lava flows especially at night that attracted the painters but also the beginning of men's curiosity in finding out about the nature of volcanoes. In fact, Hackert, the artist of Fig. 15.26, was cooperating with the great William Hamilton, father of modern scientific inquiry into the workings of volcanoes (Chap. 1). In Japan, on the other side of the globe, Hokusais Thirtysix Views of Mt. Fuji – or the later One Hundred Views of Fuji – embody the deep feeling of the Japanese people for this towering cone for more than 1 000 years (Fig. 15.27). Mt. Fuji has been the symbol for Japan and for volcanic cones par excellence not the least through the famous woodcuts of Hokusai. It matters little that the most elegant curve of the cone of the volcano that conveys an immortal sense of beauty, is rather accentuated by the artist. In reality, the slopes of the upper part of the cone are straight (Fig. 8.12). But then Hokusai was not

a scientist. And his rendition of the esthetic grandeur of Mt. Fuji has become a far more powerful symbol of the beauty of nature than any scientific paper can ever achieve.

Summary

The benefits of volcanoes have lured people to settle nearby in many parts of the world. Foremost is the importance of the fertile volcanic soils, such as in Latin America, the southwestern Pacific (Philippines, Indonesia) and Japan, but also in the young volcanic areas in Africa or the Canary Islands. This fertility is based on three properties, provided the climate is warm and wet: the ability to hold huge amounts of water on hydrated glass and secondary mineral phases, and to release it slowly; the supply of some essential plant growth elements (e.g., K, Mg, Se); and the glassy, unstable structure of the particles, which quickly release important elements, and which are rapidly transferred into crystalline and short-range-order clay minerals.

The high temperatures beneath young volcanoes, caused by underlying magma reservoirs, are used by man directly and indirectly. Geothermal energy sources are very important in some coun-

▶ Fig. 15.27. Four weather levels from fair weather with blue sky at the top to thundershowers at the foot of towering Mt. Fuji. Famous woodcut # 42 by Hokusai from "Thirty-six Views of Fuji" actually represented by 46 woodcuts

tries, such as New Zealand, Iceland or Italy, providing a major fraction of the energy supply.

Volcanic rocks have been used for tens of thousands of years as aggregates and tools because they show many advantages compared to other rock types. They are generally easy to mine. Obsidian is the perfect raw material for sharp knives, arrow points and other tools. Massive lapilli deposits or ignimbrites are splendid raw materials for lightweight and highly insulating building material because of their high porosity and soft nature.

Young volcanic landscapes provide for some of the most beautiful scenery on Earth. They represent tourist attractions of the first order because of their morphological variety, including crater lakes, geysers, and hot springs – or because volcanoes are still active or show signs of recent activity.

Volcanoes have attracted man not only for their obvious economic benefit, their soils, raw materials, the hot springs. The awe-inspiring grandeur of active and dormant volcanic edifices symbolize the power of the deep Earth like no other mountains. Quite naturally, volcanoes have played a major cultural and religious role over the millennia in many parts of the world. They have been, and commonly still are, viewed as the site of gods or demons, even in highly industrialized and secularized societies such as Japan (Fig. 1.1).

Epilogue

The fundamental motivation of scientists is the curiosity of the individual, the desire or obsession to find out. Nevertheless, the awareness that society is increasingly confronted with local, regional and global problems that have their roots in traditional attitudes of society vis-a-vis nature has confronted natural science with a whole set of "external" priorities. This awareness is reflected in the research attitudes of individuals who have accepted this new role, in large funding projects or even long-range programs such as the two decades of the 1990s: Global Change (IGBP) and International Decade of Natural Disaster Reduction (IDNDR), both continuing in the twenty-first century. Volcano-magma systems are part and parcel of many research activities in both programs.

The name of one of these priorities is climate. The continuing unwillingness of some governments to accept the near-unanimous conclusion of leading experts that man-made greenhouse gases are the major reason for global warming has generated a widespread feeling of helplessness among some scientists. Discouraged scientists should take note, however, that the concerted efforts of the scientific community to point out the fundamental role of CFCs in destroying ozone in the stratosphere has led to the Montreal Protocol. This has been a remarkable victory for scientific logic and concerted effort. The ozone layers have started to recover. In the future they will have to battle mainly with volcanogenic aerosol clouds that episodically deprive the ozone molecules of their protective cover of nitrous oxides.

In the last three chapters I have focussed on topics that are not discussed – or only partly – in classical textbooks on volcanology. Even though these themes are treated here in a mere rudimentary way, the reasons why they are taking up some 20% of this book reflect the new priorities and tasks confronting volcanologists. Some of these are easy to grasp. Despite the increasing importance of recycling – who would have thought 20 years ago that cars are now recycled part-by-part, rather than going wholesale through giant jaw crusher – the necessity to search for new deposits of all kinds of raw materials actually increases. The success will critically depend not only on employing ever new tools but on a fundamentally new approach: to understand complex systems. Attempts to adjust college curricula to educate young people with this goal in mind are still few and far between.

Political pressure is commonly felt by volcanologists monitoring a volcano that could erupt in the future – based on its past behavior – or actually threatens to erupt. It is the old story that the messenger is the culprit who brings the bad news. By developing strategies on how to bring the message to the endangered people directly and most effectively, volcanologists had to learn to better communicate and present their findings in ways that are easily understood and can be used effectively in decision making. There has been some remarkable progress. When just a couple of decades ago there were less than a handful of publicly accepted and financed hazard maps of active volcanoes, even in some technically advanced countries, there is now a considerable number available not the least due to new communicative skills adapted by volcanologists.

Some ten years ago I received anonymously a letter written by a young boy and published in a German newspaper. Never will I know which of the many exciting aspects of volcanoes inspired the boy Conny. Was it their dramatic eruptions, their beauty or their exotic flair? The letter reads as follows, translated from the German:

I want to become volcanologist.

When I grow up, I want to become a volcanologist. I will buy a jeep and a trailer. With these I will travel to the Philippines, to Lanzarote or to Sicily. Unfortunately, I cannot then marry. In no case will I smoke cigarettes, cigars or pipe. I will buy the necessary equipment from my personal savings. I will only buy superior quality. I will travel around the world, actually from volcano to volcano. This is, of course, impossible without an assistant. I will therefore employ a nice girl to accompany me on my travels from volcano to volcano. The trailer will contain a washing basin, a TV with satellite program, a completely furnished kitchen and a bed. I will also buy a dog. In the evening, after work, I will take a walk with my dog. After that I will explore the town and chat with the natives. A little bit later I will taste the local food. After dinner I will go to bed. Yes, I will certainly choose this job. Most definitely.

Conny Hulverscheidt,
9 years old, Biengen.

At that young age Conny was unlikely to be even remotely aware that volcanoes not only trigger enthusiasm and sometimes obsession but also represent unique events and monuments of deep

Earth processes whose eruptions in many countries have major impacts on social structures and whose most explosive eruptions significantly impact climate and the ozone layer. After the German edition of this book was published in 2000 I was finally able to locate Conny. We struck up a friendship and he came along with some of h[is] classmates and his teacher in a volcanological fie[ld] trip with my students to the Eifel volcanoes. H[e] will soon enter university and embark on a degre[e] in physics. Maybe some day he will rediscover h[is] love of volcanoes.

"But when I said I'd go with you
wherever your career took you,
I didn't know what a volcanologist was"

Physical Units and Abbreviations

Mass, pressure, viscosity and temperature

metric ton	= 1 t	= 10^6 g	
megaton	= 1 Mt	= 10^6 t	= 10^{12} g
gigaton	= 1 Gt	= 10^9 t	= 10^{15} g
teragram	= 1 Tg	= 10^6 t	= 10^{12} g
Pascal	= 1 Pa ($m/kg/s^2$)		
bar	= 1 bar = $10^6 g/cm/sec^2$ = 10^5 Pa		

Megapascal (Mpa) = 10^6 Pa
Gigapascal (Gpa) = 10^9 Pa = 10 kb
η = viscosity
Pa.s = pascal second
  (formerly poise;
  1 poise = 1 g/cm/sec;
  10 poise = 1 Pa s)
Kelvin (K) = °C + 273.15

Time

d = day
a = year
ka = 1000 years
Ma = 1 million years before present
m.y. = million years, in terms of duration

Abbreviations

DRE = magma volume
  (Dense Rock Equivalent)
CMB = Core mantle boundary
COSPEC = correlation spectrometer
LNB = Level of neutral bouyancy
Moho = Mohorovicic discontinuity
  (density boundary between
  crust and mantle)
MORB = Mid-ocean ridge basalt
OIB = Ocean island basalt
TOMS = total ozone mapping spectrometer
  (on satellite platform)
VEI = Volcanic Explosivity Index
T = temperature
P = pressure
X = chemical composition

Glossary

For definitions of volcanic rock names see Chapter 3

Aa lava
Hawaiian term for lava flows generally of basaltic composition, covered and underlain by loose fragments of broken crust and spiny, scoriaceous clinker that are typically decimeters in diameter

Acceptable risk → risk

Accretionary lapilli
Spherical aggregates, mostly < 20 mm in diameter, of ash particles. Coarse-grained cores are generally surrounded by thin layers of very small ash grains

Adiabatic
Commonly used for adiabatic rise of mantle material in hot plumes where temperature of rising body is maintained

Aerosol
A colloidal dispersion of fine liquid droplets or solid particles in a gas. For example, SO_2 will react with OH-radical in the atmosphere to form tiny droplets of sulfuric acid (H_2SO_4). Natural aerosols include fog, smoke and stratospheric sulfate

Agglutinate (spatter) deposit
A rock formed by hot, plastically deformed pyroclasts flattened and welded together upon landing

Ash
Small pyroclasts (glass shards, minerals, rock fragments) < 2 mm in diameter, ejected by explosive volcanic eruptions. Lithified ash is called tuff

Ash-flow tuff → ignimbrite, pyroclastic flow

Asthenosphere
A layer in the uppermost mantle beneath the lithosphere where most magmas are generated. The asthenosphere is characterized by low and attenuated seismic wave velocities and deforms by plastic flow as a result of high temperature

Autobreccia
Breccia formed by internal fragmentation within a moving lava flow or intrusion

B

Base surge
A surface-hugging, turbulent, low density volcanic current that moves outward from the base of an eruption column at high velocity. Generally applied to hydroclastic (phreatomagmatic) eruptions

Bingham plastic viscosity →viscosity

Black smoker
Hydrothermal vent on the ocean floor that emits generally hot but < 330 °C fluids producing fine, dark mineral precipitates upon mixing with cold ambient seawater

Blast
A sudden, violent, overpressured explosion projected laterally or vertically. At Mount St. Helens, the blast was directed laterally and produced a high-velocity dilute pyroclastic density current, or pyroclastic surge

Block
Fragment of solid rock > 64 mm in diameter ejected during an explosive eruption

Block-and-ash flow deposit
A small-volume, dominantly monolithologic pyroclastic flow deposit dominated by dense to moderately vesicular juvenile blocks set in a medium to coarse ash matrix of the same composition. Commonly generated by partial collapse of volcanic domes

Block lava
A lava flow whose surface is covered by angular blocks up to meters across, formed by lava fragmentation during emplacement

Bomb
A pyroclast (lava bleb), ejected hot and deformed plastically, > 64 mm in diameter. Bombs can be roundish to spindle-shaped and may be striated at the surface or may have flattened forms created upon landing (cowpatlike)

Buoyancy level, neutral
Height at which the density of a mantle plume, magma or eruption column equals that of the surrounding lithosphere or atmosphere

C

Caldera
An irregular to subspherical collapse feature several km to tens of kilometers in diameter within a volcano. Formed by roof subsidence over an evacuated magma chamber

Cinder → scoria

Climate forcing
Sum of natural and anthropogenic factors that generate climate change

Co-ignimbrite
Ash that lifts off the top of a moving pyroclastic flow and then settles; resembling ash sourced in an eruption column

Composite volcano (cone)
A relatively large, long-lived constructional volcanic edifice, consisting of intrusives overlain by interlayered lava and volcaniclastic products erupted from one or more vents, and their recycled equivalents. Flank slopes are typically > 10° → stratovolcano, stratocone

Conduit
A pipe-like pathway for rising liquid or fragmented magma

Cone
A conical hill made of volcanic material

Cone sheet swarm
Conical arrangement of dikes

Cooling unit → flow unit

COSPEC
An ultraviolet correlation spectrometer used to detect the concentration and emission rate of SO_2 from degassing volcanoes

Crust
Outermost shell of the Earth above the Mohorovičić discontinuity. The oceanic crust is generally 6 – 7 km, the continental crust 20 – 80 km thick

Cryptodome
Shallow intrusion that has uplifted the thin rock cover but did not erupt. Usually of felsic composition

Debris avalanche (deposit) (DAD)
The product of large-scale collapse of a gravitationally unstable sector of a volcanic edifice resulting in a fan-like deposit at the foot of the edifice, characterized by a hummocky surface

Debris flow (deposit)
Viscous water-saturated slurry of rock fragments, containing 10 – 25 wt % water. The solid material is carried in water, which lubricates the flow. Debris flows move downslope under the influence of gravity either as fairly

coherent "plugs", with laminar flow, or as granular flows

Dense Rock equivalent (DRE)
Volume of magma calculated after the volume of vesicles and interparticle pore space has been subtracted

Devitrification
Transformation of volcanic glass into crystalline substances such as in a cooling ignimbrite (high-T devitrification) or diagenesis (low-T devitrification)

Diatreme
Funnel-shaped breccia pipe that can reach several 1 000 m in depth. Diatremes are thought to form by hydroclastic fragmentation and wall rock collapse. They may underlie maars and grade at depth into dikes

Differentiation
Compositional modification of a cooling magma by crystal fractionation and other processes

Dike
A blade-like discordant body of (solidified) magma, cutting the country rock nearly vertically

Dome (lava dome)
A bulbous dome-shaped subvolcanic or extrusive rock body usually of felsic composition. Domes are emplaced slowly because of high magma viscosity

Elutriation
Loss of small particles by the upward flow of gas, such as in a pyroclastic flow

Epiclastics
Particles formed by the erosion or weathering of pre-existing volcanic rocks

Eruption column
A mixture of tephra particles, magmatic gases and entrained air that rises above a volcanic conduit during an explosive volcanic eruption. The lower jet (gas thrust) is transformed into the main convective column by incorporation of air and shearing against the atmosphere

Eutaxite
A historic term for welded tuff (ignimbrite) showing flame-like high-T compaction structures

Exsolution (of volatiles)
The transition of a volatile compound from solution into a gas phase as during decompression of rising magma. Also used for separation of an initially homogeneous mineral phase into two phases as during cooling

External water
Any type of water (groundwater, lake water, seawater) that is not sourced in the magma itself

Fall unit
Sedimented tephra resulting from a single fallout event

Fallout
Sedimentation of particles out of an ash cloud

Felsic
Felsic magmas are rich in alumina, alkalies, incompatible trace elements, and in some also silica (dacite and rhyolite). Felsic magmas develop from mafic magmas by differentiation or by partial melting of crustal rocks. The term felsic is preferable to silicic or acid both of which only refer to silica-rich magmas

Fiamme
Flame-like variously compacted and devitrified lapilli- to bomb-sized pumice/lava spatter

Filter pressing
The removal of an interstitial liquid from a crystal mush

Fire fountain
Fire or lava fountains are highly fluid mixtures of basaltic magma blebs in gas that characterize Hawaiian eruptions as in Kilauea volcano

Fissure eruption
An eruption that takes place from an elongate fracture in the crust

Flood basalt
Thick sequences of laterally extensive basaltic lavas that form huge lava fields typically >1 million km^3 in volume and are erupted over a short period of time, < 1 million years. Some flood basalt provinces are generated when continents break up and drift apart

Flow unit
Subdivision of a pyroclastic flow deposit (cooling unit) into separate mappable units recording a pause in eruption (source-controlled flow unit). Also applied to discontinuous subunits formed during transport (flow-controlled flow unit)

Forecast
A general description of future events, including rough estimates of time, location, and likely type of activity (in a volcano)

Fractional crystallization (fractionation)
The crystallization of minerals in a dominantly silicate liquid during cooling. Some elements are removed preferentially from the liquid, while others are concentrated in the minerals formed

Fumarole
Volcanic commonly sulfur-rich gas and steam (heated groundwater) released through a crack or vent

Gabbro
Plutonic rock formed by slow crystallization of basaltic magma

Gas thrust
Lower part of eruption column propelled by rapid expansion of magmatic gases above a vent

Geothermal reservoir
A body of hot rock containing geothermal fluid that is economically exploitable

Geothermal system
A present-day natural system heated by magma, in which hot water circulates. Surface expressions may be silica sinters, boiling pools, and geysers

Glass
A quenched silicate melt

Guyot
A flat-topped seamount

Harmonic tremor → volcanic tremor

Harzburgite
Peridotite, composed chiefly of olivine and orthopyroxene

Hawaiian eruptions
Eruptions of very low viscosity basaltic magma (10 – 100 Pa s) in the form of fire fountains. Minor ash and scoria production

Hazard (volcanic)
Eruptive and post-eruptive phenomena, such as pyroclastic flows, fallout tephra, lava flows, volcanic gases, lahars, and debris avalanches potentially damaging for population and infrastructure. Condition under which a potentially dangerous volcanic process might occur. In probabilistic assessments, hazards are the probabilities of such phenomena. Directly related to risk, where risk = (hazard) × (vulnerability), with vulnerability referring to the consequences

Heterogeneous chemistry
Reactions of gases in the atmosphere with species in solid or liquid aerosol forms

Hot spot
Confined area on the surface of the Earth characterized by abundant and often long-lived volcanism, usually in intraplate settings. Hot spots may originate from mantle plumes

Hummocky topography
Irregular hilly morphology typical of the surface of debris-avalanche deposits → DAD

Hyaloclastite
Deposit consisting of small angular fragments of volcanic glass formed by shattering, such as by flow or intrusion of lava or magma into water, ice, or water-saturated sediments. Also applied to vesiculated glassy lava fragments erupted under water

Hydroclastic (hydromagmatic, hydrovolcanic) eruption
Eruption that results from the interaction of magma and groundwater or surface water. Synonymous with phreatomagmatic eruption

Hydrothermal system
Any natural system involving the circulation of hot water heated by a hot magmatic body

Hyperconcentrated stream flow
Turbulent flow of water, mixed with enough sediment (60–75 wt%) to possess some yield strength

Ignimbrite
Pyroclastic flow deposit (ash-flow tuff), welded or unwelded, consisting dominantly of pumice and ash

Intraplate
The interior of lithospheric plates contrasted with convergent or divergent plate margins

Isopach
Line drawn (on a map) through points of equal thickness of a rock body or tephra deposit

Isopleth
Line drawn (on a map) through points of equal diameter of objects such as pumice or rock fragments

Jökulhlaup
An Icelandic term referring to both water floods and lahars generated when a volcano erupts under a glacier

K/T-boundary
The stratigraphic boundary between Cretaceous and Tertiary rocks, dated at 65 Ma

Lag breccia
Concentration of rock fragments within or at the base of pyroclastic flow deposits close to source

Lahar
An Indonesian term referring to a stream of water mixed with more volcanic rock particles than normal streams carry (→ debris flow). Many lahars form when pyroclastic flows s.l. enter rivers or when loose ash is mobilized by rain

Lapilli
Pyroclastic particles between 2 and 64 mm in diameter

Large igneous provinces (→ LIPs)

Lateral blast (→ blast)

Lava
Magma is called lava when it erupts at the surface as flows or domes

Lava delta
A delta-like rock body formed by advance of lava into the sea resulting in a lava platform overlying foreset deposits. Lava delta deposits are coarse-grained brecciated fragments of both subaerially and subaqueously deposited lava

Lava dome → dome

Lava fountain
Uprushing jet of incandescent pyroclasts and gas tens to hundreds of meters above a conduit

Lava lake
A lake of lava accumulated in a crater

Lava tube
The remainder of a roofed–over channel in which lava can flow long distances

Level of neutral buoyancy (LNB)
The level at which a rising magma or eruption column reaches the same density as the surrounding fluid

Lherzolite
Peridotite composed chiefly of olivine, orthopyroxene and clinopyroxene and minor amounts of spinel or garnet

LIPs
Massive lava piles and/or intrusions of basaltic and rarely evolved magmas. LIPs are typically localized and transient phenomena, with a lifetime of only a few million years. May result from partial melting in decompressing superplumes

Liquidus temperature
Temperature above which the system (magma) is completely liquid

Lithic fragments
Pieces of country rock that are incorporated in a lava or gas-fragment mixture

Lithosphere
The rigid outer plate of the Earth mostly 50–200 km thick, consisting of the crust and uppermost mantle above the asthenosphere

Littoral
Features that lie, or processes that occur, along the shoreline of an ocean or lake. Littoral cones form by hydroclastic eruptions where a lava flow enters a body of water

Maar
Small volcano characterized by a crater hundreds of meters to several kilometers in diameter and surrounded by a low rim. Maars form by hundreds to thousands of phreatic and phreatomagmatic explosions resulting in volcaniclastic deposits dominated by rock fragments

Magma
Dominantly silicate liquid (molten rock), consisting of melt, crystals, dissolved volatile compounds and, at low pressure, gas bubbles

Magma chamber (reservoir)
An underground reservoir in the Earth's crust filled with magma and crystals, from which volcanic materials are erupted

Magma mingling
Incomplete mixing of two or more magmas

Magma mixing
Intimate mixing of two or more magmas to form a hybrid magma. Most magmas are hybrid

Magnitude
A measure of the energy released by an earthquake, determined by measuring the highest-amplitude waves and correcting for distance and type of instrument

Mantle plume
The upwelling of anomalous mantle material through the Earth's mantle, made buoyant because of either a thermal or chemical anomaly. Mantle plumes may be the source of magmas in LIPs and hot spots

Mass extinction
The sudden simultaneous or near-simultaneous disappearance in the geological record of a large number of species and higher taxa

Mid-ocean ridge
Morphological ridge in the ocean basins formed by uplift, intrusion and extrusion along divergent plate boundaries

Mitigation
Activities toward reducing risk

Mohorovičić discontinuity (Moho)
The base of the crust is defined by a sudden increase of seismic compressional wave velocities from <7 to >8 km/s in the underlying mantle

MORB
Mid-Ocean Ridge Basalt. The dominant rock type along mid-ocean ridge spreading centers, containing on average 50 wt% SiO_2 derived from partial melting of mantle peridotite, previously depleted by partial melting

Mushroom cloud
The mushroom head-like expansion of an eruption column at the LNB

Neptunists
Geologists in the second half of the eighteenth century who believed that basalt had formed by precipitation from a primordial ocean

Network former
Elements that form strong covalent bonds with oxygen in magmatic melts

Network modifier
Elements that form weak metallic bonds with oxygen in magmas

Newtonian fluid (Newton's viscosity)
A simple fluid (such as water) in which the velocity gradient is directly proportional to the applied stress; the constant of proportionality is the viscosity

Nodule
A colloquial term for fragments of plutonic cumulate or mantle rock brought to the surface by lava flow or explosive eruptions

Nuée ardente
Synonymous with pyroclastic flow (glowing cloud)

Obsidian
Generally black, high-silica phenocryst-poor volcanic glass of felsic composition (rhyolite, dacite, phonolite)

Ophiolite
A section of sedimentary rocks, oceanic crust and uppermost mantle that has been uplifted and exposed at the Earth's surface

Overbank deposit
Thin deposit representing where pyroclastic flows or lahars have spilled out alongside a channel

Pahoehoe (lava) Hawaiian term for lava flows generally of basaltic composition with smooth, continuous, bulbous to ropy surfaces, and thin flow units

Palagonite
A mineraloid forming an intermediate product between sideromelane and layer silicate during alteration of basaltic glass, most commonly during diagenesis

Peperite
A mixture of lava and soft volcanic or nonvolcanic sediment formed by invasion of lava

Peridotite
An ultramafic rock consisting dominantly of olivine and pyroxene. Peridotite is the rock that makes up the upper mantle and is the source rock for basaltic magmas

Phenocryst
Large crystal in a fine-grained volcanic rock

Phreatic
High-level groundwater. A steam eruption erupting only country rock fragments

Phreatomagmatic → hydroclastic

Pillow lava
Interconnected, elongated lava tubes formed in a subaqueous environment. Cross sections of tubes commonly show a convex upper and flat or concave lower surface and radial fractures and resemble pillows. Pillow tubes are surrounded by a chilled glassy margin

Pillow volcano
Small conical or ridge-like edifices of pillow lava found on the seafloor and in ophiolites

Pit crater
A small crater formed by subsidence of the surface

Plinian eruption (column)
An eruption with a powerful column consisting of a basal jet (gas thrust) and upper convecting column, reaching up to 45 km height. Plinian eruptions result in widespread tephra blankets of several hundred square kilometers

Plume
Roughly cylindrical or drop-like mantle material that rises within the Earth because of its lower density and higher temperature compared to the surrounding mantle. Also applied to hot rising fluids in black smokers and to mixtures of particles and gas in eruption columns

Plutonists
Geologists in the late eighteenth and early nineteenth century who believed that basalt and granite were generated from magma that was able to extrude or intrude throughout the history of the Earth

Precursor
A geological event that occurs prior to an eruption and is related to the processes of the forthcoming eruption

Prediction
A specific description of future events, including time, size, type, location, and formal errors for each

Primary magma
A basaltic melt that is a direct partial melting product of a source rock, with which it is in equilibrium and has not been modified prior to eruption. True primary magmas are rare

Primitive magma
The most magnesian lava in a volcanic complex (field) from which more evolved magmas have been derived by differentiation

Pumice
A normally light-colored very frothy low density pyroclast mostly of felsic composition

P-wave (Primary wave)
An elastic wave with primarily compressional or longitudinal particle motion that travels faster than other waves, and is thus the first to appear on seismograms

Pyroclast
Particles produced during explosive volcanic eruptions

Pyroclastic density current
A gravity-controlled, laterally moving mixture of pyroclasts and gas. The term includes both pyroclastic flow and pyroclastic surges, but has no connotation of particle concentration or flow steadiness

Pyroclastic fall
Sedimentation of clasts through the atmosphere from an eruption jet, plume or laterally moving turbulent ash cloud during an explosive eruption

Pyroclastic flow
Flow (density current) of volcanic material consisting dominantly of vesiculated, low-density pumice and glass shards, which tends to follow topographic lows. Also includes flows consisting of poorly-vesiculated, dense lava clasts

Pyroclastic flow deposit
Predominantly massive, poorly-sorted, ash-rich deposit laid down by a pyroclastic density current. This term is mostly used for deposits from high particle concentration flows contrasting with low particle concentration pyroclastic surges

Pyroclastic surge
A turbulent, low-density, dilute high-velocity pyroclastic density current, commonly part of a pyroclastic flow. It is less constrained by topography than a pyroclastic flow

Pyroclastic surge deposit
Strongly bedded (laminated, low angle cross-bedded) deposits laid down by surges

Rarefaction wave
An expansion wave that moves into a conduit in the opposite direction to an outward expanding shock wave

Resurgent caldera
Uplift of the central collapsed block in a caldera due to subsequent high level emplacement of magma

Rheology
The physical properties of fluids

Rheomorphism
The local remobilization of a deposit e.g. of a

hot pyroclastic flow such as during the welding process

Rift zone
A roughly linear belt in a volcano characterized by linear dike swarms and scattered eruptive centers. Rifts in continents are large fault-bounded depressions caused by lithosphere rifting. Mid-ocean ridges are also referred to as rift zones

Risk (volcanic)
Hazard × vulnerability. The probability of a loss of lives, property or productive capacity within an area subject to volcanic hazard. Risks that an individual or community are willing to accept, or that the public official is prepared to allow persons in their charge to accept. Acceptable risk is a function of the benefits of risk mitigation (safety and avoided losses), and its costs (loss of jobs and business, community disruption, the costs of transportation, housing, and the food for evacuees, and costs of any structures to divert hazards)

Risk management
A chain of activities progressing through the sequence of risk identification → risk reduction → risk transfer, and aimed at minimizing the risk of loss of lives and/or damage resulting from volcanic hazards

S

S-wave (Secondary wave)
A shear elastic wave that transfers particle motion, and usually the most prominent wave on a seismogram. Travels more slowly than P-waves, and thus arrives later on seismograms; cannot pass through liquids

Scoria (cinder)
A vesicular, glassy to fine-grained pyroclast containing > 50% void space, usually of basaltic or andesitic composition and of lapilli size or larger

Scoria cone
Small volcano consisting dominantly of scoria

Seafloor spreading
Rifting apart of oceanic plates at variable velocities in different ocean basins

Seamount
Submarine volcano

Seamount asperity
Roughness of subducted oceanic lithosphere due to seamounts (also volcanic ridges and oceanic plateaus)

Sector collapse
Collapse of the flank of a volcano resulting in slumps or debris avalanches

Sheet lava
Submarine lava sheets with smooth, lineated, folded or jumbled surfaces resembling subaerial pahoehoe

Sheeted dikes
A succession of dikes lacking country rock screens. Applied to the zone between the gabbros (layer 3) and lavas (layer 2a) of the oceanic crust → (ophiolite)

Shield volcano
A broad, low-relief volcanic edifice, consisting dominantly of lava flows, usually of fluid basaltic composition. Flank slopes are typically < 5°

Sideromelane
Black volcanic glass of basaltic composition

Slab
Oceanic lithosphere and overlying sediments that are subducted

Slug flow
The upward motion of large individual bubbles in a conduit filled with magma

Solidus
The temperature at which a crystallizing magma has completely solidified

Spatter
Fluidally shaped lava fragments that are deposited hot, typically deformed, and sometimes welded together

Spatter flow
Flows of basaltic (or felsic such as dacitic or phonolitic) composition in which spatter structures can be recognized resulting from lava fountaining

Spreading rate
The rate at which oceanic lithospheric plates are moving apart

Stratocone (stratovolcano)
A large volcano with steep upper flanks made up of lava flows, pyroclastic rocks, and intrusives

Stratosphere
The upper portion of the atmosphere between about 8 km (high latitudes) and 16 km (equator) up to 50 km above ground separated from the troposphere by the tropopause

Strombolian eruptions
Discrete explosions at periodic intervals of a few seconds to minutes, or hours, named after Stromboli (Aeolian Islands, Italy). Deposits consist of basaltic lava spatter, vesicular bombs, scoriaceous lapilli, and ash

Subduction zone
Inclined region along convergent continental margins or island arcs where oceanic lithosphere plus overlying sediments sink into the mantle

Sulfuric acid aerosols
Small droplets of sulfuric acid formed in the atmosphere from oxidation of sulfur-rich gases. The aerosols can form a haze in the upper atmosphere after volcanic eruptions, which can reduce incoming sunlight and cool the planet and have a lifetime of several years

Surge (deposit)
Low density turbulent pyroclastic density current that moves outward from a pumiceous pyroclastic flow, block-and-ash-flow or at the base of an eruption column. Deposits are thin and show parallel, lensoid or cross-bedding. Base is erosional depending on substrate

Surtseyan eruptions (→ hydroclastic)
Volcanic explosions dominated by jets of wet tephra, as observed at Surtsey (1963) (Iceland), resulting in tuff cones. The term Surtseyan is mostly used for volcanoes erupting through seawater

T

Tachylite
Dark basaltic glass crowded with fine-grained Fe/Ti-oxides and variable amounts of microlitic silicates. Forms at fast cooling but not quenching

Tephra
All fragmental volcanic ejecta generated during explosive volcanic eruptions

TOMS (Total Ozone Mapping Spectrometer)
A space-based, remote sensing, ultraviolet spectrometer, which measures sulfur dioxide erupted by volcanoes into the stratosphere

Trass
An old term for ignimbrite

Tropopause
Boundary between troposphere and stratosphere

Troposphere
The lower portion of the atmosphere from ground level to about 8 to 16 km

Tsunami A large wave in the sea (or lake) generated during a sudden event such as an earthquake, sector collapse of a volcano, entry of a pyroclastic flow into the sea or submarine sediment slides

Tuff
Lithified ash

Tuff, welded
A hard pyroclastic rock, compacted by internal heat and pressure. Forms the interior of thick ash flow (ignimbrite) sheets

Tufolava
An old term for ignimbrite

U

Umbrella cloud (region) The lateral expansion of an eruption column at the level of neutral buoyancy (LNB)

Underplating
Accumulating of (basaltic) magma at the density barrier between crust and mantle (Moho)

V

VEI
Volcanic Explosivity Index, a measure of the size of an eruption, mainly based on magnitude, intensity (eruption duration, style, volume, and height of eruption column) and destructive power of an eruption. VEI has an 8-point scale

Veneer deposit
A thin sheet deposit laid down by a pyroclastic flow in the overbank region or by a high velocity pyroclastic flow

Vesicle (bubble)
A void space (bubble remnant) due to expansion of gases during decompression preserved in a solidified pyroclast or lava

Viscosity
A measure of the resistance of a material to flow in response to stress

Vitrophyre
A glassy quenched base of an ignimbrite deposited hot

Volatile
Chemical species or compound dissolved in a magma at high pressure that form bubbles during rise or crystallization of magma

Volcanic arcs
Arcuate belts of volcano chains generated above subduction zones

Volcanic field
An assemblage of volcanoes erupted in a restricted area over a limited period and characterized by specific compositional characteristics

Volcanic hazard → hazard

Volcanic risk → risk

Volcanic tremor
Continuous seismic signal with regular or irregular sine wave appearance and low frequencies (0.5 – 5 Hz). Harmonic tremor has a very uniform appearance, whereas spasmodic tremor is pulsating, and consists of higher frequencies with a more irregular appearance

Volcanists
Scientists in the second half of the eighteenth century who interpreted basalt columns as formed by cooling of hot lava

Volcaniclastics
Particles formed by the fragmentation of volcanic rocks, irrespective of the process of fragmentation or the nature of the transporting agent

Vulcanian eruption
A highly explosive event of $< 1 \, km^3$ magma volume, but with an eruption column sometimes reaching 10 – 20 km height

Vulnerability
The susceptibility of physical and human systems to be affected by hazardous natural phenomena. It can be expressed as a probability of damage

W

Welding
Softening, compaction and partial fusion of hot glass shards and pumice together under load

X

Xenolith
A rock fragment that is not part of the magma, carried to the surface by a magma. A clast in a lava flow or pyroclastic deposit

Y

Yield stress
The stress that must be exceeded for a substance to flow. True (Newtonian) liquids have zero yield stress. Also known as yield strength

References

Authors are listed alphabetically but in the text numbers are quoted instead of names of authors. Numbers in parenthesis at the end of a reference refer to the chapter an author is quoted in.

A

1 Abich H (1882) Geologische Forschungen in den Kaukasischen Ländern (Geologie des Armenischen Hochlandes). Alfred Holder, Wien, pp 1–478 (11)

2 Abbott DH, Isley AE (2002) Extraterrestrial influences on mantle plume activity. Earth Planet Sci Lett 205: 53–62 (4)

3 Ahorner L (1983) Historical seismicity and present-day microearthquake activity of the Rhenish Massif, central Europe. In: Fuchs K et al. (eds) Plateau Uplift. Springer, Berlin Heidelberg New York, pp 198–221 (11)

4 Akiyama M (1998) (ed) Dynamics of vapor explosions. Final report of several research programs. Ministry Education Science Japan, pp 1–296 (12)

5 Alidibirov M, Dingwell DB (1999) Three fragmentation mechanisms for highly viscous magma under rapid compression. J Volcanol Geotherm Res 100: 413–421 (10)

6 Allard P and 10 co-authors (1991) Eruptive and diffusive emissions of CO_2 from Mount Etna. Nature 351: 387–391 (4, 14)

7 Alley RB, Clark PU (1999) The deglaciation of the Northern Hemisphere: A global perspective. Annu Rev Earth Sci 1999: 149–182 (14)

8 Alvarado GE, Schmincke H-U (1994) Stratigraphic and sedimentological aspects of the rain-triggered lahars of the 1963–1965 Irazu eruption, Costa Rica. Zentralbl Geol Paläontol 1994: 513–530 (13)

9 Alvarez LW, Alvarez W, Asaro F, Michel HV (1980) Extraterrestrial cause for the Cretaceous-Tertiary extinction. Science 208: 1095–1108 (14)

10 Ancochea E, Braendle JL, Cubas CR, Hernan F, Huertas MJ (1996) Volcanic complexes in the eastern ridge of the Canary Islands: the Miocene activity of the island of Fuerteventura. J Volcanol Geotherm Res 70: 183–204 (6)

11 Anderson DL (1995) Lithosphere, asthenosphere, and perisphere. Rev Geophys 33: 125–149 (2, 6)

12 Anderson EM (1936) The dynamics of formation of cone-sheets, ring dykes, and cauldron-subsidences. Proc R Soc Edinbg 56: 128–157 (9)

13 Anderson T, Flett JS (1903) Report on the eruption of the Soufrière in St. Vincent in 1902 and on a visit to Montagne Pelée in Martinique, Part I. Philos Trans R Soc Lond A 200: 353–553 (11)

14 Andres RJ, Rose WI, Kyle PR, da Silva S, Francis P, Gardeweg M, Moreno R H (1991) Excessive sulfur dioxide emissions from Chilean volcanoes. J Volcanol Geotherm Res 46: 323–329 (14)

15 Andres RJ, Barquero J, Rose WI (1992) New measurements of SO_2 flux at Poás Volcano, Costa Rica. J Volcanol Geotherm Res 49: 175–177 (14)

16 Aramaki S, Ui T (1982) Japan. In: Thorpe RS (ed) Andesites. J Wiley and Sons, New York, pp 259–292 (8, 9)

17 Aramaki S, Kushiro I (eds) (1983) Arc Volcanism. Elsevier, Amsterdam, pp 1–652 (8)

B

18 Baker BH, McBirney AR (1985) Liquid fractionation. Part III: Geochemistry of zoned magmas and the compositional effects of liquid fractionation. J Volcanol Geotherm Res 24: 55–81 (3)

19 Ballard RD, Holcomb RT, van Andel TJH (1979) The Galapagos Rift at 86°W: 3. Sheet flows, collapse pits, and lava lakes of the rift valley. J Geophys Res 84: 5407–5422 (5)

20 Barberi F, Martini M, Rosi M (1990) Nevado del Ruiz volcano (Colombia). Pre-eruption observations and the November 13, 1985 catastrophic event. J Volcanol Geotherm Res 41: 1–12 (13)

21 Batiza R (1982) Abundances, distribution and sizes of volcanoes in the Pacific Ocean and implications for the origin of non-hotspot volcanoes. Earth Planet Sci Lett 60: 195–206 (6)

22 Baumann H, Illies JH (1983) Stress field and strain release in the Rhenish Massif.

In: Fuchs K et al (eds) Plateau Uplift. Springer, Berlin Heidelberg New York, pp 177–186 (7)

23 Becker A (1993) An attempt to define a "Neotectonic period" for central and northern Europe. Geol Rundsch 82: 67–83 (7)

24 Belousov A (1996) Deposits of the 30th March 1956 directed blast at Bezymianny volcano, Kamchatka, Russia. Bull Volcanol 57: 649–662 (11)

25 Bischoff JL, Rosenbauer RJ (1984) The critical point and two-phase boundary of sea-water, 200°–500°C. Earth Planet Sci Lett 68: 172–180 (15)

26 Blake S (1984) Volatile oversaturation during the evolution of silicic magma chambers as an eruption trigger. J Geophys Res 89: 8237–8244 (4)

27 Blong RJ (1982) The time of darkness; local legends and volcanic reality in Papua New Guinea. Univ Wash Press, Seattle, pp 1–257 (1)

28 Blong RJ (1984) Volcanic hazards. A source book on the effects of eruptions. Academic Press, Sydney, pp 1–424 (1)

29 Bogaard C van den, Schmincke H-U (2002) Linking the North Atlantic to central Europe: a high-resolution Holocene tephrochronological record from Northern Germany. J Quat Science 17: 3–20 (7)

30 Bogaard P van den, Schmincke H-U (1984) The eruptive center of the late Quaternary Laacher See Tephra. Geol Rundsch 73: 935–982 (11)

31 Bogaard P van den, Schmincke H-U (1985) Laacher See Tephra: A widespread isochronous late Quaternary ash layer in central and northern Europe. Geol Soc Am Bull 96: 1554–1571 (11)

32 Brasseur G (1992) Volcanic aerosols implicated. Nature 359: 275–276 (14)

33 Brauer A, Endres C, Ganter C, Litt T, Stebich M, Negendank J (1999) High resolution sediment and vegetation responses to Younger Dryas climate change in varved lake sediments from Meerfelder Maar, Germany. Quat Sci Rev: 18: 321–329(13)

34 Brayshay M, Grattan J P (1999) Environmental and social responses in Europe to the 1883 eruption of the Laki fissure vol-

cano in Iceland: a consideration of contemporary evidence. In: Firth CR, McGuire WJ (eds) Volcanoes in the Quaternary. Geol Soc London, Spec Publ 161, pp 173–187 (14)

35 Brodsky EE, Sturtevant B, Kanamori H (1998) Earthquakes, volcanoes, and rectified diffusion. J Geophys Res 1ß3: 23 827 –23 838 (4)

36 Bryan SE, Marti J, Leosson M (2002) Petrology and geochemistry of the Bandas del Sur formation, Las Cañadas Edifice, Tenerife (Canary Islands). J Petrol 43: 1 815–1 856

37 Buch CL von (1825) Physikalische Beschreibung der Canarischen Inseln. Text und Atlas, 2 vols. In: Ewald J, Roth J, Dames W (eds) Gesammelte Werke. Bd 3, pp 225–646 (1). Königl Akad Wiss, Berlin, pp 1–407

38 Buck WR (1991) Modes of continental extension. J Geophys Res 96: 20, 161–20, 178 (7)

39 Byron GG Lord (2000) Collected Poems. In: McCann JJ (ed) Oxford, pp 1–1 110 (14)

40 Caldeira K, Rampino MR (1991) The mid-Cretaceous superplume, carbon dioxide, and global warming. Geophys Res Lett 18: 987–990 (14)

41 Carey S, Sigurdsson H (1982) Influence of particle aggregation on deposition of distal tephra from the May 18, 1980 eruption of Mount St. Helens volcano. J Geophys Res 87: 7 061–7 072 (14)

42 Carey S, Sigurdsson H (1985) The May 18, 1980 eruption of Mount St. Helens II: Modeling of dynamics of the Plinian phase. J Geophys Res 90: 2 948–2 958 (10, 11)

43 Carey S, Sigursson H, Gardner JE, Criswell W (1990) Variations in column height and magma discharge during the May 18, 1980 eruption of Mount St. Helens. J Volcanol Geotherm Res 43: 99–112 (11)

44 Carmichael ISE (1979) Glass and the glassy rocks. In: Yoder HS Jr (ed) The evolution of the igneous rocks. Princeton Univ Press, Princeton, New Jersey, pp 233–244 (12)

45 Carr MJ (1977) Volcanic activity and great earthquakes at convergent plate margins. Science 197: 655–657 (13)

46 Carr MJ (1984) Symmetrical and segmented variation of physical and geochemical characteristics of the Central American volcanic front. J Volcanol Geotherm Res 20: 231–252 (8)

47 Carr MJ, Feigenson MDF, Bennett EA (1990) Incompatible element and isotopic evidence for the tectonic control of source mixing and melt extraction along the Central American Arc. Contrib Mineral Petrol 105: 369–380 (8)

48 Carroll MR, Rutherford MJ (1985) Sulfide and sulfate saturation in hydrous silicate melts. J Geophys Res 90: 601–612 (14)

49 Casadevall TJ (1993) Volcanic hazards and aviation safety: lessons of the past decade. FAA Aviation Safety J 2: 3–11 (13)

50 Cashman KV, Thornber CR, Kauahikaua JP (1999) Cooling and crystallization of lava in open channels, and the transition of pahoehoe lava to a'a. Bull Volcanol 61: 306–323 (4)

51 Chester DK, Duncan AM, Guest JE, Kilburn CRJ (1985) Mt. Etna. The anatomy of a volcano. Chapman and Hall, London New York, pp 1–405 (13)

52 Chouet BA (1996) New methods and future trends in seismological volcano monitoring. In: Scarpa R, Tilling RI (eds) Monitoring and mitigation of volcano hazards. Springer, Berlin Heidelberg New York, pp 23–97 (13)

53 Choukroune P, Francheteau J, Hekinian R (1984) Tectonics of the East Pacific Rise near 12°50'N: a submersible study. Earth Planet Sci Lett 68: 115–127 (5)

54 Christiansen RL (1983) Yellowstone magmatic evolution: its bearing on understanding large-volume explosive volcanism. In: Explosive Volcanism: Inception, Evolution and Hazard. National Res Council Washington DC: 84–95 (7)

55 Christiansen RL, Foulger GR, Evans JR (2002) Upper-mantle origin of the Yellowstone hotspot. Geol Soc Am Bull 114: 1 245–1 256 (7)

56 Clague DA, Dalrymple GB (1987) The Hawaiian-Emperor volcanic chain: Part I Geologic evolution. In: Decker RW, Wright TL, Stauffer PH (eds) Volcanism in Hawaii. US Geol Surv Prof Pap 1350, pp 5–54 (6)

57 Clague DA, Moore JG (1991) Geology and petrology of Mahukona Volcano, Hawaii. Bull Volcanol 53: 159–173 (6)

58 Clague DA, Denlinger RP (1994) The role of olivine cumulates in destabilizing the flanks of Hawaiian volcanoes. Bull Volcanol 56: 425–434 (6)

59 Clausen HB, Hammer CU (1988) The Laki and Tambora eruptions as revealed in Greenland ice cores from 11 locations. Ann Glaciol 10: 16–22 (14)

60 Cochran JR, Goff JA, Malinverno A, Fornari DJ, Keeley C, Wang X (1993) Morphology of a "superfast" mid-ocean ridge crest and flanks; the East Pacific Rise, 7 degrees – 9 degrees S. Mar Geophys Res 15: 65–75 (5)

61 Coffin MF, Eldholm O (1994) Large igneous provinces: crustal structure, dimensions, and external consequences. Rev Geophys 32: 1–36 (6, 7)

62 Coleman RG (1977) Ophiolites. Ancient oceanic lithosphere? Springer, Berlin Heidelberg New York, pp 1–229 (5)

63 Colgate SA, Sigurgeirsson T (1973) Dynamic mixing of water and lava. Nature 244: 552–555 (12)

64 Connor CB (1990) Cinder cone clustering in the Transmexican volcanic belt: implications for structural and petrologic models. J Geophys Res 95: 395–405 (7)

65 Connor CB, Stamatakos JA, Ferril DA, Hill BE, Ofoegbu GI, Conway FM (2000) Volcanic hazards at the proposed Yucca Mountain, Nevada, high-level radioactive waste repository. J Geophys Res, 105: 417– 432 (13)

66 Cordery MJ, Davis FG, Campbell IH (1997) Genesis of flood basalts from eclogite-bearing mantle plumes. J Geophys Res 102: 20 179–20 197 (7)

67 Cox KG, Bell JD, Pankhurst RJ (1979) The interpretation of igneous rocks. Allen and Unwin, London, pp 1–450 (3)

68 Crandell DR, Mullineaux DR (1978) Potential hazards from future eruptions of Mt. St. Helens volcano, Washington. US Geol Surv Bull 1 383-C: 1–25 (10)

69 Crough ST (1983) Hotspot swells. Annu Rev Earth Planet Sci 11: 165–193 (6)

70 Crutzen PJ, Birks JWI (1982) The atmosphere after a nuclear war: twilight at noon. Ambio 11: 114–125 (14)

71 Dalrymple GB, Silver EA, Jackson ED (1973) Origin of the Hawaiian Islands. Am Sci 61: 294–308 (6)

72 Dalrymple GB, Lanphere MA, Clague DA (1980) Conventional K-Ar and ^{40}Ar/^{39}Ar ages of volcanic rocks from Ojin (site 430), Nintoku (site 432), and Suiko (site

433) seamounts and the chronology of volcanic propagation along the Hawaiian-Emperor chain. Initial Rep Deep Sea Drill Project 55: 659–676 (6)

73 Dana JD (1849) Geology: United States Exploring Expedition, vol X. Sherman, Philadelphia, pp 1–756 (13)

74 Darwin CR (1842) The structure and distribution of coral reefs. Murray, London, pp 1–214 (6)

75 Darwin CR (1844) Geological observations on the volcanic islands and parts of South America visited during the voyages of the HMS Beagle, with brief notices on the geology of Australia and the Cape of Good Hope, being the second part of the voyage of the "Beagle", during 1832–36. Smith, Elder, London, pp 1–543 (6)

76 Da Silva SL, Francis PW (1991) Volcanoes of the central Andes. Springer, Berlin Heidelberg New York, pp 1–216 (8)

77 Decker R, Decker B (1997) Volcanoes. Freeman, San Francisco, pp 1–321 (1)

78 Decker R, Decker B (2002) Volcanoes in America's National Parks. Odyssey Publ, pp 1–250

79 Delaney PT (1982) Rapid intrusion of magma into wet rock: groundwater flow due to pore pressure increases. J Geophys Res 87: 7739–7756 (12)

80 De Paolo D J (1981) Trace element and isotopic effects of combined wallrock assimilation and fractional crystallization. Earth Planet Sci Lett 53: 189–202 (3)

81 Desmarest N (1771) Mémoire sur l'origine et la nature du basalt à grandes colonnes polygones, déterminés par l'histoire naturelle de cette pierre, observée en Auvergne. Mém l'Acad R Sci Paris: 705–775 (1)

82 Detrick RS, Crough ST (1978) Island subsidence, hot spots, and lithospheric thinning. J Geophys Res 83: 1236–1244 (6)

83 Devine JD, Sigurdsson H, Davis AN (1984) Estimates of sulfur and chlorine yield to the atmosphere from volcanic eruptions and potential climatic effects. J Geophys Res 89: 6309–6325 (4, 14)

84 Dieterich JH (1988) Growth and persistence of Hawaiian volcanic rift zones. J Geophys Res 93: 4258–4270 (6)

85 Dingwell DB (1998) Magma degassing and fragmentation: Recent experimental advances. In: Freundt A, Rosi M (eds) From magma to tephra: Modeling physical processes of explosive volcanic eruptions. Develop Volcanol 4. Elsevier, Amsterdam, pp 1–23 (10)

86 Duda A, Schmincke H-U (1985) Polybaric evolution of alkali basalts from the West Eifel: Evidence from green-core clinopyroxenes. Contrib Mineral Petrol 91: 340–353 (3)

87 Dullforce TA, Buchanan DJ, Peckover RS (1976) Self-triggering of small-scale fuel-coolant interactions. I. Experiments. J Phys Dev Appl Phys 9: 1295–1303 (12)

88 Duncan RA, Petersen N, Hargraves RB (1972) Mantle plumes, movement of the European plate, and polar wandering. Nature 239: 82–86 (7)

89 Dzurisin D, Koyanagi RY, English TT (1984) Magma supply and storage at Kilauea Volcano, Hawaii, 1956–1983. J Volcanol Geotherm Res 21: 177–206 (6)

90 Eaton JP, Murata DJ (1960) How volcanoes grow. Science 132: 925–938 (13)

91 Elliot T, Plank T, Zindler A, White W, Bourdon B (1997) Element transport from slab to volcanic front at the Mariana arc. J Geophys Res 102: 14991–15019 (8)

92 Endo ET, Murray T (1991) Real-time seismic amplitude measurement (RSAM): a volcano monitoring and prediction tool. Bull Volcanol 53: 533–545 (13)

93 Escher BG (1933) On a classification of central eruptions according to gas pressure of the magma and viscosity of the lava. Leidsche Geol Med 6: 45–49 (11)

94 Fenner CN (1923) The origin and mode of emplacement of the great tuff deposits in the Valley of Ten Thousand Smokes. Nat Geogr Soc Contrib Tech Pap, Katmai Ser 1: 1–74 (11)

95 Fisher RV (1979) Models for pyroclastic surges and pyroclastic flows. J Volcanol Geotherm Res 6: 305–318 (11)

96 Fisher RV, Schmincke H-U (1984) Pyroclastic rocks. Springer, Berlin Heidelberg New York, pp 1–472 (4, 9, 12)

97 Fiske RS (1983) Volcanologists, journalists, and the concerned local public: a tale of two crises in the Eastern Caribbean. In: Explosive volcanism: Inception, evolution and hazard. National Res Council Washington DC, pp 170–176 (13)

98 Forsyth DW and 15 co-authors (1998) Imaging the deep seismic structure beneath a mid-ocean ridge: the MELT experiment. Science 280: 1215–1218 (5)

99 Francis, PW (1993) Volcanoes. A planetary perspective. Oxford Univ Press, Oxford, pp 1–443 (1, 9)

100 Francis PW, Wadge G, Mouginis-Mark PJ (1996) Satellite monitoring of volcanoes. In: Scarpa R, Tilling RI (eds) Monitoring and mitigation of volcano hazards. Springer, Berlin Heidelberg New York, pp 257–298 (13)

101 Franklin B (1782) The meteorological imaginations and conceptions. Mem Lit Philos Soc Manchester 3: 373–377 (14)

102 Freundt A, Schmincke H-U (1985a) Lithic enriched segregation bodies in pyroclastic flow deposits of Laacher See Volcano (E-Eifel, Germany). J Volcanol Geotherm Res 25: 193–224 (11)

103 Freundt A, Schmincke H-U (1985b) Hierarchy of facies of pyroclastic flow deposits generated by Laacher See-type eruptions. Geology 13: 278–281 (11)

104 Freundt A, Schmincke H-U (1986) Emplacement of small-volume pyroclastic flows at Laacher See Volcano (E-Eifel, Germany). Bull Volcanol 48: 39–60 (11)

105 Freundt A, Rosi M (eds) (1998) From magma to tephra. Elsevier, Amsterdam, pp 1–318 (1)

106 Fuchs K, von Gehlen K, Mälzer H, Murawski H, Semmel A (eds) (1983) Plateau uplift, the Rhenish Shield – a case history. Springer, Berlin Heidelberg New York, pp 1–411 (7)

107 Funck T, Schmincke H-U (1998) Growth and destruction of Gran Canaria deduced from seismic reflection and bathymetric data. J Geophys Res 103: 15393–15407 (6)

108 Gansecki CA, Mahood GA, McWilliams M (1998) New ages for the climactic eruptions at Yellowstone: Single-crystal 40Ar/39Ar dating identifies contamination. Geology 26: 343–346 (7)

109 Gardner JE, Thomas RME, Jaupart C, Tait S (1996) Fragmentation of magma during a Plinian eruption. Bull Volcanol 58:144–162 (10)

110 Gerlach TM (1986) Emission of H_2O, CO_2, and S during eruptive episodes at Kilauea volcano, Hawaii. J Geophys Res 91: 12177–12185 (14)

111 Gerlach TM (1991) Present-day CO_2 emissions from volcanoes. EOS Trans Am Geophys Union 72: 249–255 (4)

112 Gerlach TM, Graeber EJ (1985) Volatile budget of Kilauea volcano. Nature 313: 213–277 (4)

113 Gerlach TM, McGee KA (1994) Total sulfur dioxide and pre-eruption vapor-saturated magma at Mount St. Helens volcano, 1980–88. Geophys Res Lett 21: 25 2833–2 836 (14)

114 Gerlach TM, Westrich HR, Symonds RB (1996) Preeruption vapor in magma of the climactic Mount Pinatubo eruption: source of the giant stratospheric sulfur dioxide cloud. In: Newhall CG, Punongbayan RS (eds) Fire and Mud – Eruptions and Lahars of Mount Pinatubo, Philippines. PHIVOLCS and Univ Wash Press, Seattle, pp 415–433 (14)

115 Gerlach TM, McGee KA, Elias T, Sutton AJ, Doukas MP (2002) Carbon dioxide emission rate of Kilauea volcano: implication for primary magma and the summit reservoir. J Geophys Res 107: ECV 3/1–15 (4)

116 Geshi N, Shimano T, Chiba T, Nakada S (2002) Caldera collapse during the 2000 eruption of Miyakejima Volcano, Japan. Bull Volcanol 64: 55–68 (9,12)

117 Gessner C (1565) De omni rerum fossilium genere, gemis, lapidibus, metallis et huiusmodi libri aliquod plerique nunc primum editi. Darin: Ex epistola joannis kentmani ad me, De Basalte lapide, qui angulis constat, minimum quarter, plurimum septer. Zürich. Das Buch über Gestaltung und ähnliche Bildungen der Mineralien, Steine und vornehmlich Edelsteine, das nicht nur für die Ärzte, sondern für alle, die sich mit den Gegenständen der Natur und Philologie beschäftigen, nützlich und willkommen sein wird. Translated by G. Fraustadt und H. Prescher (MS) Dresden (1955). In: Koch et al. (1983) (1)

118 Giggenbach WF (1996) Chemical composition of volcanic gases. In: Scarpa R, Tilling RI (eds) Monitoring and mitigation of volcano hazards. Springer, Berlin Heidelberg New York, pp 221–256 (4)

119 Giggenbach WF, Sano Y, Schmincke H-U (1991) CO_2-rich gases from Lakes Nyos and Monoun, Cameroon; Laacher See, Germany; Dieng, Indonesia, and Mt. Gambier, Australia: variations on a common theme. J Volcanol Geotherm Res 45: 311–323 (4, 11, 13)

120 Gilbert CM (1938) Welded tuff in eastern California. Geol Soc Am Bull 49: 1829–1862 (11)

121 Gill JB (1981) Orogenic andesites and plate tectonics. Springer, Berlin Heidelberg New York, pp 1–390 (8)

122 Glaser R (2001) Klimageschichte Mitteleuropas. Wissenschaftliche Buchgesellschaft, Darmstadt: pp 1–227 (14)

123 Gorshkov GS (1959) Gigantic eruption of the volcano Bezymianny. Bull Volcanol 20: 77–109 (11)

124 Graf H-F, Feichter J, Langmann B (1997) Volcanic degassing: Contribution to global sulphate burden and climate. J Geophys Res 102: 10 727–10 738 (14)

125 Graf HF, Timmrick C (2001) A general climate model simulation of the aerosol radiative effects of the Laacher See eruption (10 900 BC). J Geophys Res 106: 1 474–14 756 (14)

126 Grand SP, Hilst RD van der, Widiyantoro S (1997) Global seismic tomography: A snapshot of convection in the earth. Geol Soc Am Today 7: 1–7 (6)

127 Granet M, Wilson M, Achauer U (1995) Imaging a mantle plume beneath the Massif Central (France). Earth Planet Sci Lett 136: 281–296 (7)

128 Greeley R (1982) The Snake River Plain, Idaho: Representative of a new category of volcanism. J Geophys Res 87: 755–777 (9)

129 Griggs RE (1922) The Valley of Ten Thousand Smokes (Alaska). Washington, Nat Geogr Soc, pp 1–340 (11)

130 Gudmundsson A (1998) Magma chambers modeled as cavities explain the formation of rift zone central volcanoes and their eruption and intrusion statistics. J Geophys Res 103: 7 401–7 412 (9)

131 Gudmundsson MT, Sigmundsson F, Björnsson H (1997) Ice-volcano interaction of the 1996 Gjálp subglacial eruption, Vatnajökull, Iceland. Nature 389: 954–957 (12)

132 Guettard JE (1752) Mémoires sur quelques montagnes de la France qui ont été des volcans. Mém l'Acad R Sci Paris: 27–59 (1)

133 Guterson D (1999) East of the Mountains. Bloomsbury, London, pp 1–279 (7)

134 Gvirtzman Z, Nur A (1999) The formation of Mount Etna as the consequence of slab rollback. Nature 401: 782–785 (2)

135 Gvirtzmann Z, Nur A (2001) Residue to topography, lithospheric structure and sunken slabs in the central Mediterranien. Earth Planet Sci Lett 187: 117–13 (2)

136 Hallam A (1983) Great geological contro versies. Oxford Univ Press, Oxford, pp 1 182 (1)

137 Halmer MM, Schmincke H-U, Graf H- (2002) The annual volcanic gas inpu into the atmosphere, in particular int the stratosphere: global data set for th past 100 years. J Volcanol Geotherm Re 115: 511–528 (14)

138 Hamilton W (1779) Campi Phlegrae Observations on volcanoes of the tw Siciles. As they have been communicate to the Royal Society of London 1776 an Supplement to the Campi Phlegrae being an account of the great eruption (Mount Vesuvius with 50 plates. Naple pp 1–89 (1)

139 Hammer CH, Clausen HB, Dansgaard \ (1980) Greenland ice sheet: evidence (post-glacial volcanism and its climat impact. Nature 288: 230–235 (14)

140 Harms E, Schmincke H-U (1999) Volati composition of the Laacher See phonoli magma (12 900 yr BP): Implications fc syn-eruptive S, F, Cl and H_2O degassin Contrib Mineral Petrol 138: 84–98 (14)

141 Harms E, Gardner JE, Schmincke H-\ (2003) Phase equilibria in the Laache See Tephra (East-Eifel, Germany). Cor straints on pre-eruptive storage condit ons of a phonolitic magma reservoir. Volcanol Geotherm Res (in press)

142 Harlow DH, Power JA, Laguerta E Ambubuyog G, White RA, Hoblitt R (1996) Precursory seismicity and foreca sting of the June 15, 1991, eruption (Mount Pinatubo. In: Newhall C(Punongbayan RS (eds) Fire and Mud Eruptions and Lahars of Mount Pinatu bo, Philippines. PHIVOLCS and Uni Wash Press, Seattle, pp 285–306 (13)

143 Harris DM, Sato M, Casadevall TJ, Ros WI Jr, Bornhorst TJ (1981) Emissio rates of CO_2 from plume measurement In: Lipman PW, Mullineaux DR (ed The 1980 eruption of Mount St. Helen US Geol Surv Prof Pap 1 250, pp 201 207 (10)

144 Hauri EH, Kurz MD (1997) Melt migrat on and mantle chromatography, 2: time-series Os isotope study of Maun Loa volcano, Hawaii. Earth Planet S Lett 153: 21–36 (6)

145 Hauri EH, Whitehead JA, Hart SR (1994) Fluid dynamic and geochemical aspects of entrainment in mantle plumes. J Geophys Res 99: 24 275 – 24 300 (6)

146 Heilprin A (1904) The tower of Pelee. Lippincott, Philadelphia, pp 1 – 59 (11)

147 Herd D, Comité de Estudios Vulcanologicos (1986) The 1985 Ruiz Volcano Disaster. EOS, Trans Am Geophys Union 67 (19): 457 – 460 (13)

148 Hervig RL, Dunbar N, Westrich HR, Kyle P (1989) Pre-eruptive water content of rhyolitic magmas as determined by ion microprobe analyses of melt inclusions in phenocrysts. J Volcanol Geotherm Res 36: 293 – 302 (4)

149 Herzog M, Graf H-F, Textor C, Oberhuber JM (1998) The effect of phase changes of water on the development of volcanic plumes. J Volcanol Geotherm Res 270: 1 – 31 (10)

150 Hildreth W (1981) Gradients in silicic magma chambers: implications for lithospheric magmatism. J Geophys Res 86: 10 153 – 10 192 (3, 8, 11)

151 Hildreth W (1983) The compositionally zoned eruption of 1912 in the Valley of Ten Thousand Smokes, Katmai National Park, Alaska. J Volcanol Geotherm Res 18: 1 – 56 (11)

152 Hildreth W, Christiansen RL, O'Neil JR (1984) Catastrophic isotopic modification of rhyolitic magma at times of caldera subsidence, Yellowstone Plateau volcanic field. J Geophys Res 89: 8 339 – 8 369 (4)

153 Hildreth W, Fierstein J (2000) Katmai volcanic cluster and the great eruption of 1912. Geol Soc Am Bull 112: 1 594 – 1 620 (10)

154 Hildreth W, Moorbath S (1988) Crustal contributions to arc magmatism in the Andes of Central Chile. Contrib Mineral Petrol 98: 455-489

155 Hill B, Connor CB, Jarzemba MS, La Femina PC, Navarro M, Strauch W (1998) 1995 eruptions of Cerro Negro volcano, Nicaragua, and risk assessment for future eruptions. Geol Soc Am Bull 10: 1 231 – 1 241 (7)

156 Hill DP, Pollitz F, Newhall C (2002) Earthquake-volcano interaction. Physics Today Nov 2002: 41 – 47 (4)

157 Hoernle KA, Schmincke H-U (1993) The role of partial melting in the 15-Ma geochemical evolution of Gran Canaria: A blob model for the Canary Hotspot. J Petrol 34: 599 – 627 (6)

158 Hoernle KA, Zhang YS, Graham D (1995) Seismic and geochemical evidence for large-scale mantle upwelling beneath the eastern Atlantic and western and central Europe. Nature 374: 34 – 39 (6)

159 Hofman DJ, Oltmans SJ, Harris JM, Solomon S, Deshler T, Johnson BJ (1992) Observation and possible causes of new ozone depletion in Antarctica in 1991. Nature 359: 283 – 287 (14)

160 Hofmann AW (1997) Mantle geochemistry: the message from oceanic volcanism. Nature 385: 219 – 229 (6)

161 Holmes A (1931) Radioactivity and the Earth's thermal history. Geol Mag 62: 102 – 112 (2)

162 Hulme G (1974) The interpretation of lava flow morphology. Geophys J Royal Astron Soc 39: 361 – 383 (9)

163 Humboldt A von (1823) Über den Bau und die Wirksamkeit der Vulkane in verschiedenen Erdstrichen. In: Kosmos, Stuttgart 1847, Leonhards Taschenb Min 18: 1 – 24 (2, 12)

164 Hutton J (1785) Theory of the Earth. Abstract of a dissertation. Roy Soc Edinburgh: 1 – 30. Reprint Scottish Academic Press, Edinburgh 1987 (1)

165 Hutton J (1795) Theory of the earth with proofs and illustrations. 2 vols. WM Creech, Edinburgh. Reprinted 1960, Wheldon and Wesley, London (1)

166 Iverson RM (1997) The physics of debris flows. Rev Geophys 35: 245 – 296 (13)

167 Jull M, McKenzie D (1996) The effect of deglaciation on mantle melting beneath Iceland. J Geophys Res 101: 21 815 – 21 828 (14)

168 Kaiser KF (1993) Klimageschichte vom späten Hochglazial bis ins frühe Holozän, rekonstruiert mit Jahrringen und Molluskenschalen aus verschiedenen Vereisungsgebieten. Eidgen Forsch Wald Schnee Land, Zürich, pp 1 – 203 (14)

169 Karato S, Jung H (1998) Water, partial melting and the origin of the seismic low velocity and high attenuation zone in the upper mantle. Earth Planet Sci Lett 157: 193 – 207 (2)

170 Kellogg LH, Hager BH, van der Hilst RD (1999) Compositional stratification in the deep mantle. Science 283: 1 881 – 1 884 (2)

171 Kennett JP (1981) Marine tephrochronology. In: Emiliani C (ed) The Sea 7. Wiley, New York, pp 1 313 – 1 416 (14)

172 Kieffer SW (1977) Sound speed in liquid-gas mixtures: water-air and water-steam. J Geophys Res 82: 2 895 – 2 904 (10)

173 Kieffer SW (1981) Blast dynamics at Mount St. Helens on 18 May 1980. Nature 291: 568 – 570 (10, 12)

174 Kienle J, Kyle PR, Self S, Motyka RJ, Lorenz V (1980) Ukinrek Maars, Alaska, I. April 1977 eruption sequence, petrology and tectonic setting. J Volcanol Geotherm Res 7: 11 – 37 (12)

175 Kilburn CRJ (1981) Pahoehoe and aa lavas: a discussion and continuation of the model of Peterson and Tilling. J Volcanol Geotherm Res 11: 373 – 382 (4)

176 Kircher A (1664) Mundus Subterraneus in XII Libros Digestus; quo Divinum Subterrestris Mundi Opificium, 2 vols Amsterdam, pp 1 – 194 (1)

177 Kling GW and 9 co-authors (1987) The 1986 Lake Nyos gas disaster in Cameroon, West Africa. Science 236: 169 – 175 (13)

178 Kokelaar BP (1983) The mechanism of Surtseyan volcanism. J Geol Soc Lond 140: 939 – 944 (12)

179 Koyaguchi T, Woods AW (1996) On the formation of eruption columns following explosive mixing of magma and surface-water. J Geophys Res 101: 5 561 – 5 574 (12)

180 Koyanagi RY, Swanson DA, Endo ET (1972) Distribution of earthquakes related to the mobility of the south flank of Kilauea Volcano, Hawaii. US Geol Surv Prof Pap 800 D: 89 – 97 (6)

181 Krafft M (1993) Volcanoes: Fire from the Earth. Harry N Abrams, pp 1 – 223 (1)

182 Krastel S, Schmincke H-U, Jacobs CL, Rihm R, LeBas TT, Alibés B (1999) Submarine landslides around the Canary Islands. J Geophys Res 106: 3 977 – 3 998 (6)

183 Krueger AJ, Walter LS, Bhartia PK, Schnetzler CC, Krotkov NA, Sprod I, Bluth GJS (1995) Volcanic sulfur dioxide measurements from the total ozone mapping spectrometer instruments. J Geophys Res 100: 14 057 – 14 076 (14)

184 Kuno H (1966) Lateral variation of basalt magma across continental margins and island arcs. Bull Volcanol 29: 195 – 222 (8)

185 Kushiro I (1983) On the lateral variations in chemical composition and volume of Quaternary volcanic rocks across Japanese arcs. J Volcanol Geothermal Res 18: 435–447 (8)

186 Kushiro I (1990) Partial melting of mantle wedge and evolution of island arc crust. J Geophys Res 95: 15 929–15 939 (8)

187 Kyser TK, O'Neil JR (1984) Hydrogen isotope systematics of submarine basalts. Geochim Cosmochim Acta 48: 2 123–2 133 (4)

188 Lacroix A (1904) La Montagne Pelée et ses eruptions. Masson et Cie, Paris, pp 1–662 (11)

189 Lamb HH (1970) Volcanic dust in the atmosphere, with a chronology and assessment of its meteorological significance. Philos Trans R Soc Lond A 266: 425–533 (14)

190 Larson RL (1997) Superplumes and ridge interactions between Ontong Java and Manihiki Plateaus and the Nova-Canton trough. Geology 25: 779–782 (6)

191 LeGuern F, Tazieff H, Faivre-Pierret R (1982) An example of health hazard: people killed by gas during a phreatic eruption: Dieng Plateau (Java, Indonesia), February 20th, 1979. Bull Volcanol 45: 153–156 (13)

192 Linde AT, Sacks IS (1998) Triggering of volcanic eruptions. Nature 395: 88–890 (4)

193 Lipman PW (1984) The roots of ash flow calderas in western North America: windows into the tops of granitic batholiths. J Geophys Res 89: 8 801–8 841 (9)

194 Lipman PW (1997) Subsidence of ash-flow calderas: relation to caldera size and magma-chamber geometry. Bull Volcanol 59: 198–218 (9)

195 Lipman PW, Mullineaux DR (eds) (1981) The 1980 eruption of Mount St. Helens. US Geol Surv Prof Pap 1250: 1–844 (4, 10)

196 Lister CRB (1980) Heat flow and hydrothermal circulation. Ann Rev Earth Planet Sci 8: 95–117 (15)

197 Litt T, Schmincke H-U, Kromer B (2003) Environmental response to climatic and volcanic events in central Europe during the Weichselian Late glacial. Quat Sci Rev 22: 7–32 (11)

198 Lockwood JP, Koyanagi RY, Tilling RI, Holcomb RT, Peterson DW (1976) Mauna Loa threatening. Geotimes 21:12–15 (13)

199 Lockwood JP and 9 co-authors (1985) The 1984 eruption of Mauna Loa Volcano, Hawaii. EOS 66: 169–171 (13)

200 Lorenz V (1974) On the formation of maars. Bull Volcanol 37: 183–204 (12)

201 Luhr JF, Simkin T (eds) (1993) Paricutin, the volcano born in a Mexican cornfield. Geoscience Press, Phoenix, pp 1–427 (9)

202 Luhr JF, Carmichael ISE, Varekamp JC (1984) The 1982 eruptions of El Chichón Volcano, Chiapas, Mexico: mineralogy and petrology of the anhydrite-bearing pumices. J Volcanol Geotherm Res 23: 69–108 (14)

203 Macdonald GA (1972) Volcanoes. Prentice-Hall, Englewood Cliffs, pp 1–510 (1)

204 Macdonald GA, Katsura T (1964) Chemical compositions of Hawaiian lavas. J Petrol 5: 82–133 (6)

205 Macdonald KC (1982) Mid-ocean ridges: Fine-scale tectonic, volcanic and hydrothermal processes within the plate boundary zone. Annu Rev Earth Planet Sci 10: 155–190 (5)

206 Macdonald KC, Becker K, Spiess FN, Ballard RD (1980) Hydrothermal heat flux of the 'black smoker' vents on the East Pacific Rise. Earth Planet Sci Lett 48: 1–7 (15)

207 Major JJ, Newhall CG (1989) Snow and ice perturbation during historical volcanic eruptions and the formation of lahars. A global review. Bull Volcanol 52: 1–27 (13)

208 Marsh BD (1996) Solidification fronts and magmatic evolution. The 1995 Hallimond Lecture. Mineral Mag 60: 5–40 (3)

209 Marshall P (1935) Acid rocks of the Taupo-Rotorua volcanic district. Trans R Soc N Z 64: 323–366 (11)

210 Marti J, Mitjavila J, Araña V (1994) Stratigraphy, structure and geochronology of the Las Cañadas caldera (Tenerife, Canary Islands). Geol Mag 131: 715–727 (9)

211 Masson DG (1996) Catastrophic collapse of the volcanic island of Hierro 15 Ka ago and the history of landslides in the Canary Islands. Geology 24: 231–234 (6)

212 Mastin LG, Witter JB (2000) The hazards of eruptions through lakes and seawater. J Volcanol Geotherm Res 97: 197–214 (12)

213 McCormick PM, Thomason LW, Trepte CR (1995) Atmospheric effects of the Mt. Pinatubo eruption. Nature 373: 399–404 (14)

214 McDougall I (1964) Potassium-argon ages from lavas of the Hawaiian Islands. Geol Soc Am Bull 75: 107–128 (6)

215 McGetchin TR, Settle M, Chouet BA (1974) Cinder cone growth modeled after northeast crater, Mount Etna, Sicily. J Geophys Res 79: 3 257–3 272 (9)

216 McGuire WJ, Kilburn CRJ, Murray JB (1995) Monitoring active volcanoes: strategies, procedures and techniques. Univ College London Press, London, pp 1–420 (1, 14)

217 McLean DM (1988) K-T transition into chaos. J Geol Ed 36: 237–243 (14)

218 McNutt SR (1996) Seismic monitoring and eruption forecasting of volcanoes: a review of the state-of-the-art and case histories. In: Scarpa R, Tilling RI (eds) Monitoring and mitigation of volcano hazards. Springer, Berlin Heidelberg New York, pp 99–146 (13)

219 Mercalli G (1907) I volcani attivi della Terra. Ulrico Hoepli, Milano, pp 1– 421 (4)

220 Mertes H, Schmincke H-U (1985) Petrology of potassic mafic magmas of the Westeifel volcanic field. Major and trace elements. Contrib Mineral Petrol 89: 330–345 (7)

221 Merzbacher C, Eggler DH (1984) A magmatic geohygrometer: application to Mount St. Helens and other dacitic magmas. Geology 12: 587–590 (4)

222 Metrich N, Clocchiati R (1989) Melt inclusion investigation of the volatile behaviour in historic alkali basaltic magmas of Etna. Bull Volcanol 51: 185–198 (14)

223 Metrich N, Sigurdsson H, Meyer PS, Devine J (1991) The 1783 Lakagigar eruption in Iceland: geochemistry, CO_2 and sulfur degassing. Contrib Mineral Petrol 107: 435–447 (14)

224 Minakami T (1974) Seismology of volcanoes in Japan. In: Civetta L, Gasparini P, Luongo G, Rapolla A (eds) Physical volcanology. Developments in solid earth geophysics 6. Elsevier, Amsterdam, pp 1–27 (13)

225 Miura D, Niida K (2002) Two-stage growth model of cryptodome by shallow intrusions, the 2000 eruption of Usu volcano, northern Japan. In Japanese with English abstract. Bull Volcanol Soc Jpn 47: 119–130 (4)

226 Moore GW, Moore JG (1988) Large-scale bedforms in boulder gravel produced by giant waves in Hawaii. Geol Soc Am Spec Pap 229: 101–109 (13)

227 Moore JG (1975) Mechanism of formation of pillows. Am Sci 63: 269–277 (12)

228 Moore JG (1979) Vesicularity and CO_2 in mid-ocean ridge basalts. Nature 282: 250–253 (4)

229 Moore JG (1985) Structure and mechanism at Surtsey volcano, Iceland. Geol Mag 122: 649–661 (12)

230 Moore JG (1987) Subsidence of the Hawaiian Ridge. US Geol Surv Prof Pap 1350: 85–100 (6)

231 Moore JG, Fiske RS (1969) Volcanic substructure inferred from dredge samples and ocean-bottom photographs, Hawaii. Geol Soc Am Bull 80: 1 191–1 202 (6)

232 Moore JG, Melson WG (1969) Nuées ardentes of the 1968 eruption of Mayon Volcano, Philippines. Bull Volcanol 33: 600–620 (13)

233 Moore JG, Schilling JG (1973) Vesicles, water, and sulfur in Reykjanes Ridge basalts. Contrib Mineral Petrol 41: 105–118 (4, 12)

234 Moore JG, Albee WC (1981) Topographic and structural changes, March-July 1980 – photogrammetric data. US Geol Surv Prof Pap 1250: 123–134 (10)

235 Moore JG, Clague DA (1992) Volcano growth and evolution of the island of Hawaii. Geol Soc Am Bull 104: 1 471–1 484 (6)

236 Moore JG, Nakamura K, Alcaraz A (1966) The 1965 eruption of Taal Volcano. Science 151: 955–960 (11)

237 Moore JG, Phillips RL, Grigg RW, Peterson DW, Swanson DA (1973) Flow of lava into the sea 1969–1971, Kilauea Volcano, Hawaii. Geol Soc Am Bull 84: 537–546 (6)

238 Moore JG, Clague DA, Holcomb RT, Lipman PW, Normark WR, Torresan ME (1989) Prodigious submarine landslides on the Hawaiian Ridge. J Geophys Res 94: 17 465–17 484 (6)

239 Moore JG, Normark WR, Holcomb RT (1994) Giant Hawaiian landslides. Annu Rev Earth Planet Sci 22: 119–144 (6, 9, 13)

240 Moore JG, Bryan WB, Beeson, MH (1995) Giant blocks in the South Kona Landslide, Hawaii. Geology 23: 125–128 (6)

241 Morgan WJ (1972) Plate motions and deep mantle convection. Geol Soc Am Mem 132: 7–22 (2, 6)

242 Morrissey MM, Mastin LG (1999) Vulcanian eruptions. In: Sigurdsson H et al. (eds) Encyclopedia of Volcanology, Academic Press, San Diego ,pp 463–475 (12)

243 Muenow DW, Garcia MO, Aggrey KE, Bednarz U, Schmincke H-U (1990) Volatiles in submarine glasses as a discriminant of tectonic origin: application to the Troodos ophiolite. Nature 343: 159–161 (4)

244 Muffler LJP, White DE, Truesdell AH (1971) Hydrothermal explosion craters in Yellowstone National Park. Geol Soc Am Bull 82: 723–740 (12)

245 Mutch TA, Arvidson RE, Head JW III, Jones KL, Saunders RS (1976) The geology of Mars. Princeton Univ Press, Princeton, pp 1–400 (6)

N

246 Nairn IA, Wiradiradja S (1980) Late Quaternary hydrothermal explosions at Kawerau geothermal field, New Zealand. Bull Volcanol 43: 1–13 (12)

247 Nakada S, Fujii T (1993) Preliminary report on the activity at Unzen volcano (Japan), November 1990-November 1991: dacite lava domes and pyroclastic flows. J Volcanol Geotherm Res 54: 319–333 (11)

248 Neri A, Dobran F (1994) Influence of eruption parameters on the thermofluid dynamics of collapsing volcanic columns. J Geophys Res 99: 11 833–11 857 (11)

249 Neumann van Padang M (1931) Der Ausbruch des Merapi (Mittel Java) im Jahre 1930. Z Vulkanol 14: 135–148 (11)

250 Newhall CH, Self S (1982) The volcanic explosivity index (VEI): an estimate of explosive magnitude for historical volcanism. J Geophys Res 87: 1 231–1 238 (13)

251 Newhall CH, Punongbayan RS (1996) Fire and mud: eruptions and lahars of Mount Pinatubo, Philippines. PHIVOLCS and Univ Wash Press, Seattle, pp 1–1 126 (8, 13)

O

252 Oskarsson N (1980) The interaction of volcanic gases and tephra: fluorine adhering to tephra of the 1970 Hekla eruption. J Volcanol Geotherm Res 8: 251–266 (4)

P

253 Palais JM, Sigurdsson H (1989) Petrologic evidence of volatile emissions from major historic and pre-historic volcanic

eruptions. In: Berger A, Dickinson RE, Kidson JW (eds) Understanding climate change. Am Geophys Union Geophys Monogr 52: 31–53 (4)

254 Panza GF, Müller ST, Calcagnile G (1980) The stress features of the lithosphere-asthenosphere system in Europe from seismic surface waves and body waves. Pure Appl Geophys 118: 1 209–1 213 (7)

255 Park C, Schmincke H-U (1997) Lake formation and catastrophic dam burst during the late Pleistocene Laacher See eruption (Germany). Naturwissenschaften 84: 521–525 (11, 13)

256 Parra E, Cepeda H (1990) Volcanic hazard maps of the Nevado del Ruiz volcano, Colombia. J Volcanol Geotherm Res 42: 117–127 (13)

257 Pearce JA, Lippert SJ Roberts S (1984) Characteristics and tectonic significance of supra-subduction zone ophiolites. In: Kokelaar BP, Howells MF (eds) Marginal basin geology. Geol Soc Spec Publ. Blackwell, Oxford, pp 77–94 (5)

258 Peck D, Wright TL, Moore JG (1966) Crystallization of tholeiitic basalt in Alae lava lake, Hawaii. Bull Volcanol 29: 629–656 (3)

259 Peck D, Wright TL, Decker RW (1979) The lava lakes of Kilauea. Sci Am 241: 114–128 (3)

260 Perret FA (1937) The eruption of Mt. Pelée: 1929–1932. Carnegie Inst Wash Publ 458: 1–125 (11)

261 Peterson DW, Tilling RI (1980) Transition of basaltic lava from pahoehoe to aa, Kilauea Volcano, Hawaii: field observation and new factors. J Volcanol Geotherm Res 7: 271–293 (4, 9)

262 Phillips WJ (1974) The dynamic emplacement of cone sheets. Tectonophysics 24: 69–84 (9)

263 Phipps Morgan J, Morgan WJ, Price E (1995) Hotspot melting generates both hotspot volcanism and a hotspot swell? J Geophys Res 100: 8 045–8 062 (6)

264 Pierson TC, Janda RJ, Umbal JV, Daag AS (1992) Immediate and long-term hazards from lahars and excess sedimentation in rivers draining Mt. Pinatubo, Philippines. US Geol Surv, Water-Resources Invest Rep 9–4 039, pp 1–35 (13)

265 Pierson TC, Daag AS, Delos Reyes PJ, Regalado Ma TM, Solidum RU, Tubianosa BS (1996) Flow and deposition of posteruption hot lahars on the east side of Mount Pinatubo, July-October 1991.

In: Newhall CG, Punongbayan RS (eds) Fire and Mud – Eruptions and Lahars of Mount Pinatubo, Philippines. PHI-VOLCS and Univ Wash Press, Seattle, pp 921–950 (13)

266 Pope KO (2002) Impact dust not the cause of the Cretaceous-Tertiary mass extinction. Geology 30: 99–102 (14)

267 Pope KO, Baines KH, Ocampo AC, Ivanov BA (1997) Energy, volatile production, and climatic effects of the Chicxolub Cretaceous/Tertiary impact. J Geophys Res 102: 21 645 – 21 664 (14)

268 Porter SC (1972) Distribution, morphology, and size frequency of cinder cones on Mauna Kea volcano, Hawaii. Geol Soc Am Bull 83: 3 607–3 612 (9)

269 Press F, Siever R (1982) Earth, 3rd edn. Freeman, San Francisco, pp 1–613 (1, 3)

270 Proussevitch AA, Sahagian DL (1996) Dynamics of coupled diffusive and decompressive bubble growth in magmatic systems. J Geophys Res 101: 17 447–17 455 (4)

271 Proussevitch AA, Sahagian DL (1998) Dynamics and energetics of bubble growth in magmas: analytical formulation and numerical modeling. J Geophys Res 103: 223–251 (10)

272 Przedpelski Z, Casadevall TJ (1994) Impact of volcanic ash from 15 December 1989 Redoubt volcano eruption on GE CF6–80C2 turbofan engines. US Geol Surv Bull 2047: 129–135 (13)

273 Pyle DM, Beattie PD, Bluth GJS (1996) Sulphur emissions to the stratosphere from explosive volcanic eruptions. Bull Volcanol 57: 663–671 (14)

274 Raikes S, Bonjer KP (1983) Large-scale mantle heterogeneity beneath the Rhenish Massif and its vicinity from teleseismic residuals measurements. In: Fuchs K et al. (eds) Plateau Uplift – The Rhenish Shield – A Case History. Springer, Berlin Heidelberg New York, pp 315–331 (6, 7)

275 Raitt RW (1963) The crustal rocks. In: Hill MN (ed) The Sea, vol 3. Wiley, New York, pp 85–102 (5)

276 Rampino MR (1991) Volcanism, climatic change, and the geological record. Sedimentation in Volcanic Settings. Soc Sediment Geol Spec Publ 45: 9–18 (14)

277 Rampino MR, Self S, Fairbridge RW (1979) Can rapid climatic change cause volcanic eruptions? Science 206: 826–829 (14)

278 Reid RG (1976) Superheated liquids. Am Sci 64: 146–156 (12)

279 Ribe NM, Christensen UR (1994) Three-dimensional modeling of plume-lithosphere interaction. J Geophys Res 99: 669–682 (6)

280 Ritter JR, Jordan M, Christensen U, Achauer U (2001) A mantle plume below the Eifel volcanic fields, Germany. Earth Planet Sci Lett 186: 7–14 (7)

281 Robock, A (2002) Pinatubo Eruption: The climatic aftermath. Science 295: 1 242–1 244 (14)

282 Rogers R, Ka´rason H, van der Hilst R (2002) Epeirogenic uplift above a detached slab in northern central America. Geology 30: 1 031–1 034 (8)

283 Romanowicz B, Gung Y (2002) Superplumes from the core-mantle boundary to the lithosphere: implications for heat flux. Science 296: 513–515

284 Rona PA (1984) Hydrothermal mineralization at seafloor spreading centers. Earth Sci Rev 20: 1–104 (15)

285 Ross CS, Smith RL (1961) Ash-flow tuffs: their origin, geologic relations and identification. US Geol Surv Prof Pap 366: 1–77 (11)

286 Rüpke L, Phipps Morgan J, Hort M, Connolly JAD (2002) Are the regional variations in Central America arc lavas due to differing basaltic versus peridotitic slab sources of fluids? Geology 30: 1 035–1 038 (8)

287 Rutherford MJ, Sigurdsson H, Carey S (1985) The May 18, 1980 eruption of Mount St. Helens. I. Melt composition and experimental phase equilibria. J Geophys Res 90: 2929–2947 (10)

288 Ryan MP (1987) Neutral buoyancy and the mechanical evolution of magmatic systems. In: Mysen BO (ed) Magmatic processes: Physicochemical principles. Geochem Soc Spec Publ 1, Univ Park, PA, pp 259–287 (6)

289 Ryan MPR, Sammis CHA (1981) The glass transition in basalt. J Geophys Res 86: 9 519–9 535 (12)

290 Saito G, Uto K, Kazahaya K, Satoh H, Kawanabe Y, Shinohara H (2005) Petrological characteristic and volatile contents of magma of August 18, 2000 eruption of Miyakejima volcano: magma ascent, phreatomagmatic eruption and magma degassing. Bull Volcanol 67: 268–280 (4)

291 Sapper K (1927) Vulkankunde. J Engelhorn's Nachf, Stuttgart, pp 1–424 (4)

292 Sarna-Wojcicki AM, Meyer CE, Woodward MJ, Lamothe PJ (1981) Areal distribution, thickness, mass, volume and grain size of air-fall ash from the six major eruptions of 1980. In: Lipman PW, Mullineaux DR (eds) The 1980 eruptions of Mount St. Helens, Washington. US Geol Surv Prof Pap 1250, pp 577–600 (10)

293 Sasaki M, Fujimoto K, Sawaki T, Tsukamoto H, Kato O, Komatsu R, Doi N (2003) Petrographic features of a high-temperature granite just newly solidified magma at the Kakkonda geothermal field, Japan. J Volcanol Geotherm Res 121: 247–269

294 Sato H, Fujii T, Nakada S (1992) Crumbling of dacite dome lava and generation of pyroclastic flows at Unzen volcano. Nature 360: 664–666 (11)

295 Sato H (2002) Comparative petrological study of the 864 effusive and 1707 AD explosive eruption of Mt. Fuji volcano, Japan (abstract). Mt. Pelee IAVCEI symposium Martinique

296 Sawada Y (1996) Detection of explosive eruptions and regional tracking of volcanic ash clouds with geostationary meteorological satellite (GUS) In: Scarpa R, Tilling RI (eds) Monitoring and mitigation of volcano hazards. Springer, Berlin Heidelberg New York, pp 299–314 (13)

297 Scandone R, Malone SD (1985) Magma supply, magma discharge and readjustment of the feeding system of Mount St. Helens during 1980. J Volcanol Geotherm Res 23: 239–262 (10)

298 Scarpa R, Tilling RI (1996) Monitoring and mitigation of volcano hazards. Springer, Berlin Heidelberg New York, pp 1–841 (1, 13)

299 Schirnick C, Bogaard P van den, Schmincke H-U (1999) Cone sheet formation and intrusive growth of an oceanic island – The Miocene Tejeda complex on Gran Canaria (Canary Islands). Geology 27: 207–210 (9)

300 Schmidt R, Schmincke H-U (1999) Seamounts and island building. In: Sigurdsson H et al. (eds) Encyclopedia of Volcanoes. Academic Press, San Diego, pp 383–402 (6)

301 Schmincke H-U (1967) Flow directions in Columbia River Basalt flows and paleocurrents of interbedded sedimenta-

ry rocks, south-central Washington. Geol Rundsch 56: 992–1020 (9)

302 Schmincke H-U (1977) Phreatomagmatische Phasen in quartären Vulkanen der Osteifel. Geol Jahrb 39: 3–45 (9, 12)

303 Schmincke H-U (1982a) Vulkane und ihre Wurzeln. Rhein Westf Akad Wiss Westd Verlag Opladen, Vorträge N315: 35–78 (1, 7)

304 Schmincke H-U (1982b) Volcanic and chemical evolution of the Canary Islands. In: von Rad U, Hinz K, Sarnthein M, Seibold E (eds) Geology of the Northwest African Continental Margin. Springer, Berlin Heidelberg New York, pp 273–308 (6)

305 Schmincke H-U (1998) Geological field guide Gran Canaria. Pluto Press, Ascheberg, pp 1–270 (6)

306 Schmincke H-U (2002) Tanz auf dem Vulkan. Goethe und die Entfaltung der Vulkanologie als Wissenschaft. Grafik M Sumita. In: Kossatz A, Steininger I (eds) "Senckenberg Kleine Reihe". Schweizerbart'sche Verlagsbuchhandlung, Stuttgart, pp 119–176 (1)

307 Schmincke H-U, Bednarz U (1990) Pillow-, sheet flow- and breccia-volcanoes and volcano-tectonic-hydrothermal cycles in the Extrusive Series of the northwestern Troodos Ophiolite (Cyprus). In: Malpas J, Moores EM, Panayiotou A, Xenophontas C (eds) Symposium Troodos 87 – Ophiolites and Oceanic Lithosphere. Nicosia, The Geological Survey Dept, pp 185–207 (5)

308 Schmincke H-U, Sumita M (1998) Volcanic evolution of Gran Canaria reconstructed from apron sediments: Synthesis of VICAP project drilling (ODP Leg 157). In: Weaver PPE, Schmincke H-U, Firth JV, Duffield WA (eds) Proc ODP, Sci Results, 157: College Station, TX (Ocean Drilling Program), pp 443–469 (6, 13)

309 Schmincke H-U, Fisher RV, Waters AC (1973) Antidune and chute and pool structures in the base surge deposits of the Laacher See area, Germany. Sedimentology 20: 553–574 (11, 12)

310 Schmincke H-U, Bogaard P van den, Freundt A (1990) Quaternary Eifel Volcanism. Excursion guide, workshop in explosive volcanism, IAVCEI Intern Volcanol Congr Mainz (FRG). Pluto Press, Witten, pp 1–188 (11, 14)

311 Schmincke H-U, Klügel A, Hansteen TH, Hoernle K, Bogaard P van den (1998) Samples from the Jurassic ocean crust beneath Gran Canaria, La Palma and Lanzarote (Canary Islands). Earth Planet Sci Lett 163: 343–360 (6)

312 Schmincke H-U, Graf G, (1999a) Meteor – DECOS 7 OMEX II Cruise No. 43. Inst Meeresk Hamburg , pp 1–78 (6, 11)

313 Schmincke H-U, Park C, Harms E (1999) Evolution and environmental impacts of the eruption of Laacher See volcano (Germany) 12 900 a BP. Quat Int 61: 61–72

314 Schumacher R, Schmincke H-U (1990) The lateral facies of ignimbrites at Laacher See volcano. Bull Volcanol 52: 271–285 (11)

315 Schumacher R, Schmincke H-U (1995) Models for the origin of accretionary lapilli. Bull Volcanol 56: 626–639 (11)

316 Schwarzkopf L, Schmincke H-U (2000) Die Vulkanwelt Indonesiens. Geogr Rdsch 52: 4-12 (11)

317 Schwarzkopf L, Schmincke H-U, Cronin S (2003) A conceptual model for block-and-ash flow basal avalanche transport and deposition, based on deposit architecture of 1998 and 1994 Merapi flows. J Volcanol Geotherm Res (in press)

318 Scott SD (1983) Basalt and sedimentary-hosted seafloor polymetallic sulfide deposits and their ancient analogues. MTS-IEEE conference 'oceans 83', San Francisco, pp 1–6 (15)

319 Scrope GP (1825) Considerations on volcanoes. The probable causes of their phenomena, the laws which determine their march, the disposition of their products, and their connexion with the present state and past history of the globe; leading to the establishment of a new theory of the Earth. W Phillips, London, pp 1–270 (4)

320 Seck HA, Wedepohl KH (1983) Mantle xenoliths in the Rhenish Massif and the northern Hessian Depression. In: Fuchs K et al. (eds) Plateau Uplift. The Rhenish shield – a case history. Springer, Berlin Heidelberg New York, pp 343–351 (7)

321 Seifert F (1985) Struktur und Eigenschaften magmatischer Schmelzen. Rhein Westf Akad Wiss, Westd Verlag, Opladen, Vorträge N 341: 31–64 (4)

322 Seyfried R, Freundt A (1999) Analog experiments on conduit flow, eruption behavior, and tremor of basaltic volcanic eruptions. J Geophys Res 105: 23 727–23 740 (10)

323 Shelly M W (1831) Frankenstein, or the modern Prometheus. Colburn and Bentley, London. New edition: Shelley M, Hindle M (1992) Frankenstein. Penguin Books, pp 1–320

324 Shepherd JB, Sigurdsson H (1982) Mechanism of the 1979 explosive eruption of Soufrière Volcano, St. Vincent. J Volcanol Geotherm Res 13: 119–130 (12)

325 Sheridan MF, Wohletz KH (1983) Hydrovolcanism: basic considerations and review. J Volcanol Geotherm Res 17: 1–29 (12)

326 Shimozuru D, Kubo N (1983) Volcano spacing and subduction. In: Shimozuru D, Yokoyama I (eds) Arc Volcanism: Physics and Tectonics. Reidel, Dordrecht, pp 141–152 (8)

327 Shinohara H, Fujui K, Kazahaya K, Saito G (2003) Degassing process of Miyakejima volcano: implications of gas emission rate and melt inclusion data. In: Bodnar RJ, de Vivo B (eds) Melt inclusions involcanic systems: Methods, applications and problems. Developm Volcanol Elsevier, Amsterdam, pp 147-161 (4)

328 Siebert L (1984) Large volcanic debris avalanches: characteristics of source areas, deposits, and associated eruptions. J Volcanol Geotherm Res 22: 163–197 (9)

329 Siebert L (1996) Hazards of large volcanic debris avalanches and associated eruptive phenomena. In: Scarpa R, Tilling RI (eds) (1996) Monitoring and mitigation of volcano hazards. Springer, Berlin Heidelberg New York, pp 1–20 (9)

330 Siebert L, Glicken H, Ui T (1987) Volcanic hazards from Bezymianny- and Bandai-type eruptions. Bull Volcanol 49: 435–459 (13)

331 Sigurdsson H (1982) Volcanic pollution and climate: The 1783 Laki Eruption. EOS 63: 601–602 (14)

332 Sigurdsson H (1989) Evidence of volcanic loading of the atmosphere and climate response. Palaegeogr Palaeoclim Palaeoecol 89: 277–289 (14)

333 Sigurdsson H (1999) Melting the Earth. The history of ideas of volcanic eruptions. Oxford Univ Press, Oxford, pp 1–260 (1)

334 Sigurdsson H, Carey SN (1989) Plinian and co-ignimbrite tephra fall from the 1815 eruption of Tambora Volcano. Bull Volcanol 51: 243–270 (13, 14)

335 Sigurdsson H, Carey SN, Palais JM, Devine J (1990) Pre-eruption compositional gra-

dients and mixing of andesite and dacite magma erupted from Nevado del Ruiz volcano, Colombia in 1985. J Volcanol Geotherm Res 41: 127–151 (14)

336 Sigurdsson H, Houghton BF, McNutt S, Rymer H, Stix J (2000) Encyclopedia of Volcanology. Academic Press, San Diego, pp 1–1417 (1)

337 Simkin T, Fiske RS (1983) Krakatau 1883: The volcanic eruption and its effects. Smithsonian Inst Press, Washington, DC, pp 1–464 (13, 14)

338 Simkin T, Siebert L (1984) Explosive eruptions in space and time: durations, intervals, and a comparison of the world's active volcanic belts. In: Explosive volcanism: inception, evolution, and hazards. National Res Council Washington DC: 110–121 (2, 13)

339 Simkin T, Siebert L (1994) Volcanoes of the World. 2nd edn. Geoscience Press, Missoula, pp 1–368 (8)

340 Simkin T, Siebert L, McClelland L, Bridge D, Newhall C, Latter JH (1981) Volcanoes of the world. Smithsonian Inst Hutchinson Ross Publ, Stroudsberg, pp 1–232 (9, 13, 14)

341 Simkin T, Unger JD, Tilling RI, Vogt PR, Spall H (1994) This dynamic planet: World map of volcanoes, earthquakes, impact craters, and plate tectonics. US Geological Surv, Reston (2)

342 Sinton JM, Detrick RS (1992) Mid-ocean ridge magma chambers. J Geophys Res 97: 197–216 (5)

343 Sleep NH (1997) Lateral flow and ponding of starting plume material. J Geophys Res 102: 10001–10012 (6)

344 Smith DK, Cann JR (1992) The role of seamount volcanoes in crustal construction at the Mid-Atlantic Ridge. J Geophys Res 97: 1645–1658 (6)

345 Smith RB, Braile LW (1984) Crustal structure and evolution of an explosive silicic volcanic system at Yellowstone National Park. In: Explosive volcanism: inception, evolution, and hazards. National Res Council, Washington DC, pp 96–109 (7)

346 Smith RL (1960) Ash flows. Geol Soc Am Bull 71: 795–842 (11)

347 Smith RL (1979) Ash-flow magmatism. In: Chapin CE, Elston WE (eds) Ash-flow tuffs. Geol Soc Am Spec Pap 180: pp 5–27 (11)

348 Smith RL, Bailey RA (1968) Resurgent cauldrons. Geol Soc Am Mem 116: 613–662 (11)

349 Sobolev AV, Hofmann AW, Nikogosian IK (2000) Recycled oceanic crust observed in "ghost plagioclase" within the source of Mauna Loa lavas. Nature 404: 986–990 (5)

350 Sontag S (1993) The volcano lover. Doubleday, New York, pp 1–419 (1,15)

351 Sparks RSJ (1978) The dynamics of bubble formation and growth in magmas: A review and analysis. J Volcanol Geotherm Res 3: 1–37 (10)

352 Sparks RSJ (1986) The dimensions and dynamics of volcanic eruption columns. Bull Volcanol 48: 3-15 (10)

353 Sparks RSJ, Wilson L (1976) A model for the formation of ignimbrite by gravitational column collapse. J Geol Soc London 132: 441–451 (10, 11)

354 Sparks RSJ, Self S, Walker GPL (1973) Products of ignimbrite eruption. Geology 1: 115–118 (11)

355 Sparks RSJ, Huppert HE, Turner JS (1984) The fluid dynamics of evolving magma chambers. Philos Trans R Soc Lond A 310: 511–534 (3)

356 Sparks RSJ, Bursik ME, Carey SN, Gilbert JS, Glaze LS, Sigurdsson H, Woods AW (1997) Volcanic plumes. Wiley, New York, pp 1–574 (1, 10)

357 Sparks RSJ, Young SR (2002) The eruption of Soufrière Hills Volcano, Montserrat (1995–1999): overview of scientific results. In: Druitt TH, Kokelaar BP (eds) The eruption of Soufrière Hills Volcano, Montserrat, from 1995 to 1999. Geol Soc Lond Mem 21, pp 45–69 (8, 13)

358 Spörli KB, Eastwood VR (1997) Elliptical boundary of an intraplate volcanic field, Auckland, New Zealand. J Volcanol Geotherm Res 79: 169–179 (7)

359 Staudigel H, Schmincke H-U (1984) The Pliocene seamount series of La Palma (Canary Islands). J Geophys Res 89: 11195–11215 (6)

360 Stephens CD, Chouet BA, Page RA, Lahr JC, Power JA (1994) Seismological aspects of the 1989–1990 eruptions at Redoubt Volcano, Alaska: The SSAM perspective. J Volcanol Geotherm Res 62: 153–182 (13)

361 Stern RJ, Arima M (1998) Introduction to geophysical and geochemical studies of the Izu-Bonin-Mariana Arc System. Island Arc 7: 295–300 (8)

362 Stoiber RE, Williams SN (1990) Monitoring active volcanoes and mitigating volcanic hazards; the case for including sim-

ple approaches. J Volcanol Geotherm Res 42: 129–149 (13)

363 Stommel H, Stommel E (1983) Volcano weather. The story of the year without summer: 1816. Seven Seas Press, Newport, RI, pp 1–177 (14)

364 Stott PA, Tett SFB, Jones GS, Allen MR, Mitchell JFB, Jenkins GJ (2000) External control of twentieth century temperature by natural and anthropogenic forcings. Science 290: 2133–2137 (14)

365 Stott PA, Kettleborough JA (2002) Origins and estimates of uncertainty in predictions of twenty first century temperature rise. Nature 416: 723–726

366 Swanson DA (1972) Magma supply rate at Kilauea volcano, 1952–1971. Science 175: 169–170 (6)

367 Swanson DA, Holcomb RT (1990) Regularities in growth of the Mount St. Helens dacite dome 1980–1986. In: Fink JH (ed) Lava flows and domes. Springer, Berlin Heidelberg New York, pp 3–24 (9)

368 Swanson DA, Wright TL, Helz RT (1975) Linear vent systems and estimated rates of magma production and eruption for the Yakima Basalt on the Columbia Plateau. Am J Sci 275: 877–905 (7)

369 Swanson DA, Duffield WA, Jackson DB, Peterson DW (1979) Chronological narrative of the 1969–71 Mauna Ulu eruption of Kilauea Volcano, Hawaii. US Geol Surv Prof Pap 1056: 1–55 (1)

370 Swanson DA, Casadevall DJ, Dzurisin D, Malone SD, Newhall CG, Weaver CS (1983) Predicting eruptions at Mount St. Helens, June 1980 through December 1982. Science 221: 1369–1376 (13)

371 Tabazadeh A, Turco RP (1993) Stratospheric chlorine injection by volcanic eruptions: HCl scavenging and implications for ozone. Science 260: 1082–1085 (4)

372 Tait SR (1988) Samples from the crystallizing boundary layer of a zoned magma chamber. Contrib Mineral Petrol 100: 470–483 (3)

373 Tait SR, Wörner G, Bogaard P van den, Schmincke H-U (1989) Cumulate nodules as evidence for convective fractionation in a phonolite magma chamber. J Volcanol Geotherm Res 37: 21–37 (3)

374 Taylor GAM (1958) The 1951 eruption of Mount Lamington volcano, Papua. Aust Bur Miner Res Geol Geophys Bull 38: 1–117 (11, 13)

375 Thompson GA (1998) Deep mantle plumes and geoscience vision. Geol Soc Am Today 4: 17–25 (7)

376 Thorarinsson S (1967) Surtsey. The new island in the North Atlantic. Viking, New York, pp 1–68 (1)

377 Thorarinsson S (1979) On the damage caused by volcanic eruptions with special references to tephra and gases. In: Sheets PA, Grayson DK (eds) Volcanic activity and human ecology. Academic Press, New York, pp 125–159 (14)

378 Thorarinsson S (1981) Tephra studies and tephrochronology: A historical review with special reference to Iceland. In: Self S, Sparks RSJ (eds) Tephra studies. Reidel, Dordrecht, pp 1–12 (10)

379 Thordarson T, Self S, Óskarsson N, Hulsebosch T (1996) Sulfur, chlorine, and fluorine degassing and atmospheric loading by the 1783–1784 AD Laki (Skaftár Fires) eruption in Iceland. Bull Volcanol 58: 205–225 (14)

380 Thorpe RS (ed) (1982) Andesites. Orogenic andesites and related rocks. Wiley, New York, pp 1–724 (8)

381 Tilling RI (1984) Eruptions of Mt. St. Helens: past, present and future. US Geol Surv General Int Publ, pp 1–46 (10)

382 Tilling RI (1989) Volcanic hazards and their mitigation: progress and problems. Rev Geophys 27: 237–269 (13)

383 Tilling RI (1996) Hazards and climatic impact of subduction-zone volcanism: A global and historical perspective. In: Bebout GE, Scholl DW, Kirby SH, Platt JP (eds) Subduction top to bottom. Am Geophys Union, Geophys Monogr 96, pp 331–335 (2, 13)

384 Tilling RI, Dvorak JJ (1993) Anatomy of a basaltic volcano. Nature 363: 125–133 (6, 13)

385 Tilling RI, Lipman PW (1993) Lessons in reducing volcano risk. Nature 364: 277–280 (13)

386 Tilling RI, Heliker C, Wright TL (1987) Eruptions of Hawaiian volcanoes: past, present, and future. US Geol Surv General Int Publ, pp 1–54 (6)

387 Tilling RI, Topinka L, Swanson DA (1990) Eruptions of Mt. St. Helens: past, present, and future. US Geol Surv General Int Publ, pp 1–56 (10)

388 Toon OB (1984) Sudden changes in atmospheric composition and climate. In: Holland HD, Trendall AF (eds) Patterns of change in earth evolution. Sprin-

ger, Berlin Heidelberg New York, pp 41–62 (14)

389 Toon OB, Pollack JB (1982) Stratospheric aerosols and climate. In: Whitten RC (ed) The stratospheric aerosol layer: topics in current physics. Springer, Berlin Heidelberg New York, pp 121–147 (14)

390 Turco RP, Toon OB, Arkermann T, Pollack JB, Sagan C (1983) Global atmospheric consequences of nuclear war. Science 222: 1 283–1 292 (14)

391 Turner S, Hawkesworth CH, Gallagher K, Stewart K, Perte D, Mantovani M (1996) Mantle plumes, flood basalts, and several models for magma generation beneath continents: assessment of a conductive heating model and application to the Paraná. J Geophys Res 101:11 503–11 518 (7)

U

392 Ui T, Matsuwo N, Sumita M, Fujinawa A (1999) Generation of block and ash flows during the 1990–1995 eruption of Unzen Volcano, Japan. J Volcanol Geotherm Res 89: 123–137 (9)

393 Urgeles R, Canals M, Baraza J, Alonso B, Masson D (1997) The most recent megalandslides of the Canary Islands: El Golfo debris avalanche and Canary debris flow, west El Hierro Island. J Geophys Res 102: 20 305–20 323 (13)

394 Uyeda S (1984) Subduction zones: their diversity, mechanism and human impact. Geojournal 8: 381–406 (8)

V

395 van der Hilst R, Widiyantoro S, Engdahl ER (1997) Evidence for deep mantle circulation from global tomography. Nature 386: 578–584 (6)

396 Varekamp JC, Kreulen R, Poorter RPE, van Bergen MJ (1992) Carbon sources in arc volcanism, with implications for the carbon cycle. Terra Nova 4: 363–373 (4)

397 Verne J (1867) Voyage au centre de la terre. Hetzel, Paris, pp 1–220 (1)

398 Vine FJ, Matthews DH (1963) Magnetic anomalies over oceanic ridges. Nature 199: 947–949 (2)

399 Vitaliano DB (1973) Legends of the earth: their geological origins. Indiana Univ Press, Bloomington, pp 1–305 (1)

400 Völzing K (1907) Der Trass des Brohltales. Jb Preuss Geol LA 28: 1–56 (11)

401 Voight B (1988) A method for predicting volcanic eruptions. Nature 332: 125–130 (13)

402 Voight B (1990) The 1985 Nevado del Ruiz catastrophe: anatomy and retrospection. J Volcanol Geotherm Res 42: 151–188 (13)

403 Voight B (1996) The management of volcano emergencies: Nevado del Ruiz. In: Scarpa R, Tilling RI (eds) Monitoring and mitigation of volcano hazards. Springer, Berlin Heidelberg New York, pp 718–769 (13)

404 Voight B, Janda RJ, Glicken H, Douglass PM (1983) Nature and mechanics of the Mount St. Helens rockslide-avalanche of 18 May 1980. Geotechnique 33: 243–273 (10)

405 von Engelhardt W (1982) Neptunismus und Plutonismus. Fortschr Miner 60: 21–43 (1)

406 von Fritsch R, Reiss W (1868) Geologische Beschreibung der Insel Tenerife. Wurster und Co, Winterthur, pp 1–494 (11)

407 von Huene R, Scholl DW (1991) Observations at convergent margins concerning sediment subduction, subduction erosion, and the growth of the continental crust. Rev Geophys 29: 279–316 (8)

W

408 Walker GPL (1973) Explosive volcanic eruptions – a new classification scheme. Geol Rundsch 62: 431–446 (4)

409 Walker GPL, Wilson CJN, Froggat PC (1981) An ignimbrite veneer deposit: the trailmarker of a pyroclastic flow. J Volcanol Geotherm Res 9: 409–421 (11)

410 Wallace PJ (2001) Volcanic SO_2 emissions and the abundance and distribution of exsolved gas in magma bodies. J Volcanol Geotherm Res 108: 85–106 (14)

411 Wallace PJ, Carmichael ISE (1999) Quaternary volcanism near the Valley of Mexico: implications for subduction zone magmatism and the effects of crustal thickness variations on primitive magma compositions. Contrib Mineral Petrol 135: 291–314 (7, 8)

412 Walter TR, Schmincke H-U (2002) Rifting, recurrent landsliding and structural reorganization on NW Tenerife, Canary Islands. Int J Earth Sci 91: 615–628 (6)

413 Watts AB, Masson DG (1995) A giant landslide on the north flank of Tenerife, Canary Islands. J Geophys Res 100: 24 487–24 498 (6, 13)

414 Weaver PPE and 6 co-authors (1998) Neogene turbidite sequence on the

Madeira Abyssal Plain: basin filling and diagenesis in the deep ocean. In: Weaver PPE, Schmincke H-U, Firth JV, Duffield WA (eds) Proc ODP, Sci Results, 157: College Station, TX (Ocean Drilling Program), pp 619–634 (6)

415 Wegener A (1912) Die Entstehung der Kontinente. Geol Rundsch 3: 276–292 (2)

416 Werner AG (1774) Von den äusserlichen Kennzeichen der Fossilien. SL Crusius, Leipzig (1)

417 Werner R, Schmincke H-U (1999) Englacial vs lacustrine origin of volcanic table mountains: evidence from Iceland. Bull Volcanol 60: 335–354 (6, 12)

418 Williams H, McBirney A (1979) Volcanology. Freeman, Cooper and Co, San Francisco, pp 1–391 (1,12)

419 Williams RS, Moore JG (1973) Iceland chills a lava flow. Geotimes 18: 14–17 (13)

420 Wilson JT (1963) A possible origin of the Hawaiian islands. Can J Phys 41: 863–870 (6)

421 Wilson L (1980) Relationships between pressure, volatile content, and ejecta velocity in three types of volcanic explosions. J Volcanol Geotherm Res 8: 297–313 (10)

422 Wilson L, Head JW III (1981) Ascent and eruption of basaltic magma on the earth and moon. J Geophys Res 86: 2 971–3 001 (10)

423 Wilson L, Head JW III (1983) A comparison of volcanic eruption processes on Earth, Moon, Mars, Io and Venus. Nature 302: 663–669 (10)

424 Wilson L, Sparks RSJ, Huang TC, Watkins ND (1978) The control of volcanic column eruption heights by eruption energetics and dynamics. J Geophys Res 83: 1 829–1 836 (10, 11)

425 Wilson L, Sparks RSJ, Walker GPL (1980) Explosive volcanic eruptions, IV. The control of magma properties and conduit geometry on eruption column behavior. Geophys J R Astron Soc 63: 117–148 (10, 11)

426 Wörner G, Schmincke H-U (1984) Mineralogical and chemical zonation of the Laacher See tephra (East Eifel, Germany). J Petrol 25: 836–851 (3, 11)

427 Wörner G, Staudigel H, Zindler A (1985) Isotopic constraints on open system evolution of the Laacher See magma chamber (Eifel, West Germany). Earth Planet Sci Lett 75: 37–49 (3)

428 Wohletz K, Heiken G (1992) Volcanology and geothermal energy. Univ California Press, Berkeley, pp 1–432 (1)

429 Wohletz KH (1983) Mechanism of hydrovolcanic pyroclast formation: grain-size, scanning electron microscopy, and experimental studies. J Volcanol Geotherm Res 17: 31–63 (12)

430 Wohletz KH, McGetchin TR, Sanford II MT, Jones EM (1984) Hydrodynamic aspects of caldera-forming eruptions: numerical models. J Geophys Res 89: 8 269–8 285 (11)

431 Wolf T (1878) Der Cotopaxi und seine letzte Eruption am 26. Juni 1877. N Jahrb Miner Geol: 113–167 (11)

432 Wolfe CJ, Bjarnasson IT, VanDecar JC, Solomon SC (1997) Seismic structure of the Icelandic mantle plume. Nature 385: 245–247 (6)

433 Wolfe E, Okamura A, Dvorak J, Koyanagi R, Greenland P, Garcia M, Neal C, Jackson D (1987) The first 20 phases of the Pu'u 'O'o eruption on the middle East Rift of Kilauea Volcano, Hawaii. US Geol Surv Prof Pap 1 350: 471–508 (6)

434 Wood CA (1980) Morphometric evolution of cinder cones. J Volcanol Geotherm Res 7: 387–414 (9)

435 Woods AW (1995) The dynamics of explosive volcanic eruptions. Rev Geophys 33: 495–530 (11)

436 Wright TL, Peck D, Shaw HR (1976) Kilauea lava lakes: natural laboratories for study of cooling, crystallization, and differentiation of basaltic magma. In: The Geophysics of the Pacific Ocean Basin and its Margins. Am Geophys

Union Geophys Monogr 19: pp 315–392 (3)

437 Wyllie PJ (1976) The way the earth works: an introduction to the new global geology and its revolutionary developments. Wiley, New York, pp 1–296 (2)

438 Yoder HS (1976) Generation of basaltic magma. National Acad Press, Washington, DC, pp 1–265 (3)

439 Yogodzinski GM, Lees JM, Churikova TG, Dorendorf F, Wörner G, Volynets ON (2001) Geochemical evidence for the melting of subducting oceanic lithosphere at plate edges. Nature 409: 500–504 (8)

440 Young SR and 6 co-authors (1998) Overview of the eruption of Soufriere Hills volcano, Montserrat, 18 July 1995 to December 1997. Geophys Res Lett 25: 3 389–3 392 (13)

441 Zen MT, Hadikusumo D (1965) The future danger of Mt. Kelut (Eastern Java-Indonesia). Bull Volcanol 28: 275–282 (13)

442 Zeyen H, Novak O, Landes M, Prodehl C, Driad L, Hirn A (1997) Refraction-seismic investigations of the northern Massif Central (France). Tectonophysics 178: 329–352 (7)

443 Zielinski GA and 8 co-authors (1994) Record of volcanism since 7000 BC from the GISP 2 Greenland ice core and implications for the volcano-climate system. Science 264: 948–952 (14)

444 Zielinski GA, Mayewski PA, Meeker LD, Whitlow S, Twickler MS (1996) A 110 000-yr record of explosive volcanism from the GISP 2 (Greenland) ice core. Quat Res 45: 109–118 (14)

445 Zimanowski B, Büttner R, Lorenz V, Häfele HG (1997) Fragmentation of basaltic melt in the course of explosive volcanism. J Geophys Res 102 B1: 803–814 (12)

Subject Index

Index of Geographical Names

Place names referred to in the text are summarized under regional headings

Index of Names